BETA TITANIUM ALLOYS IN THE 1990'S

BETA TITANIUM ALLOYS IN THE 1990'S

Proceedings of a Symposium on Beta
Titanium Alloys sponsored by the Titanium
Committee of TMS, held at the 1993 Annual TMS
Meeting in Denver, Colorado, February 22-24, 1993

Edited by

Daniel Eylon
Graduate Materials Engineering
University of Dayton
Dayton, OH 45469-0240

Rodney R. Boyer
Boeing Commercial Airplane Company
Seattle, WA 98124-2207

Donald A. Koss
Department of Metals Science Engineering
Penn State University
University Park, PA 16802-7501

A Publication of

TMS
Minerals • Metals • Materials

A Publication of The Minerals, Metals & Materials Society
420 Commonwealth Drive
Warrendale, Pennsylvania 15086
(412) 776-9000

Printed in the United States of America
Library of Congress Catalog Number 93-79546
ISBN Number 0-87339-200-0

TMS
Minerals • Metals • Materials

If you are interested in purchasing a copy of this book, or if you would like to receive the latest TMS publications catalog, please telephone 1-800-759-4867.

FOREWORD

Structural Titanium Alloys are generally divided into **near alpha, alpha+beta, beta** and **ordered** alloys. The wider definition of **beta alloys** used in this includes stable beta alloys (i.e. Ti-30Mo), metastable beta alloys (i.e. Ti-15V-3Cr-3Sn-3Al) and beta-rich alpha+beta alloys (i.e. SP-700). As a group, beta alloys provide wider range of processing, physical, chemical and mechanical properties than any of the other groups. Beta alloys can be cold or hot worked; can have low or high elastic moduli; can have high environmental resistance; and can have low or very high combination of strength and fracture resistance. In spite of this wide range of attributes, these alloys account for only 1% of the titanium market in the USA. The editors of this book felt that further research and development of beta alloys can bring about additional applications for titanium alloys as a replacement for high strength and corrosion resistant steels.

This book is based on the **Beta Titanium Alloy Symposium** which was held in February, 1993 in Denver, Colorado as part of the 1993 Annual TMS Meeting. This symposium was held exactly 10 years after the first Beta Titanium Alloy Symposium that took place in Atlanta, Georgia in March, 1983. When comparing the proceedings of both conferences, it is interesting to note that after 10 years, some alloys (such as Transage 134 and 175 or Beta III) are no longer the subject of research, while new alloys (such as Beta 21S, Beta-CEZ or SP-700) were introduced. Another interesting fact is the first large scale implementation of beta alloys for applications such as in aircraft landing gears, airframe parts and gas turbine engine vectored nozzles, and an expected implementation in orthopedic implants, sport instruments and power generation turbine blades. The editors of this book hope that the papers presented will inspire additional developments and new applications for beta alloys.

The book has been divided into 5 major sections:

- **Alloy Development**
- **Environmental Aspects**
- **Microstructure-Property Relationships**
- **Applications**
- **Data Sheets**

Each of the first 4 sections is opened by an invited review article. The 39 papers were contributed by researchers from the USA, Japan, France and Germany and therefore provide a good summary of the international state-of-technology of beta alloys. The section on the Beta Alloy Data Sheets has been added after the symposium for the benefit of the readers.

The organizers wish to thank the Titanium Committee of TMS and TMS for sponsoring this symposium. The help of Mr. John Porter of the University of Dayton in organizing the manuscripts and indexing the texts is greatly appreciated. All articles published were thoroughly reviewed by the three editors, and the reviewed drafts were corrected and resubmitted by the authors.

Danny Eylon
University of Dayton, *Dayton, OH*

Rod Boyer
Boeing Commercial Airplane Company, *Seattle, WA*

Don Koss
Penn State University, *University Park, PA*   June, 1993

TABLE OF CONTENTS

DEVELOPMENT OF BETA TITANIUM ALLOYS

ENVIRONMENTAL ASPECTS OF BETA TITANIUM ALLOYS

MICROSTRUCTURE-PROPERTY RELATIONSHIPS
IN BETA TITANIUM ALLOYS

APPLICATIONS OF BETA TITANIUM ALLOYS

DATA SHEETS OF BETA TITANIUM ALLOYS

DEVELOPMENT OF
BETA TITANIUM ALLOYS

BETA TITANIUM ALLOYS AND THEIR ROLE IN THE TITANIUM INDUSTRY

--Keynote Lecture--

Paul J. Bania

TIMET
P.O. Box 2128
Henderson, NV 89009

Abstract

The class of titanium alloys generically referred to as "the beta alloys" is arguably the most versatile in the titanium family. Since they offer the highest strength-to-weight ratio and deepest hardenability of all titanium alloys, one might expect them to compete favorably for a variety of aerospace applications. To the contrary, however, with the exception of one very successful application (Ti-13V-11Cr-3Al on the SR-71), the beta alloys have remained a very small segment of the industry. This paper will review some of the alloys and applications of the past and present. It will also examine some unique new alloys and applications which promise to reverse the trends of the past.

BACKGROUND: What is a Beta Alloy

The term "beta alloy" is used rather broadly. As shown in Figure 1, for the purposes of this discussion we will define a beta alloy to mean any alloy with enough total beta stabilizer content to retain 100% beta upon quenching from above the beta transus (the boundary between the single phase beta region and the two-phase alpha-beta region). This infers that there is enough beta stabilizer content to avoid passing through the martensite start (M_s) upon quenching, thus precluding the formation of martensite. Alloys that lie between this critical minimum level of beta stabilizer content (β_c) and the β_s point of Figure 1 are still within the two-phase region. Thus, even though one can quench to retain 100% beta in such alloys, the beta phase is metastable and will precipitate a second phase (usually alpha) upon aging. Alloys to the right of β_s are considered stable beta alloys and, theoretically, no precipitation occurs. Thus, stable beta alloys should not be hardenable.

Table I lists the more frequently used beta stabilizers and their respective β_c values. It also shows that there are two types of beta stabilizers - isomorphous and eutectoid. In general, the isomorphous beta stabilizers:

  a)   Have high β_c values.

  b)   Don't decompose the beta phase to form alpha + compound at equilibrium.

  c)   Must be added as aluminum master alloys due to their high melting points and densities.

  d)   Are relatively high cost additions.

Beta Titanium Alloys in the 1990's
Edited by D. Eylon, R.R. Boyer and D.A. Koss
The Minerals, Metals & Materials Society, 1993

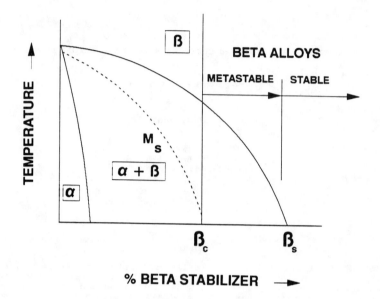

Figure 1 - Pseudo binary phase diagram of titanium and a beta stabilizer.

On the other hand, the eutectoid beta stabilizers generally:

a)   Have lower β_c values and are more potent in suppressing the beta transus.

b)   Will tend to form compounds.

c)   Can be added elementally.

d)   Are relatively low cost additions.

Note: (Although Sn and Zr act to lower the beta transus, they are considered neutral additions. Aluminum is an alpha stabilizer.)

There is a convenient manner in which to consider the overall beta stability of an alloy with various alloying additions. By arbitrarily using molybdenum as a baseline, we can define a "moly equivalent" (Mo. Eq.) as follows:

Mo. Eq. = 1.0 (Wt. % Mo) + .67 (Wt. % V) + .44 (Wt. % W) + .28 (Wt. % Nb) + .22 (Wt. % Ta) + 2.9 (Wt. % Fe) + 1.6 (Wt. % Cr) ... ... − 1.0   (Wt. % Al).

In the above equation, the constant before each alloying element reflects the ratio of the β_c for the moly baseline (i.e. 10.0) divided by the β_c for the specific element. Note also that aluminum is added as a negative value to reflect it's opposite tendency to stabilize alpha.

In general, a Mo. Eq. value of about 10.0 is required to stabilize beta upon quenching. It is not clear what Mo. Eq. is required to produce a truly stable beta alloy, since even Ti-30Mo (Mo. Eq. = 30) is unstable at elevated temperatures under applied stress.

Table I. Beta Stabilizing Elements

β-Stabilizer	Type	β_c (Wt %)[1]	β_t Suppression (°F)[2]
Mo	Isomorphous	10.0	17
V	"	15.0	22
W	"	22.5	7
Cb	"	36.0	13
Ta	"	45.0	4
Fe	Eutectoid	3.5	32
Cr	"	6.5	27
Cu	"	13.0	22
Ni	"	9.0	40
Co	"	7.0	38
Mn	"	6.5	40
Si	"	-	70

[1]Approximate wt. % needed to retain 100% beta upon quenching.
[2]Approximate amount of beta transus reduction per wt. % of addition.

Figure 2 provides a comparison of the calculated Mo. Eq. values for various commercial alloys. Note that Ti-10V-2Fe-3Al (Ti-10-2-3) is very marginal whereas alloys such as *TIMETAL®15-3*, *TIMETAL®21S* and **Ti-38-6-44** are comfortably within the metastable region. The Ti-13V-11Cr-3Al alloy is quite heavily stabilized while Alloy C (Ti-35V-15Cr) is very heavily stabilized. Even Alloy C, at such a high Mo. Eq., is not a truly stable alloy.

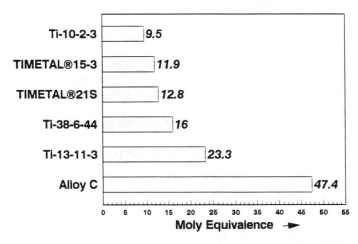

Figure 2 - Comparison of Mo. Eq. values for various commercial alloys. In general, a Mo. Eq. above about 10.0 is required to retain 100% beta upon quenching from above the beta transus.

5

A further comment should be noted regarding the various beta stabilizers. As shown in Figure 3, eutectoid beta stabilizers such as iron tend to have a wide freezing range (spread between liquidus and solidus at a given composition) and hence segregate strongly. Compounded by the fact that they also have a relatively large effect on beta transus temperature, the segregation can result in regions with significantly lower beta transus values than the bulk. These are known as "beta flecks". On the other hand, isomorphous stabilizers such as molybdenum tend to have narrower freezing ranges and hence segregate less. Note also that isomorphous stabilizers tend to segregate in the opposite direction of eutectoid stabilizers. It is primarily because of the smaller segregation tendency, coupled with the freedom from compound formation, that most beta alloys rely primarily on isomorphous beta stabilizers.

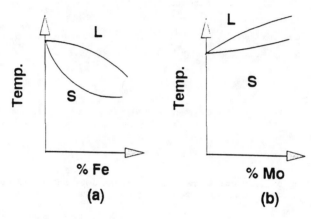

Figure 3 - Schematic comparison of the binary phase diagram freezing ranges of typical eutectoid (Fe) and isomorphous (Mo) beta stabilizers.

GENERAL ATTRIBUTES OF BETA ALLOYS

Table II provides a comparison of the advantages and disadvantages of the beta alloys:

Table II. General Attributes of Beta Alloys

Advantages	Disadvantages
• High strength-to-density	• High density
• Low modulus	• Low modulus
• Strip producible (low cost TMP)	• High formulation cost
• Good weldability characteristics	• Tend to segregate
• Very cold formable	• High springback
• Easy to heat treat	• Oxygen/hydrogen pick-up
• Some have good castability	• Segregation
• Excellent corrosion properties for some compositions	• Poor low & E.T. props.

- While they offer the highest strength to weight ratio's of any titanium alloy, they are nonetheless higher in density compared to other alloys. Densities on the order of (4.70 to 4.98 g/cc) .17 to .18 lb/cu-in are typical, with ultimate strength levels in the 220 to 230 ksi (1517 to 1586 MPa) range.

- In the solution treated condition, modulus values of (69 to 76 GPa) 10-11 M-Psi are typical whereas in the aged condition (103 to 110 GPa) 15-16 M-Psi are common. In some cases, such as medical implants where matching the low modulus of bone is important, low modulus is a plus. However, in most aircraft structure, high modulus is desired.

- Although beta alloys are expensive to formulate, the strip rollable alloys can take advantage of low cost/high yield cold strip processing vs. hot band mill operations.

- Although segregation can be a problem during melting (solidification), the rapid cooling rate obtained in most weldments often prevents transformations, hence improving weldability.

- Beta alloys typically exhibit outstanding cold formability. However, their low modulus/high strength can often result in a high springback which must be accounted for in tooling design.

- Because of the high degree of stability of the beta phase, a water quench is often not required for solution treating; air cooling greatly reduces distortion, as does the low solutionizing temperature. However, beta alloys are notorious for surface oxygen contamination, even in supposed "vacuum" heat treatments, so care must be exercised to remove such contamination, usually by pickling.

- While beta alloys usually exhibit excellent fill characteristics during casting, segregation may limit the utility of some alloys.

- Although beta alloys are generally not considered for elevated or cryogenic temperature service, some exceptions exist. Also, many beta alloys have outstanding corrosion resistance. Generally, corrosion resistance increases with moly content.

On balance, the advantages listed in Table II would seem to easily outweigh the disadvantages, most of which can be fairly easily overcome. Yet, as shown in Figure 4, beta alloys have garnered only about 1% of the total U.S. market for titanium mill products. The reasons for the limited success are probably complex and quite varied. The higher cost of beta alloys in most product forms has been a deterrent, as has the cost of qualifying such new alloys. For example, high strength beta alloys are designed to compete with precipitation hardened stainless steels which sell for considerably less (on the order of one-fourth to one-eight) than most beta alloys. On a performance basis, beta alloys not only compete with (lower cost) aluminum and iron-based alloys, but also with other conventional alpha-beta titanium alloys such as Ti-6Al-4V. For example, superplastic forming of complex sheet metal details from alpha beta alloys competes with cold forming of beta alloys. (Beta alloy are not considered SPF able.)

In order to obtain a better perspective of the beta alloy family, it is instructive to look at its history.

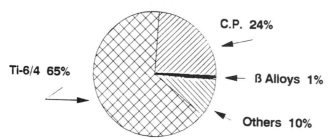

Approx. U.S. Market by Alloy Type

C.P. 24%

Ti-6/4 65%

ß Alloys 1%

Others 10%

Figure 4 - Approximate breakdown of the U.S. titanium market by alloy type.

7

THE EARLY DAYS OF BETA ALLOYS

Beta alloys are certainly not a new or recent development in the titanium industry. Even though the commercial titanium industry began in the 1949-1950 timeframe, one of the early alloy patents[1] filed in 1949 and issued in 1952, covered a corrosion resistant beta alloy Ti-30Mo to Ti-40Mo. Numerous other patents[2-6] filed in the 1949-1952 period also covered a variety of beta alloys.

• <u>Ti-13V-11Cr-3Al (B120 VCA)</u>: Without question, the most successful of the early beta alloys, and the only one to reach significant commercial production, was the Ti-13V-11Cr-3Al alloy (also designated B120 VCA) introduced by Crucible Steel Corporation in the mid-1950's. As shown in Figure 2, the alloy has a high Mo. Eq. and hence would be expected to be rather stable. In fact, the alloy is so stable that in the fully recrystallized condition aging times up to 100 hours were required.[7] One of the "tricks" to using the alloy was to not fully recrystallize the structure thereby leaving residual hot/cold work within the bulk to assist aging response. Typically, the material was left in a polygonized state after solution-treating above the beta transus so that more reasonable aging cycles (8-12 hours) could be employed.

While the alloy found various uses such as springs (cold worked + aged condition), by far the biggest application for the alloy was as a primary structural material for the Mach 3+ SR-71 Blackbird.[8] While the alloy offered many advantages such as high strength-to-weight and cold/hot formability, perhaps the key to its widespread use was the alloys high degree of stability. Even though aged sheet was used, a great deal of the structure was spot welded. Since it was impractical to age the large spot welded structures, the spot welds had to be left in the solution-treated (as-welded) condition. However, since the hot structure was going to see extended service up to 550°F (288°C), the solution-treated spot weld had to be stable - ie. resist aging - with extended exposure at these service temperatures. Only Ti-13-11-3, with its high Mo. Eq., could resist aging at these temperatures upon prolonged exposures in the as-welded condition. Most beta alloys with lower Mo. Eq.'s would eventually age to a brittle state upon extended exposure to such temperatures.

While the Ti-13-11-3 alloy was the most successful of the early beta alloys, a number of other beta alloys were also being offered in the 1950's and 1960's. Examples include:

• <u>Ti-8V-8Mo-2Fe-3Al (8-8-2-3)</u>: Developed under Army contracts through TIMET[9], this alloy was primarily intended as a high strength forging alloy. There were some melting problems with the alloy (suitable master alloys; segregation due to high iron content) and it was never truly commercialized. It is interesting to note the similarity to Ti-10V-2Fe-3Al introduced a decade later, through which melting segregation problems at this iron level were eventually solved.

• <u>Ti-11.5Mo-6Zr-4.5Sn (Beta III)</u> : This alloy was originally developed by Crucible and later supported through Air Force funding. Note that there are no eutectoid beta stabilizers used, thereby avoiding compound formation problems. Also note that there is no aluminum used, thus precluding the use of a lower melting point Mo-Al master alloy. This ultimately led to melting problems when the pure moly alloying addition used was shown to result in unmelted, pure moly high density conclusions (HDI's).

• <u>Ti-1Al-8V-5Fe (1-8-5)</u>: This alloy was developed by RMI primarily as a fastener alloy. It was capable of very high strength/ductility combinations. Solution treatments were generally carried out below the beta transus.[10]

• <u>Ti-30Mo (and Various Modifications)</u>: As noted earlier, such alloys were developed primarily for applications requiring extremely high corrosion resistance. The obvious VAR melting problems with such a high requirement for a pure moly addition rendered this an impractical alloy. However, the advent of today's hearth melting technology could provide sufficient justification to revisit such alloys.

TODAY'S BETA ALLOYS

The prominent beta alloys of today are listed below, along with their approximate year of introduction.

Alloy	When Introduced
Ti-3Al-8V-6Cr-4Zr-4Mo (38-6-44; Beta C)	1969
Ti-10V-2Fe-3Al (10-2-3)	1971
Ti-15V-3Cr-3Sn-3Al (15-3)	1978
Ti-15Mo-2.7Nb-3Al-.2Si (*TIMETAL*®21S)	1989

While all of the above alloys are currently specified for a variety of applications, Figure 4 shows that none of these alloys represent a very substantial portion of the overall titanium market. As a general rule-of-thumb, a new alloy usually takes between 10 to 15 years to establish itself in the marketplace. Once established, it can enjoy a relatively long life as it is often difficult to displace an established alloy . However, if an alloy is not firmly established in this timeframe, one would expect that the technology would advance to a point wherein newer more attractive alloys become available. In such a scenario, it would appear to be a critical time for the first three alloys listed above.

On the other hand, *TIMETAL*®21S is an interesting exception to the conventional rule. It has progressed to a production alloy in less than four years. In this case, the fortuitous timing of the alloy introduction coupled with the weight reduction needs of the Boeing 777 allowed the alloy "short circuit" the normal evolution cycle.

• Ti-3Al-8V-6Cr-4Zr-4Mo (38-6-44; Beta C): This alloy was developed by RMI in the late 1960's. It is known to precipitate a second phase at high temperatures[11] (possibly silicon rich phospho-sulfides). The zirconium addition is known[11] to promote uniform precipitation of these precipitates thus enhancing ductility and retarding grain growth. The alloy is generally used in the cold-worked + aged condition in order to promote uniform aging response. The primary commercial use for the alloy is for aerospace springs made from rod stock, in many cases displacing the Ti-13-11-3 in this application. It is also being used as a high strength fastener on the Boeing 777, although there is a slight reduction in properties associated with the hot-headed region of the fastener (ie. not taking advantage of the cold work + age process). Interestingly, the alloy has also been promoted heavily in a non-aerospace application for "downhole" oil drilling/exploration equipment. In this case, it's high-strength/density coupled with excellent corrosion resistance which drives this application. For improved corrosion resistance, a modified version with roughly .04 to .05 Pd has been NACE qualified for downhole service.

• Ti-10V-2Fe-3Al (10-2-3, Ti-10-2-3): This alloy was developed by TIMET as a high strength, high toughness alloy to compete with steels in airframe applications. Because of its low flow stress at high temperature vs. alloys such as Ti-6Al-4V, it is an excellent alloy for isothermal forging practices. In the early days of the alloy (early 1980's) there were melting problems associated primarily with iron segregation which led to "beta-flecks - areas enriched in iron with a significantly low beta transus than the matrix which caused problems when heat treating just below the matrix transus. However, these issues have been solved by proper melt practice and subsequent homogenization.

When processed properly, 10-2-3 offers the best combination of strength, toughness and high cycle fatigue resistance of any titanium alloy. However, the processing window is tight[12], generally requiring beta forging + quenching, controlled alpha-beta forging (15-25%) and sub-transus solution heat treatment (sometimes 2-step) followed by aging. The reason for the sub-transus heat treatment is shown in Figure 5. As Figure 2 shows, this alloy is marginal in terms of beta stability (Mo. Eq.) Thus, by heat treating below the transus as shown in Figure 5, the beta phase which exists at the solutioning temperature is enriched and can be retained. If solutioned above the beta transus, the alloy might retain a 100% beta structure but upon any deformation (as little as a quenching strain) the alloy will undergo a shear

9

transformation[13] to an orthorhombic martensite.

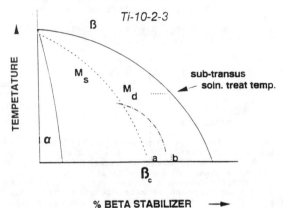

Figure 5 - Schematic diagram showing the reason for solutioning Ti-10-2-3 below its beta transus thereby enriching the beta phase.

The Ti-10-2-3 alloy has been used on airframe forgings by Boeing and is scheduled to be used for landing gear truck beams on the Boeing 777 - the first use of titanium in commercial landing gear. It is also scheduled for various forging applications on the C-17 and as a rotor head on the Westlands Lynx helicopter. The alloy has also been shown to exhibit a shape-memory effect.[14]

• Ti-15V-3Cr-3Sn-3Al (15-3; Ti-15-3): This alloy was developed under U.S. Air Force sponsorship. While the initial contract which identified the composition was conducted by Lockheed[15], a series of follow-on studies leading to production scale-up were performed by TIMET.[16,17] The alloy was developed primarily as a strip-producible, cold formable, age hardenable and weldable alloy. The alloy has also been shown to exhibit excellent castability, hydrogen tolerance[17] and damage tolerance.[17] In terms of formulation cost, 15-3 is a more expensive alloy than, for example, Ti-6Al-4V. However, since it is strip producible (vs. hot pack rolling for alpha-beta alloys such as Ti-6/4) it can be less costly than Ti-6/4 in sheet form. At roughly .080" thickness, the costs of both alloys in sheet form are roughly comparable. For thinner gages, the lower processing costs and higher yields of cold strip rolling gradually render Ti-15-3 more and more cost competitive than Ti-6/4. At the extreme, Ti-15-3 foil is roughly one tenth the cost of Ti-6/4 foil at the same gage.

Ti-15-3 is emerging as a viable aerospace alloy. Following application in over 250 details on the B-1, it is now slated for various sheet metal applications on the C-17 and the Boeing 777.[18] The 777 applications include the ECS (Environmental Control System) ducts as well as fire extinguisher bottles. It is also used in cast form on the 777.[18] Pratt & Whitney has also committed to use the alloy for all of its engine brackets in service below about 550°F. Judging by the success of some of these programs, it seems safe to conclude that such applications will transfer to other airframe and engine models.

Perhaps the most exciting new application for this alloy is the Ti-15-3 softball bat introduced by Easton. This is a significant development in that it represents the use of an aerospace titanium alloy in a consumer market.

• Ti-15Mo-2.7Nb-3Al-.2Si (*TIMETAL* 21S): This alloy was developed by TIMET in 1989 in order to provide an oxidation resistant foil product for use by McDonnell Douglas as a matrix for titanium metal-matrix composites (MMC's) on the N.A.S.P. (National AeroSpace Plane). Compared to Ti-15-3, this alloy provides roughly a 100-fold improvement in oxidation.[20] The comparison is made to Ti-15-3 because McDonnell Douglas had decided to work with a production producible foil product - hence a cold rollable strip alloy. Ti-15-3 had been tested as a MMC as high as 815°C (1500°F) and found adequate except for oxidation resistance. The 21S alloy is also strip rollable to foil but offers the 100-fold reduction in oxidation weight gain vs. 15-3. Primarily due to the NASP studies, *TIMETAL*21S has established itself as the primary titanium MMC alloy being studied today. However, because of its exceptionally good high temperature strength and creep resistance (for a beta alloy), it is also finding application in warm to hot structures such as ducts and engine plugs and nozzles. Boeing has selected *TIMETAL*21S for 777 plugs and nozzles, as has P&W for the PW 4168 engine. Also, Garrett is specifying it for the Garrett 331-500 APU eductor. In each case, the max service temperature is over 540°C (1000°F) but stresses are quite low. Often, such substitutions are in favor of *TIMETAL*21S vs. a nickel-based alloy such as Inconel 625. *TIMETAL*21S is also being considered for higher stress "warm" applications in the 260°C (500°F) to 425°C (800°F) ranges.

Another attribute of the alloy is its exceptional corrosion resistance,[20] particularly against hydraulic fluid. This has led to a variety of nacelle applications on the Boeing 777[18] wherein nickel-based alloys would normally be used.

As in the Ti-15-3 case, many of these applications would be expected to eventually translate to other airframe and engine models.

It is no doubt clear to the reader by now that the Boeing 777 will be an important user of these advanced beta alloys. Figure 6[21] shows the evolution of titanium usage on Boeing aircraft highlighting the impact of such alloys. Roughly 10 % of the 777 airframe weight is titanium, much of which is represented by Ti-10-2-3 on the landing gear as well as the above cited applications of *TIMETAL*15-3 and *TIMETAL*21S.

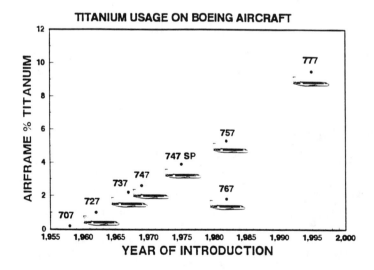

Figure 6 - Progression of titanium usage on Boeing airframes.[21]

TOMORROW'S BETA ALLOYS

Virtually all of the beta alloys developed to date have been performance driven - either by mechanical/physical properties, fabricability (forming, forging ...) corrosion resistance, oxidation resistance or some combination thereof. While such efforts will undoubtedly continue, it is also becoming apparent that a new driver will prompt a special class of beta alloys in the future. That driver will be COST.

It is recognized that in order to expand the overall titanium market, new applications involving substitution of titanium for other materials will have to take place. Many of these new applications will be driven primarily by cost, especially those in non-traditional titanium markets. An excellent example is that of springs for automotive use (valve springs and coil suspension springs). Titanium offers nearly a 25% space savings and a 50% weight savings compared to steel springs. Indeed, trial programs have been successful in demonstrating the advantages of titanium for springs and titanium has been used for several years as a spring material for high performance racing cars. However, the beta alloys used for such springs are invariably too costly for general commercial use.

The cost of such alloys begins at the high formulation cost precipitated by the need for a high concentration of expensive beta stabilizers such as vanadium, molybdenum and niobium. Such additions must be added as V/Al, Mo/Al or Nb/Al master alloys, because of the high melting point and high density of the pure elements. The other key contributors to cost are the conversion process used and the associated product yield. Thus, a low cost beta alloy will require:

- low formulation cost
- low cost conversion processing
- high product yield

This section will briefly highlight two recently introduced alloys - one developed primarily as a high performance alloy, the other a low cost alloy.

- Ti-6V-6Mo-5.7Fe-2.7Al (*TIMETAL*®125): This alloy was developed by TIMET as a competitor to high strength steels and nickel-based alloys (H-11 A-286, IN 718) used as fasteners in airframes and engines. In order to provide a 1:1 fastener substitution and hence save about 42% of the weight, the alloy has to be capable of (860 MPa) 125 ksi shear strength and about (1585 MPa) 230 ksi tensile strength at 7-8% elongation. While cold-worked plus aged beta alloys such as Ti-38-6-44 and *TIMETAL*®21S have been shown capable of these targets, the problem lies in the fact that the fasteners need to be hot-headed, thus dissipating the cold work effect. The *TIMETAL*®125 alloy is designed to offer the target properties in a solution-treated and aged condition. It should be noted that for this beta alloy the strength/ ductility relationship is significantly better when solutioning below the beta transus.[22]

Although the target properties have been demonstrated on section sizes below 12.7mm (.5") diameter, there is still some development required for larger sizes. This is a clear example of the importance of thermomechanical processing in determining final properties.

- Ti-4.5Fe-6.8Mo-1.5Al (*TIMETAL*®LCB): This LCB (Low Cost Beta) alloy takes advantage of ferro-moly as a low cost beta stabilizer addition, thereby lowering its formulation cost to roughly the same as commercially pure titanium. Depending on the alloy, this can represent a reduction of over 50% in formulation cost alone vs. the more expensive beta alloys. Preliminary lab heat results suggest that the alloy also has excellent forging and rolling characteristics, thus additional cost savings vs. other beta alloys are also possible.

Table III provides some preliminary tensile data on the LCB alloy. These values are clearly competitive with any other beta alloy, particularly considering the fact that scale-up to larger heats usually improves strength/ductility relationships due to added thermomechanical processing. The initial examination of this alloy is focusing on non-aerospace applications such as automotive/Army suspension systems (springs, torsion bars...) and high strength corrosion resistant applications where existing CP or alpha-beta alloys are either too low in strength or too costly. However, it is easy to understand the wide potential utility of such a unique alloy.

Table III. Example of **TIMETAL®LCB** Tensile Properties

Condition	YS, MPa (ksi)		UTS, MPa (ksi)		% El	% RA
Solution Treated	1034	(150)	1061	(154)	16	43
ST + 950°F Age	1503	(218)	1545	(224)	5	14
ST + 1000°F Age	1365	(198)	1400	(203)	8	25
ST + 1100°F Age	1145	(166)	1172	(170)	15	44
ST + 1200°F Age	1048	(152)	1076	(156)	17	37

Properties from 12.7mm (.5") dia. rod rolled from 22.7 kg (50-lb.) heat

SUMMARY

Beta alloys have had an important but relatively small role in the titanium industry thus far. However, with the advent of significant new applications in the aerospace sector, coupled with some very promising applications and opportunities in the non-aerospace consumer sector, the future of this class of titanium alloy seems brighter than ever.

REFERENCES

1. U.S. Patent No. 2,614,041, issued Oct. 14, 1952; W. L. Finlay inventor; Assigned to Rem-Cru Titanium Inc.

2. U.S. Patent No. 2,645,575, issued July 14, 1953; S. A. Herres and T. K. Redden inventors; assigned to Allegheny Ludlum Steel Corp.

3. U.S. Patent No. 2,718,465, issued Sept. 20, 1955; S. A. Herres and T. K. Redden inventors; assigned to Allegheny Ludlum Steel Corp.

4. U.S. Patent No. 2,740,711, issued Apr. 3, 1956; S. A. Herres and T. K. Redden inventors; assigned to Allegheny Ludlum Steel Corp.

5. U.S. Patent No. 2,739,887, issued May 27, 1956; J. O. Brittain and P. D. Frost inventors; assigned to Battelle Development Corp.

6. U.S. Patent No. 2,640,773, issued June 2, 1953; R. K. Pitler inventor; assigned to Allegheny Ludlum Steel Corp.

7. W. M. Parris and H. W. Rosenberg, "Producing Ti-13V-11Cr-3Al Mill Product at TMCA", Beta Titanium Alloys in the 1980's, ed. by Boyer and Rosenberg, published by TMS-AIME, (1984) 9-15.

8. R. R.Boyer and H.W. Rosenberg, "Beta Titanium on the SR-71", Ibid, (1-8).

9. U. S. Army Contract DA-30-069-ORD-3743.

10. Titanium Alloys Handbook, MCIC-HB-02, published by Battelle Metals and Ceramics Center, Dec. 1972, 1-20:72-17 through 1-20:72-19.

11. Beta Titanium Alloys, ed. by R. A. Wood, MCIC 72-11, Published by Battelle Metals and Ceramics Information Center, (1972) 168.

12. G. W. Kuhlman, Alcoa Green Letter #224, published by Alcoa, Aug. 1987.

13. T. W. Duerig, et.al., Metallurgical Transactions A. Vol. 11A, Dec. 1980, 1987-1998.

14. T. W. Duerig, et.al., Acta Metallurgical, Vol. 30, (1982) 2161-2171.

15. H. W. Stemme, "Development of a Formable Sheet Titanium Alloy", AFML-TR-73-49.

16. T. L. Wardlaw, et.al., "Development of An Economical Sheet Titanium Alloy", AFML-TR-73-296, Jan. 1974.

17. G. A. Lenning, et.al., "Cold Formable Titanium Sheet", AFML-TR-82-4174, Dec. 1982.

18. R. R. Boyer, this symposium.

19. P. J. Bania and W. M. Parris, "Beta-21S: A High Temperature Metastable Beta Titanium Alloy", Presented at the 7th World Conference on Titanium, San Diego, June 1992.

20. J. S. Grauman, "Beta-21S: A New High Strength Corrosion Resistant Titanium Alloy", Presented at the 7th World Conference on Titanium, San Diego, June 1992.

21. R. R. Boyer, "New Titanium Applications on the Boeing 777 Airplane", Journal of Metals, May 1992, 23-25.

22. P. J. Bania, et.al., "Ultra High Strength Titanium Alloy for Fasteners", Presented at the 7th World Conference on Titanium, San Diego, June 1992.

DEVELOPMENT AND APPLICATIONS OF BETA AND

NEAR BETA TITANIUM ALLOYS

A.Takemura*, H.Ohyama*, T.Abumiya**, T.Nishimura*

*   Iron and Steel Research Laboratories, Kobe Steel,LTD.
    2222-1,Ikeda,Onoe-cho,Kakogawa,Hyogo,675,JAPAN
** Titanium Technology Department, Kobe Steel,LTD.
    1-8-2,Marunouchi,Chiyoda-ku,Tokyo,100,JAPAN

Abstract

In this report we introduced application of beta and near beta titanium alloys ,also development and processing of these alloys at Kobe Steel LTD .
Ti-15Mo-5Zr-3Al is an alloy developed by Kobe Steel which has been applied for variety of sporting goods, also used as an erosion shield of steam turbine blades. Ti-15Mo-5Zr-3Al high strength wire for valve springs is under development.
New beta alloys(Ti-V-Nb-Sn-Al) are under developement which have lower flow stress at room temperature than Ti-15V-3Cr-3Sn-3Al, expected to improve productivity of cold forging.
NNS forging and thermo mechanical treatment of Ti-10V-2Fe-3Al were studied. Ti-10V-2Fe-3Al steam turbine blades and structural parts for aircraft were developed.
Fine grain cold strips of Ti-15V-3Cr-3Sn-3Al are produced by annealing and pickling process. These cold strips are used for parts of a fishing rod .

Beta Titanium Alloys in the 1990's
Edited by D. Eylon, R.R. Boyer and D.A. Koss
The Minerals, Metals & Materials Society, 1993

15

Introduction

Beta and near beta titanium alloys are most active field of research and development of titanium alloys and the production of these titanium alloys has been growing steadily in recent years in Japan.

The applications are expanding to many areas, for example, sporting goods, automobile use and aerospace use.

In this report we introduced a few research,development and application topics concerning beta and near beta titanium alloys at Kobe Steel Ltd.

As the examples of alloy development,Ti-Mo based high strength and corrosion resistant alloy (Ti-15Mo-5Zr-3Al) and low flow stress cold forgeable alloy(Ti-V-Nb-Sn-Al) are outlined.

As process development,NNS forging and thermo mechanical treatment of near beta alloy and grain size controle of beta alloy cold strip are illustrated.

Ti-15Mo-5Zr-3Al high strength corrosion resistant alloy

Ti-15Mo-5Zr-3Al, developed by Kobe Steel, has high strength and better corrosion resistance in nonoxidizing enviroments than commercially pure titanium (1).

Ti-Mo base alloys are expected to have superior corrosion resistance in nonoxidizing enviroments such as sulfuric acid or hydrochloric acid.

Ti-15Mo is used as a base alloy which allows beta stability sufficient to retain beta phase and also show good age hardenability. Zr minimizes the effects of the solution treatment cooling rate, while having little effect of suppression of alpha precipitation during aging (2). Figure 1 shows the effects of Zr content and the cooling rate after beta treating on the tensile properties of Ti-15Mo-x Zr alloys. When the alloys containing less than 5% Zr are slow-cooled from the beta region, hardening and embrittlement due to the decomposition of beta is remarkable,and when quenched, a low strength was shown because of martensitic strain transformation.

On the other hand,the mechanical properties of the alloys containing more than 5 % Zr show only a modest dependance on the rate of cooling from the beta region, and the alloys are stable mechanicaly and thermally.

Therefore Ti-15Mo-5Zr was considered as a candidate corrosion resistant alloy.

Figure 2 shows the relative intensity of the X-ray diffraction of the omega and alpha phases during aging for Ti-15Mo-5Zr. Figure 3 shows the isochronal curve of tensile properties at various aging temperatures for Ti-15Mo-5Zr (1) . In the Ti-15Mo-5Zr the maximum age hardening and the minimum value in ductility are observed concurrently. It is found from the X-ray diffraction results that these phenomena are related to the omega phase precipitation. The alpha phase precipitation led to a decrease in tensile strength and an increase in ductility.Therefore,the mechanical properties are controlled by the omega and the alpha in the Ti-15Mo-5Zr.But omega is thermally unstable and is not appropriate as a strengthening phase.

Al suppresses omega formation and solution hardenes alpha precipitation with which age hardenability is enhanced.

Figure 4 shows the relative intensity of X-ray diffraction peaks of the omega and alpha phase during aging for Ti-15Mo-5Zr-3Al. Figure 5 shows the isochronal curve of tensile properties at various aging temperatures for Ti-15Mo-5Zr-3Al (1). In this alloy the omega phase precipitation during aging is suppressed by Al addition and the age hardening is related to the amount of the alpha phase, strengthened by Al. Therefore, the mechanical properties of the Ti-15Mo-5Zr-3Al are controlled by the alpha. This alloy has a good combination of strength and ductility, and has better stability than Ti-15Mo-5Zr.

Ti-15Mo-5Zr-3Al high strength wire for valve springs is under development. Its mechanical properties are tensile strength 180kgf/mm^2 (1760 MP), elongation 5%, fatigue strength 70 kgf/mm^2(680MPa).Figure 6 shows a prototype of the valve spring (with a courtesy of NHK Spring Co Ltd.).

Ti-15Mo-5Zr-3Al is used as an erosion shield for steam turbine blades,because it has better erosion resistance than Ti-6Al-4V. Figure 7 shows a Ti-6Al-4Vsteam turbine blade with a Ti-15Mo-5Zr-3Al erosion shield on the blade edge. These turbine blades are used in the last stage of steam turbine in Japanease power plant.Ti-15Mo-5Zr-3Al also has been applied to a variety of sporting goods, such as fishing equipment, diver knives and golf club shafts.

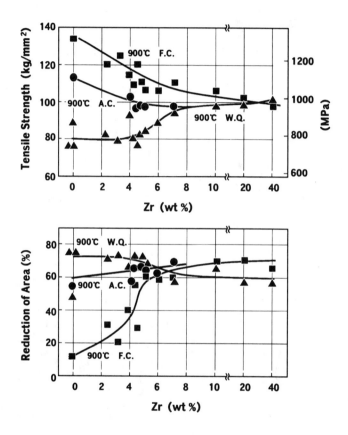

Fig.1 Effect of Zr content and cooling rate
after beta treating on the tensile properties
of Ti-15Mo-Zr alloy

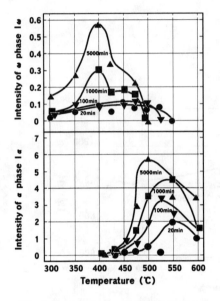

Fig.2 Amount of ω and α phases
during aging treatment of
Ti-15Mo-5Zr.

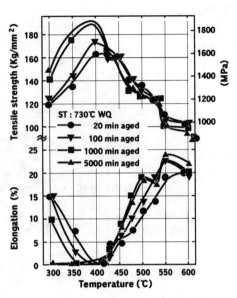

Fig.3 Isochronal curve of tensile
properties at various aging
temperatures for Ti-15Mo-5Zr.

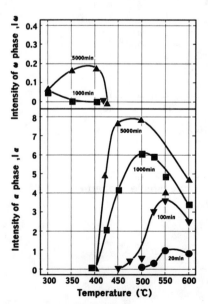

Fig.4 Amount of ω and α phases
during aging treatment of
Ti-15Mo-5Zr-3Al.

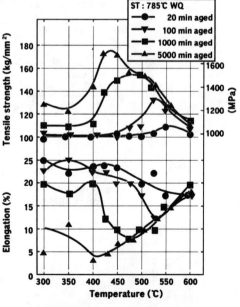

Fig.5 Isochronal curve of tensile
properties at various aging
temperatures for Ti-15Mo-5Zr-3Al.

Fig.6   Beta Titanium Alloy Valve Springs.

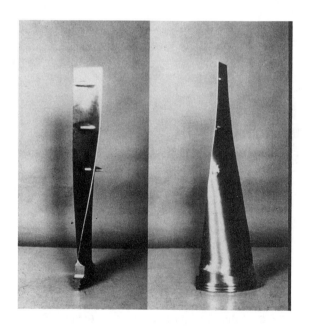

Fig.7   Ti-6Al-4V 40 inch Turbine Blade with
Ti-15Mo-5Zr-3Al Erosion Shield.

Ti-16V-4Sn-3Al-3Nb cold forgeable alloy

It is important to reduce the cost of secondary working of titanium alloys by the user, as it is a large share of the final product cost. Cold forging, a very efficient process, is applied as secondary working to steel, copper and aluminum products. But the flow stress of the current beta titanium alloy, such as Ti-15V-3Sn-3Al-3Cr or BetaIII, is not low enough to be efficiently cold forged.   A high flow stress requires a large die forging press and causes heavy die wear.
A new beta alloy is required which has low flow stress, good workability at room temperature, and is suitable for cold working .
The effects of the alloying elements on flow stress were investigated (3)(4). It was revealed that the addition of Nb led to a new alloy, Ti-16V-4Sn-3Al-3Nb now under development , which has 10% lower flow stress than Ti-15V-3Sn-3Al-3Cr.   Figure 8 shows a comparison of the cold forging flow stress of Ti-16V-4Sn-3Al-3Nb and Ti-15V-3Sn-3Al-3Cr. The flow stress of Ti-16V-4Sn-3Al-3Nb is lower than that of Ti-15V-3Sn-3Al-3Cr throughout the tested strain range. Figure 9 shows cold forged bolts of Ti-16V-4Sn-3Al-3Nb in an as forged condition. No crack or any other defect was observed.This alloy has improved the productivity of cold forging and will be applied to bolts, nuts and valve retainers for automobile engines.

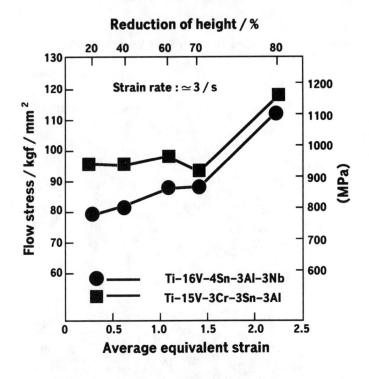

Fig.8   Relationship between upset forging ratio and flow
stress of Ti-16V-4Sn-3Al-3Nb

Fig.9 Ti-16V-4Sn-3Al-3Nb. Cold Forged Bolts.

NNS forging of Ti-10V-2Fe-3Al

The application of Ti-6Al-4V for steam turbine blades has increased in recent years. While Ti-6Al-4V is a widely used and reliable titanium alloy,higher strength and better forgeability are required to improve the performance and cost efficiency of turbine blades. Ti-10V-2Fe-3Al near beta titanium alloy,known to have a high strength per weight ratio and a low hot flow stress,are expected to replace Ti-6Al-4V in aerospace use.A 20-inch Ti-10V-2Fe-3Al near net shape model blade was made by isothermal forging.
Billet forging is done in the alpha + beta region to obtain forging stocks with uniform structure. It is expected that fluctuation in volume fraction and morphology of primary-alpha are relatively small in the case of alpha + beta forging. Although the alpha + beta forging structure which consists of fine equiaxed alpha grain,has lower fracture toughness than beta forged structure,it is possible to improve the combination of fracture toughness and strength by solution treatment near the beta transus temperature(5). Figure 10 shows an increase in transformed beta volume fraction with solution treatment temperature.   Improvement of fracture toughness and strength is expected from this structural change, because transformed beta has relatively high fracture toughness and strength. The small volume fraction of primary alpha prevents corsening of beta subgrains,and minimizes the loss of ductility .
Figure 11 shows a coarsening of precipitated alpha in transformed beta with an increase in solution treatment temperature. Improvement of fracture toughness is also expected from this structural change .
Figure 12 shows an increase in strength and fracture toughness with an increase in solution treatment temperature.This high temperature solution treatment can't be applied unless beta transus is uniform throughout the material. It is necessary for the ingot to be uniform in composition, especialy the iron content. The melting condition is carefully controlled so that no beta flecks are formed even if solution treatment is done near the beta transus temperature.
Figure 13 shows the isothermally forged Ti-10V-2Fe-3Al near net shape model blade.This blade is 500mm (20inch) in length (6). Near net shape blades were formed in a single heat cycle of isothermal forging. No crack or any other defect was observed after shot and pickling. This blade has uniform mechanical properties and micro structure from root to airfoil.This research and development was performed under the sponsorship of the Advanced Nuclear Equipment Research Institute (ANERI) under contract with the Agency of Natural Resources and Energy, Ministry of International Trade and Industry, Japan .

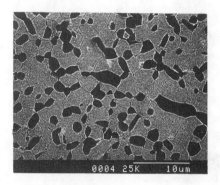 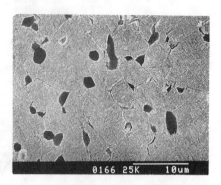

ST Temp. 1013K   **ST Temp. 1053K**

Fig.10 Changes in P-α Volume Fraction in Ti-10V-2Fe-3Al with
Solution Treatment Temperature.

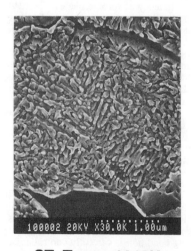 

ST Temp. 1013K   **ST Temp. 1053K**

Fig.11 Changes in Precipitated α Morphology in Ti-10V-2Fe-3Al
with Solution Treatment Temperature.

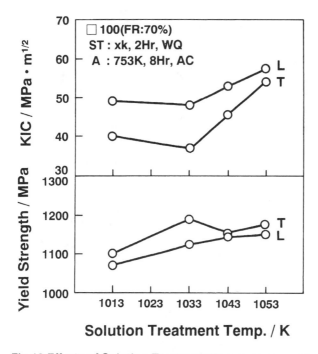

Fig.12 Effects of Solution Treatment Temperature on Fracture Toughness and Yield Strength of Ti-10V-2Fe-3Al

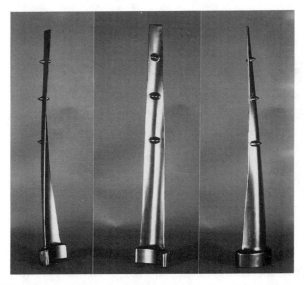

Fig.13 Isothermaly forged 500mm (20inch) Ti-10V-2Fe-3Al model blade.

Fine grain cold strip of Ti-15V-3Sn-3Al-3Cr

As for sheet products of titanium alloys, cold strips of beta titanium alloys are the most cost-effective in terms of both primary and secondary fabrication costs. However, inappropriate heat treatments often lead to grain coasening and result in problems such as orange peel after bending,and low ductility. In order to get good cold formability, it is very important to control the grain size after annealing. For producing fine-grained and fully recrystallized beta titanium sheets, it is necessary to take into account the effect of heating time on recrystallization and grain growth behaviors.
We have investigated the behavior of the recrystallization and the grain growth of a cold-rolled Ti-15V-3Sn-3Al-3Cr and proposed a method for estimating the grain size after recrystallization during a continuous annealing process(7). Figure14 shows a schematic concept of the model for grain growth estimation when the material temperature is continuously changing.The estimation of the grain size recrystallized under continuous annealing has been made by sequentially integrating,step by step,each increment of the isothermal grain growth at each temperature. Figure 15 shows the correlation between grain size estimated by the model and the measured grain size when the 80% cold rolled materials are heated from room temperature to 1073K at various rates.The estimation of the grain size recrystallized under a continuous heating process is in good agreement with the actual grain size.
Figure 16 shows Ti-15V-3Sn-3Al-3Cr cold strip produced in an annealing and pickling process. The thickness is 1.0mm and the width is 950mm . The grain size of this cold strip is almost the same as estimated ,that is, 21 μm. Figure17 shows parts of a fishing rod (with a courtesy of Fuji Kogyo Co.,Ltd.).The surface of these parts is very smooth because of fine grain size.

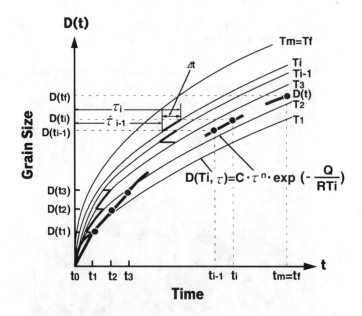

Fig.14 Schematic concept of the model for grain growth estimation when the material temperature is continuously changing.

24

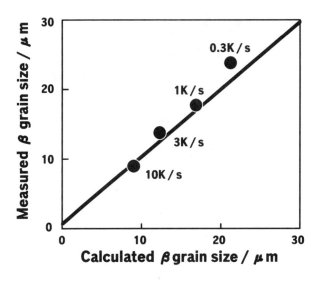

Fig.15 Graphic correlation between grain size estimated by the model and measured grain size when the 80% cold rolled materials are heated from room　temperature to 1073K at various rates.

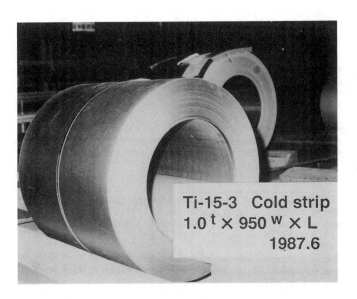

Ti-15-3　Cold strip
1.0^t × 950^w × L
1987.6

Fig.16 Ti-15V-3Cr-3Sn-3Al cold strip produced in an annealing and pickling process.

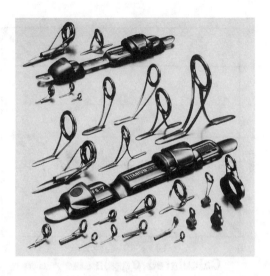

**Fig.17 Beta Titanium Alloy Parts of Fishing Rod.
(with a courtesy of Fuji Kogyo Co., Ltd.)**

Summary

Our experiences mentioned above and in other instances, confirm the beta and near beta titanium alloys'advantages in properties and fabrication. But,it is necessary to study the characteristics of transformation under dynamic condition for better control of thermo mechanical treatment.
New alloy design is also required which has wider range of permissible processing conditions suitable for commercial production.
Although beta and near beta titanium alloys still have many problems to solve,they will be applied for a wider range of promising applications than Ti-6Al-4V.We are convinced that beta and near beta titanium alloys are becoming more available and indispensable .

References

1)T.Nishimura, M.Nishigaki and S.Ohtani: J.Japan Inst.Metals, 40(1976),219.
2)S.Ohtani and M.Nishigaki: J.Japan Inst.Metals, 35(1971),97.
3)H.Ohyama,A.Takemura,T.Nishimura and Y.Ashida: CAMP-ISIJ, 5(1992),761.
4)T.Abumiya,K.Yasui,K.Suzuki,H.Ohyama and H.Nishimoto:R&D Kobe Steel Engineering Reports, 43(1992),64.
5)A.Takemura,T.Nishimura,S.Ishigai and T.Matsumoto: CAMP-ISIJ, 5(1992),759.
6)A.Takemura,Y.Ashida,H.Ohyama,A.Morita,S,Hattori,K.Tani,A.Hasegawa and T.Matsumoto: Proc.7th Int.Conf. on Titanium, in press.
7)H.Ohyama and Y.Ashida: ISIJ Int., 31(1991),799.

PROCESSING, PROPERTIES AND APPLICATIONS

OF THE ß-CEZ ALLOY

Y. Combres and B. Champin

Centre de Recherches de CEZUS
BP 33
73400 Ugine Cedex France

ABSTRACT

The desire to acheive a good strength-toughness-medium temperature creep resistance balance on titanium products provided the driving force for the development of a new alloy in CEZUS. Besides its intrinsic mechanical properties, multifunctionality of the alloy as well as enhanced formability has been sought from the beginning. It was thought that good combinations of properties associated with easy processing would get sustained interest from industries, where the demand for improved materials is still strong, such as the aeronautic and aerospace fields. This led to the design of the ß-CEZ alloy at the end of the 80's. This alloy can be defined as a multipurpose alloy exhibiting high strengh and toughness and intermediate temperature creep resistance. Its large processing flexibility makes it suitable for a wide range of applications.

The paper will first review the alloy design background and history and present some of the thermal properties. Then, the microstructures after heat treatment (solution treatment and/or ageing, continuous cooling...) will be detailed. The fabrication characteristics during hot working, either on the producer practice (ß, or α+ß, or "through the transus" conversions...), or the user viewpoint (ß processing, "trough the transus" forging, superplastic forming...), will be then discussed. At last, the relevant mechanical properties (tensile, toughness, fatigue, creep...) on semi finished products (forged and rolled bars, plates and sheets), and those of final products (heavy section forgings, fasteners...) will be presented.

The authors acknowledge with gratitude the french DoD (DGA/DRET-STPAé) financial support, for the ß-CEZ alloy development. They also thank all the CEZUS researchers who contributed to this work over the last past decade : Mrs Rizzi, Gigante, Chittaro, Prandi, Dumas and Grandemange. They would like also to underline the work of Mr. Pichol, who mastered its technical development, from the elaboration to the final products, and gathered and compiled most of the data available to date. Fruitfull discussions with industrial partners : SNECMA, Aérospatiale, Boeing, SEP, Fortech, GE... have been appreciated.

Beta Titanium Alloys in the 1990's
Edited by D. Eylon, R.R. Boyer and D.A. Koss
The Minerals, Metals & Materials Society, 1993

INTRODUCTION

Almost ten years elapsed since the last symposium on ß titanium alloys held by the US Titanium Committee of AIME (Annual Meeting of TMS, in Atlanta, Georgia). This conference focused on the state of art of the alloys available at this period : Ti-13V-11Cr-3Al [1], ß III [2], Ti-15-3 [3-5], ßC (Ti-38644) [6,7], Ti 17 [8]... From that time, design requirements have become more stringent. Higher properties are saught to improve the titanium parts efficiency, and, nowadays, final price considerations motivate series of studies such as near net shape forming for instance. The use of ß titanium alloys provide an answer in these fields.

Over the last past five years, the development of ß metastable titanium alloys has then been stressed by the need of ever improved materials, either on the viewpoint of strength, toughness, creep resistance, formability, or corrosion resistance. Some of the newly designed ß metastable or ß rich alloys are refered to as SP 700 [9,10], a high strength-low temperature superplastic material, or TIMETAL 21S [11], a high strength-corrosion resistant alloy, or ß-CEZ [12,13], a high strength-high toughness-medium temperature creep resistant grade developed by CEZUS.

First, the alloy design background and history will be reviewed as well as some of the thermal properties. In a second part, the characteristics of the alloy, which may be attractive, will be be described such as :

> - the variety of microstructures after heat treatment (solution treatment and/or ageing, continuous cooling...) or thermo-mechanical treatment,
> - the relevant mechanical properties (tensile, toughness, fatigue, creep...) on semi finished products (forged and rolled bars, plates and sheets), and those of final products (heavy section forgings, fasteners...),
> - the fabrication characteristics during hot working, either on the producer practice (ß, or α+ß, or "through the transus" conversions...), or the user viewpoint (ß processing, "through the transus" forging, superplastic forming...).

The last part of this paper will be devoted to the applications that are considered for this type of alloy.

BACKGROUND OF THE DEVELOPMENT AND ALLOY DESIGN

Development outlines

In the early 70's, among several titanium compositions developed by CEZUS, the Zr modified 662 alloy (labeled hereinafter 662-Zr in short) had been designed to provide an improvement over the conventional Ti-64 and 662 grades [14-16]. It was the result of an investigation dealing with thirty experimental alloys. The aim of the program was a material with UTS > 1300 MPa, 0.2%YS > 1150 MPa, with sufficient ductility : El > 6% and R.A. > 15%, and notch insensitivity at this level of strength : TS(notched with kt = 3.8)/UTS(unnotched) > 1. It has been found that, with 6% of Zr addition, i.e. the composition of the alloy being Ti-6%Al-6%V-2%Sn-6%Zr, this target was reached (UTS =1280-1380 MPa, 0.2%YS = 1175-1350 MPa, El = 8-14%, R.A. = 20-44%, TS(kt = 3.8)/UTS(kt = 1) = 1.4). For similar ductility (10 %) and toughness (50 MPa√m), the 662-Zr was 100 MPa (or 200 MPa) stronger than 662 (or Ti-64). In 1983, the need of improved material for compressor disks at intermediate temperatures has emerged with engine makers. It was thought that, based on its 662-Zr experience, CEZUS could respond favourably to this challenge, keeping in mind the high strength and creep resistance at medium temperatures (as for instance in the Ti 17) [17].

Another concept was also to obtain strength-ductility-toughness balance through microstructure control [17] ; this can be readily achieved with ß metastable alloys, which exhibit several variants of morphologies and fineness of structures, through heat treatment response (such as 10-2-3 for instance).

Alloy design

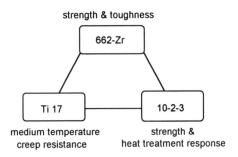

Figure 1 - Alloy design philosophy after [17]

Several guidelines were followed for the choice of the alloy composition :

    ① limitation of the exploration to the known alloy systems (fig. 1):
        - V strengthened alloys (such as 662-Zr),
        - Mo strengthened alloys (such as Ti 17),
        - ß rich alloys (such as 10-2-3) ;
    ② definition of Aleq-Moeq range where :
        - density < 4.7 g/cm^3 (to get high specific properties),
        - low ß transus or Tß (to increase the dies life by reducing
         the forming temperature),
        - not too much ß (toughness and creep resistance) ;
    ③ replacement of V by Mo :
        - to increase heat treatment response,
        - to prevent the oxydation [18].

An evaluation of 28 different compositions tested on 80 kg small laboratory ingots in tensile, creep and fracture toughness was performed. The following ranges were investigated (table I).

Element	Al	Sn	Zr	Mo	V	Fe	Cr	Si
min	4	0	0	2	0	0	0	0
max	6	3	5	4	4	2	4	0.4

compositions in wt %

Table I - Ranges of compositions investigated for the alloy design

It has been found that [19]:
- Al should be kept in the 4-5 % range to get maximum hardening effect without impairing ductility, and to prevent massive Ti3Al precipitation after long time exposure at elevated temperatures ;
- 2-3 % Sn and 3-5 % Zr are necessary for creep strength ;
- V has been ruled out to improve oxydation resistance ;

- addition of Mo improves both tensile properties and creep and the content has been set to the maximum permitted value which accounts of density limitations, i.e. 4 % ;
- 1 % Fe is equivalent to 2 % Cr with respect to tensile and creep strengths ; reduction of ß transus, Tß, and heat treatment response is by far stronger in the case of Fe additions ; however, it has been limited to 1 % to avoid ß flecks ;
This led to the definition of the ß-CEZ composition (table II) :

Element	wt %	effect
Al	5	high strength
Sn	2	high strength
Zr	4	strength + creep resistance
Mo	4	ß phase stabilizer
Cr	2	creep strength + ß stabilizer
Fe	1	creep strength + strong ß stabilizer
O	800-1200 ppm	hardness/toughness compromise

Table II - ß-CEZ composition

ß-CEZ is a ß metastable titanium alloy with 18 % addition elements. Mo, Cr and Fe strongly stabilize the ß phase insuring a large heat treatment response and improved formability. For this composition, it should be noticed that Si has no effect on creep resistance, since the silicides solvus appears to be above Tß [20].

Thermal properties

With this composition, the alloy exhibits the following thermal properties (table III) :

Tliquidus	Tsolidus	Tß	density	thermal expension*	specific heat*	thermal conductivity*	emissivity*
1602 °C	1552 °C	890 °C	4.69 g/cm³	9 µm/m/K	580 J/kg/K	6.7 W/m/K	0.7

* thermal properties at 20 °C

Table III - Thermal properties of ß-CEZ alloy

Product development

Industrial scale up took advantage of the modelling of fusion and ingot conversion. Especially, the best combinations of melting rate and stirring mode have been chosen according to theory [21]. Both triple VAR (Vacuum Arc Remelting) or VAR+EBCHR(Electron Beam Cold Earth Remelting)+VAR melting routes have been tested each of them having its limitations with respect of inclusion elimination or chemical composition control [22]. Ø 660 mm (dia 26") ingots are converted using a 2500 ton fully computerized forging press facility, which insures the microstructure homogeneity along the bars length and in the cross section and the reproducibility from one batch to another one. The alloy can be easily hot forged in bars or hot rolled in bars or plates.

PRINCIPAL CHARACTERISTICS OF THE ß-CEZ ALLOY

Besides the alloy design, the mechanical properties are strongly dependant on the microstructures. For ß-CEZ, several microstructure sizes and morphologies allow covering a wide choice of corresponding properties. These structural characteristics can be obtained through appropriate thermomechanical and heat treatments. In this part, the different type of microstructures will be first reviewed. Then the mechanical properties associated with them will be presented. At last, the forming processes which allow to get them will be detailed.

Varieties of morphologies and sizes

ß-CEZ microstructures consist of primary alpha (αI) present at high temperatures during thermomechanical treatment or solution treatment (ST), secondary alpha (αII), which precipitates out upon ageing (A), stable ß (ßs) present at high temperatures, metastable ß (ßm) retained at room temperature by rapid quenching and transformed ß (ßt), a mixture of α+ß due to slow decomposition of high temperature ß phase.

The morpohologies can be divided depending on the αI shape. They can be referred to as equiaxed, or lamellar, or necklaced microstructures (fig. 2).

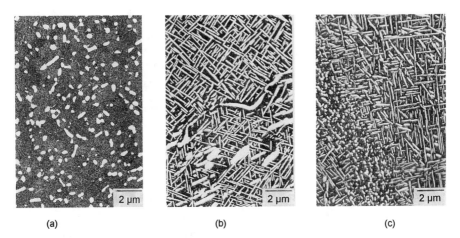

(a)   (b)   (c)

Figure 2 - Different types of morphologies : (a) equiaxed, (b) lamellar, (c) necklaced

To quantify these microstructures, Image Analysis techniques have been developed [23]. Both αI and αII sizes can be determined. Table IV show the results after ST+A treatments on samples from a Ø 150 mm (dia. 6") forged bar, and the wide range of αII size.

Treatment	αI size (μm)	αII size (μm)	%αI	% ß
830 °C/2h/WQ	3.2	-	25	75
830 °C/2h/WQ + 550 °C/8H/AC	3.2	0.03	25	19
830 °C/2h/WQ + 600 °C/8H/AC	3.2	0.08	25	27
830 °C/2h/WQ + 650 °C/8H/AC	3.2	0.16	25	36
860 °C/2h/WQ	3.1	-	4	96
860 °C/2h/WQ + 550 °C/8H/AC	3.1	0.04	4	21
860 °C/2h/WQ + 600 °C/8H/AC	3.1	0.07	4	31
860 °C/2h/WQ + 650 °C/8H/AC	3.1	0.19	4	45
910 °C/2h/WQ	-	-	0	100
910 °C/2h/WQ + 650 °C/8H/AC	-	0.18	0	32

Table IV - Microstructure quantification after ST+A treatments on samples from a Ø 150 mm (dia. 6") forged bar.

Figure 3 presents the CCT diagram from the ß field. The behaviour of ß-CEZ is typical of that kind of alloy (10-2-3, Ti-15-3...). a precipitation occurs, at first, at the ß grain boundaries (GB)

and, secondly, inside the grains. For instance, at 1 °C/min continuous cooling rate, the time difference between grain boundary and intragranular precipitation is about 1 h.

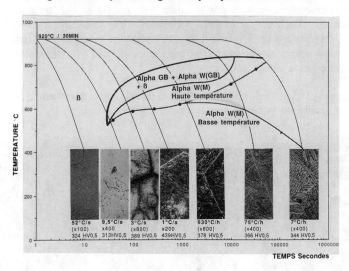

Figure 3 - CCT diagram from ß field (920 °C) ; Haute≡high, basse≡low.

TTT diagram for specimens heat treated above Tß presents three regions : for Tß > T > 700°C, a rather coarse α precipitation occurs (this is the range of solution treatments and αI precipitation), for 700 °C > T > 450 °C, very fine αII particles precipitate (this is the corresponding domain of ageing treatments), for T < 400 °C, isothermal ω phase precipitates out. It should be noticed that when the stability of ß phase is increased (i.e. the solution treatment temperature is lowered), α or ω phases precipitations are delayed and, the ω stability domain is reduced (T < 200 °C) [24]. On the other hand, neither athermal or isothermal ω precipitation is to be feared in the conventional industrial heating or cooling rates [25].

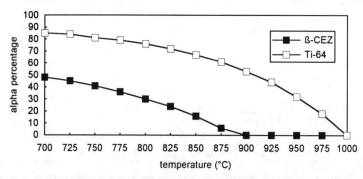

Figure 4 - α phase percentage variations vs. T in the case of ß-CEZ and Ti-64

In figure 4, α percentage variations with temperature is given. It is compared to that of Ti-64 alloy. When cooled from the ß range, α precipitation rate appears to be lower in the case of ß-CEZ, certainly because of its ß metastable nature. The maximum content is about 20-30 % at room temperature. It should be noticed that the temperature corresponding to 50 % α-50 % ß is nearly 700 °C, and, then, 200 °C below the one of Ti-64.

Mechanical properties

Product	Heat treatment	UTS (MPa)	0.2% YS (MPa)	El (%)
	as forged	<u>1040</u>	<u>960</u>	<u>18</u>
Ø 150 mm	830 °C/1 h/WQ + 550 °C/8 h/AC	1601	1518	2
forged bar	830 °C/1 h/WQ + 600 °C/8 h/AC	1283	1208	11
(dia. 6")	860 °C/1 h/WQ + 550 °C/8 h/AC	1557	1478	2
	860 °C/1 h/WQ + 600 °C/8 h/AC	1370	1304	5
	as rolled	<u>1490</u>	<u>1345</u>	<u>11</u>
Ø 12.8 mm	830 °C/1 h/WQ + 550 °C/8 h/AC	<u>1506</u>	<u>1460</u>	<u>13</u>
rolled bar	830 °C/1 h/WQ + 600 °C/8 h/AC	1373	1349	15
(dia. 0.5")	860 °C/1 h/WQ + 550 °C/8 h/AC	<u>1723</u>	<u>1683</u>	<u>7</u>
	860 °C/1 h/WQ + 600 °C/8 h/AC	1540	1485	9
	as rolled L	1222	1124	15
	as rolled T	1260	1163	11
25^t mm	830 °C/1 h/WQ+600 °C/8 h/AC L	1334	1287	13
rolled plate	830 °C/1 h/WQ+600 °C/8 h/AC T	1351	1300	12
(1" thick)	860 °C/1 h/WQ+600 °C/8 h/AC L	<u>1405</u>	<u>1338</u>	<u>10</u>
	860 °C/1 h/WQ+600 °C/8 h/AC T	1418	1340	6
Ø 300 mm	600 °C/8 h/AC	1608	1472	2
ß processed	830 °C/1 h/WQ + 570 °C/8 h/AC	1357	1171	5
pancake (dia. 12")	830 °C/1 h/WQ + 600 °C/8 h/AC	1326	1188	6
Ø 300 mm	600 °C/8 h/AC	1227	1138	10
through the	830 °C/1 h/WQ + 570 °C/8 h/AC	<u>1314</u>	<u>1200</u>	<u>10</u>
transus proc. (dia. 12")	830 °C/1 h/WQ + 600 °C/8 h/AC	1263	1170	11

Table V - Selection of tensile properties

In table V, a selection of some tensile properties is given. Forging and rolling has been performed at 850 °C, whereas ß process and through the transus process took place at 920 °C. Some data are underlined which enlighten the range of properties that can be get with ß-CEZ. In figure 5, the strength-ductility-toughness balance is shown for both ß-CEZ and Ti-64, either for equiaxed or lamellar/necklaced microstructures.

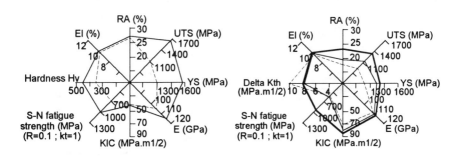

(a)   (b)

Figure 5 - Strength-ductility-toughness balance : (a) for 3-5 µm equiaxed morphology, (b) for lamellar/necklaced morphologies ; broken line is Ti-64 (equiaxed or lamellar) ; solid line is ß-CEZ (equiaxed or lamellar) ; strong solid line is ß-CEZ necklaced microstructure.

It can be seen that ß-CEZ provides a better compromise than Ti-64. In figure 6, the strength-ductility-toughness-creep balance is illustrated by comparison with Ti 17.

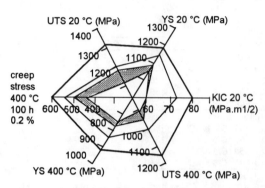

Figure 6 - Strength-ductility-toughness-creep balance solid line is lamellar ß-CEZ ; strong solid line is necklaced ß-CEZ ; greyed zone is Ti 17 (literature data).

Here again, there is a better properties combination in the case of ß-CEZ. It can be said that, for 10 % elongation :
- for structural applications, the alloy is 200-400 MPa stronger than Ti-64 depending on the morphology ; the equiaxed microstructures exhibit a fatigue limit 300 MPa higher ; the necklaced microstructures give also interesting balance of properties with high fatigue limit and toughness ;
- for high temperatures applications, the alloy keeps its high strength until 400 °C ; ß-CEZ is 100 MPa stronger than Ti 17, and creep stress is 20 % larger at 400 °C.

Formability

As said in the beginning of this paper, ß-CEZ offers attractive forming characteristics due to the choice and level of all the ß stabilizers. Figure 7 presents the formability characteristics of the alloy. One can find the rheological parameters, such as the strain rate sensitivity exponent m, which characterizes the material flow stability during hot working, and the apparent activation energies Q, which measure the temperature dependance of the material flow stress. Compared with Ti-64 :

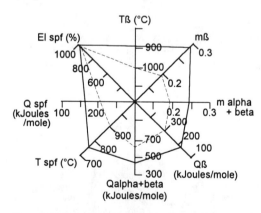

Figure 7 - Characteristics of the formability of ß-CEZ (solid line) at elevated temperature (T>700 °C) ; comparison with Ti-64 (broken line).

- the ß transus is 100 °C lower and the superplastic temperature is 150-200 °C lower too ; dies for hot forging or superplastic forming can be then heated 100 °C below the conventional temperature, and their life increase ;
- whatever the temperature range (ß or α+ß), the strain rate sensitivity exponent is larger, insuring stability and preventing flow localization ; in the case of the superplastic range, its value is about 0.5 (for a 3 μm a grain size) ;
- irrespective of the phase field (ß or α+ß), the activation energies are lower, making the alloy less sensitive to temperature variations during the process ; this also applies in the case of superplasticity.

By hot working, the α grains can be more easily refined. Especially, compared to Ti-64, the microstructures stability is enhanced during hot working. Series of deformation-reheating cycles can be performed without degradation of the microstructure by grain coarsening during reheating. Figure 8 shows both ß and α grain growth kinetics.

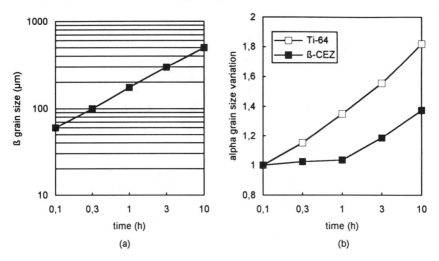

(a) (b)

Figure 8 - Grain size variations : (a) in the ß field at 920 °C, ß grain coarsening ; (b) in the α+ß field, comparison of the α grain size variations for heat treatment at Tß-40 °C between ß-CEZ and Ti-64.

Classically, equiaxed microstructures are obtained by hot forming below Tß or by a series of cold working/recrystallization annealing. Lamellar morphologies are produced by ß process ; equiaxed+lamellar structures can be achieved by processing just below Tß and by controlling the cooling rate in order to get the desired lamellae thickness. The special necklaced microstructure, which exhibits such good combinations of mechanical properties, is obtained by working the material upon cooling from the ß field (labeled through the transus, "tTß" process).

APPLICATIONS

Types of applications

Over the past 3 years, several applications have been considered for ß-CEZ. They concern mostly the aeronautic and space fields. They can be divided into propulsion and structure.

In the case of propulsion, medium temperature strength-ductility-toughness-creep balance is attractive for compressor disk at 400 °C [26]. The strength-toughness-fatigue compromise is used at lower temperatures for impellers or blisks ; the fabrication of wide cord hollow fan blades is of interest, which associates the high mechanical properties with the good formability of the alloy. Figure 9 shows a forged disk preform.

In the case of structure, the specific properties are being evaluated in the case of landing gears, rotor hub shaft for helicopters, rudder axis for missiles. The strength-ductility-fatigue balance is used for fasteners (class UTS=1500 MPa-El=15 %). In the table VI, the specific tensile properties on massive forged parts are presented and a comparison is made between 300 M steel, 7010 aluminum alloy, and other titanium alloys (Ti-64, 10-2-3) [27] ; ß-CEZ data are added. Comparison is made for 4 % elongation and 10 % elongation. Of course, because of their strength and density, the titanium alloys offer significant improvement of the specific properties. Among them, and especially compared with 10-2-3, ß-CEZ gives the best strength (at least +30 MPa.dm³/kg). It should be noticed that, with the necklaced microstructures, the ductility has been found always larger than 6 %.

Figure 9 - Disk preform forged by SNECMA. Courtesy of SNECMA.

Alloy	density d (kg/dm³)	Specific tensile properties El≈4%				KIc (MPa√m)	fatigue lim.
		UTS/d (MPa/d)	YS/d (MPa/d)	El (%)	E/d (GPa/d)		σ_lim/d (MPa/d) #
10-2-3 @	4.65	258	237	4	23.7	44	127
ß-CEZ ß §		288	253	3.85	24.1	90	-
" tTß §§	4.69	305	294	6	< 24	55	-
" α+ß §§§		292	278	4	23.4	40	-
300 M	7.83	246	203	4	26.1	45	102
7010	2.80	171	150	4	25.4	22	60.7
Alloy	density d (kg/dm³)	Specific tensile properties El≈10%				KIc (MPa√m)	fatigue lim.
		UTS/d (MPa/d)	YS/d (MPa/d)	El (%)	E/d (GPa/d)		σ_lim/d (MPa/d)
TA6V @@	4.45	202	187	10	25.8	50	110 #
ß-CEZ ß *		244	232	10.7	23.5	90	-
" tTß *	4.69	280	256	10.1	24.3	90	170 ##
" α+ß *		273	258	8	23.8	50	170 ##

@ ST+A : 760 °C (Tß-30)/1 h/WQ + 520 °C (Tß-290)/8 h/AC
§ ST+A : 830 °C (Tß-60)/1 h/WQ + 570 °C (Tß-320)/8 h/AC
§§ trough transus forging octo 80 mm ; ST+A : 875 °C (Tß-15)/1 h/WQ + 580 °C (Tß-310)/8 h/AC
§§§ α+ß process Ø 150 mm bar ; ST+A : 830 °C (Tß-60)/1 h/WQ + 550 °C (Tß-340)/8 h/AC
# R = -1, kt = 1.035 et N = 10⁷ cycles
@@ ST+Annealing : 940 °C (Tß-50)/1 h/WQ + 730 °C (Tß-260)/2 h/AC
*ß process pancake ; ST+A : 830 °C (Tß-60)/1 h/WQ + 570 °C (Tß-320)/8 h/AC
** trough transus forging pancake ; ST+A : 830 °C (Tß-60)/1 h/WQ + 570 °C (Tß-320)/8 h/AC
*** α+ß process barre Ø 150 mm ; ST+A : 830 °C (Tß-60)/1 h/TE + 600 °C (Tß-290)/8 h/TA
R = 0.1, kt = 1.035 et N = 10⁵-10⁷ cycles

Table VI - Comparison of the specific properties of ß-CEZ with other alloys

Effect of temperature in service

In table VII, it is shown that ß-CEZ alloy behaves rather well at cryogenic temperatures.

Alloy	Treatment	Temperature (K)	UTS (MPa)	0.2%YS (MPa)	EI (%)
ß-CEZ	None	300	1030	950	18
		77	1680	1560	7
		20	1915	1780	5
	860°C/1h/WQ + 600°C/8h/AC	300	1325	1280	5
		20	2200	2100	2
IMI 834		300	1060	940	12
		20	1800	1530	5
Ti-6-4	730°C/2h/AC	300	960	920	13
		77	1490	1425	10
		20	1610	1560	10
10-2-3	760°C /1h/WQ + 520°C/8h/AC	300	1170	1110	10
		77	1860	1785	4
		20	2110	2090	1

Table VII - Effect of cryogenic temperatures on ß-CEZ and other titanium alloys

ß-CEZ is also likely to keep its strength and ductility balance at elevated temperature. For long time exposure, there is no embrittlement due to Ti_3Al precipitation for instance [20]. These results are presented in table VIII.

Temp. (°C)	State (h/MPa)	E (GPa)	0.2%YS (MPa)	UTS (MPa)	EI (%)	RA (%)	KIc (MPa√m)
R.T. (ref.)	-	123	1195	1270	15	23	58
425	300/-	127	1210	1280	14	22	56
	1000/-	122	1230	1300	13	18	53
	1000/250	126	1185	1265	12	19	-
	5000/-	125	1233	1277	14	22	-
475	100/-	127	1210	1270	14	27	57
	100/250	123	1200	1260	11	17	-
	500/-	124	1226	1274	14	25	-

Table VIII - Effect of long time exposure at elevated temperature on ß-CEZ properties with and without applied stress.

CONCLUSIONS

As a matter of conclusion, it can be said that ß-CEZ :
- ♦ has a high level of strength-ductility-toughness balance and an excellent creep resistance up to 450 °C,
- ♦ has enhanced hot working, low temperature superplastic behaviour, and enlarged processing window,
- ♦ exhibits different microstructures that can be obtained through thermomechanical properties and heat treatment.

Then, the ß-CEZ alloy has a very good prospect for applications :
- in first stages of compressor disks,
- in structural parts for aircraft, aerospace, automotive industry...(as landing gear, rotors, impellers, fasteners, spring...)

References

[1] W.M. Parris and H.W. Rosenberg, Proc. "ß Titanium Alloys in the 80's", Boyer & Rosenberg eds, TMS AIME Warrendale, Pennsylvania, (1984) 9-15.

[2] F.H. Froes, C.F. Yolto and J.P. Hirth, Proc. "ß Titanium Alloys in the 80's", Boyer & Rosenberg eds, TMS AIME Warrendale, Pennsylvania, (1984) 185-208.

[3] P.J. Bania, G.A. Lenning and J.A. Hall, Proc. "ß Titanium Alloys in the 80's", Boyer & Rosenberg eds, TMS AIME Warrendale, Pennsylvania, (1984) 209-229.

[4] A.G. Hicks and H.W. Rosenberg, Proc. "ß Titanium Alloys in the 80's", Boyer & Rosenberg eds, TMS AIME Warrendale, Pennsylvania, (1984) 231-237.

[5] A.E. Leach, Proc. "ß Titanium Alloys in the 80's", Boyer & Rosenberg eds, TMS AIME Warrendale, Pennsylvania, (1984) 331-347.

[6] A.M. Sherman and S.R. Seagle, Proc. "ß Titanium Alloys in the 80's", Boyer & Rosenberg eds, TMS AIME Warrendale, Pennsylvania, (1984) 281-293.

[7] R.R. Boyer, R. Bajoraitis, D.N. Greenwood and E.E. Mill, Proc. "ß Titanium Alloys in the 80's", Boyer & Rosenberg eds, TMS AIME Warrendale, Pennsylvania, (1984) 295-305.

[8] T.K. Redden, Proc. "ß Titanium Alloys in the 80's", Boyer & Rosenberg eds, TMS AIME Warrendale, Pennsylvania, (1984) 239-254.

[9] K. Takahashi, A. Ogawa and K. Minakawa, Proc. 1990 TDA Int. Conf., TDA Ed. Dayton, OH, (1990) 755-769.

[10] C. Ouchi, Tetsu to Hagane, vol. 39, (1991) 134-137.

[11] J.S. Grauman, Proc. 1990 TDA Int. Conf., TDA Ed. Dayton, OH, (1990) 290-299.

[12] B. Prandi, J.-F. Wadier, F. Schwartz, P.-E. Mosser and A. vassel, Proc. 1990 TDA Int. Conf., TDA Ed. Dayton, OH, (1990) 150-159.

[13] B. Champin, B. Prandi, P.-E. Mosser and Y. Honorat, Proc. 1991 AAAF Meeting, (1991) 55-77.

[14] R. Tricot, L. Seraphin, R. Syre, R. Molinier and J.-M. Logerot, La Revue de Métallurgie, Jan. (1970) 43-54.

[15] R. Molinier, L Seraphin, R. Tricot and R. Castro, La Revue de Métallurgie, Jan. (1974) 37-49.

[16] B. de Gélas, R. Molinier, L. Seraphin, M. Armand and R. Tricot, Proc. 2nd World Conf. on Titanium, Williams & Belov Eds, Plenum Publ. Co., (1982) 2121-2129.

[17] B. Champin, CEZUS internal report, (1983).

[18] C. Coddet, K. Ramoul, A.-M. Chaze, G. Beranger, B. Champin, L. Graff and M. Armand, Proc. 4th World Conf. on Titanium, Kimura & Izumi Eds, TMS AIME Warrendale, Pennsylvania, (1980) 2755-2764.

[19] B. Prandi, E. Alheritiere, F. Schwarz and M. Thomas, Proc. 6th World Conf. on Titanium, Lacombe et al. Eds., Ed. Phys. Les Ulis, (1989) 811-816.

[20] A. Henri and A. Vassel, ONERA technical report n° 22/3578 M, (1992).

[21] Y. Combres, CEZUS technical report STPA contract n° 90-96-023, (1992).

[22] P. Paillere, CEZUS technical report STPA contract n° 90-96-014, (1991).

[23] G. Dumas, Y. Combres, B. Champin and J.-F. Müller, Proc. GS Ti meeting, CNRS Ed, 143-152.

[24] Y. Combres, J. Bechet and A. Vassel, to be published.

[25] A. Henri and A. Vassel, Proc. 7th World Conf. on Titanium, to be published.

[26] P.-E. Mosser, N. Marnier, and Y. Honorat., Proc. 7th World Conf. on Titanium, to be published.

[27] Taylor and Huet, MESSIER-BUGATTI n° PV-DB/RM n° 557 report (1992).

ULTRA HIGH STRENGTH BETA TITANIUM ALLOY FOR FASTENERS

P. J. Bania, A. J. Hutt, and R. E. Adams

TIMET
P.O. Box 2128
Henderson, NV 89009

Abstract

A new high strength titanium alloy has been developed primarily intended for fastener applications. While Ti-6Al-4V is used extensively as a fastener alloy in the aerospace industry, its shear strength allowable is limited to 655 MPa (95 ksi). For higher shear strength requirements, various steels or nickel-based alloys are used (up to 860 MPa (125 ksi)), but with the attendant density penalty. This new alloy is intended to provide the 860 MPa shear strength at roughly a 40% weight savings. After screening various alloy systems, the optimum chemistry has been selected as follows: Ti-6.0V-6.2Mo-5.7Fe-3Al. In light of its 125 ksi shear strength goal, the alloy has been designated *TIMETAL*®125.

Beta Titanium Alloys in the 1990's
Edited by D. Eylon, R.R. Boyer and D.A. Koss
The Minerals, Metals & Materials Society, 1993

Introduction

Titanium alloys have been used extensively in the aerospace industry as fasteners, with Ti-6Al-4V being the most commonly used alloy. There are, however, two primary limitations suffered by Ti-6Al-4V which led to the development of an improved alloy. First, Ti-6Al-4V is limited to a guaranteed shear strength level of 655 MPa (95 ksi shear) which translates into a tensile strength minimum of about 1100 MPa (160 ksi). Secondly, because of its limited hardenability, it is often difficult or not possible to obtain such properties in fasteners on the order of 19mm (.75") diameter and beyond. When higher strength or larger diameter fasteners are required, occasionally other high strength titanium alloys such as Ti-6Al-6V-2Sn are employed. More often, however, other iron or nickel based alloys such as A-286, H-11 or IN718 are required. While these fastener alloys provide the desired strength and section size requirements, albeit with difficulty at times, they do so at an accompanying weight penalty of about 40% over the titanium. This weight penalty could be recouped with a titanium alloy capable of providing 860 MPa (125 ksi) shear at section sizes up to roughly 32mm (1.25") in diameter. This translates to roughly 1586 MPa (230 ksi) average tensile strength. The goal ductility is at least 7 to 8% elongation, which is what fastener producers require in order to be able to roll threads onto a fastener _after_ solution treating and aging. Thus, the first tier typical properties for the goal alloy were established as follows:

> Shear Strength: ≥ 860 MPa (125 ksi)
> Yield Strength: ≥ 1520 MPa (220 ksi)
> Tensile Strength: ≥ 1586 MPa (230 ksi)
> Ductility (% El): ≥ 7%

Technical Approach

The first phase of the program concentrated on producing small heats of existing high strength alloys such as Ti-6Al-6V-2Sn (Ti-6/6/2), Ti-10V-2Fe-3Al (Ti-10/2/3), Ti-15V-3Cr-3Sn-3Al (Ti-15-3), Ti-3Al-8V-6Cr-4Zr-4Mo (Ti-3/8/6/4/4, Beta C) and Ti-6Al-2Sn-4Zr-6Mo (Ti-6/2/4/6) as well as various modifications thereof. Each heat was processed to 12.5mm (.5") dia. bar, solution treated plus aged to a wide range of strengths and then tensile tested at room temperature.

The resultant tensile data was then analyzed by regression analysis to establish the following general equation for each alloy (ie, baseline alloy or modification thereof):

$$\% \text{ El} = A + B \times (UTS) \tag{1}$$

This equation was then used to calculate the expected ductility at 1586 MPa (230 ksi) tensile strength. Table 1 provides a summary of the expected "best" alloy from each group. Considering these results, it was clear that a new alloy base chemistry was required.

Table I   Expected Ductility (% El) From Best Composition at 1586 MPa

Alloy Group	Elongation[1]
Ti-6/2/4/6	5%
Ti-15/3	3.5%
Ti-10/2/3	4%
Ti-38644	4%
Ti-6/6/2	5%

[1]Expected values are from "best" modification of given alloy group and not necessarily from the specific alloy cited.

Based on some earlier work at TIMET, an alloy based on Ti-V-Mo-Fe-Al was selected for the next phase. While each of the beta stabilizers (V, Mo, Fe) was originally felt to be optimized at roughly the 5% level, little work had been conducted to define optimum levels. Thus, a series of 18-kg (40-lb) heats were melted as outlined in Table II. Alloys A and F had one beta stabilizer at the "low" level (~ 4.5%) while Alloys B, C, and D had at least two of the three beta stabilizers "low". Alloy E, however, had all three beta stabilizers at the "high" (~ 6%) level.

Table II Ti-V-Mo-Fe-Al Series With High and Low Levels of Beta Stabilizers[1]

Alloy	V	Mo	Fe	Al	O$_2$	Ti
A	4.5	5.8	5.7	3.0	.13	Bal
B	4.5	5.8	4.5	3.0	.13	Bal
C	4.3	4.8	5.7	2.7	.13	Bal
D	4.3	4.8	4.5	2.7	.13	Bal
E	6.2	6.2	5.7	2.7	.13	Bal
F	6.2	6.0	4.5	2.7	.13	Bal

[1]All chemistries cited in Wt. %.

The Table II alloys were processed to 12.5mm dia. rod and solution treated plus aged to various strength levels. Again, the data was analyzed by regression analysis and the results are given in Table III. This data clearly shows that the best ductility is obtained when all three beta stabilizers are at the high (> 5%) level. It also shows that when at least two of the three beta stabilizers are low (< 5%) the expected ductility at the 1586 MPa strength level is no better than the initial series of alloys (Table I). A graphic illustration of the advantage of Alloy E over the other formulations is shown in Figure 1.

Table III   Results of Regression Analysis of Tensile data from Table II Alloys[1]

Alloy	Regression Coefficient[2] A	B	Calculated % El At 1586 MPa UTS[2]
A	31.55	− .0161	6.03
B	42.64	− .0243	4.09
C	38.21	− .0211	4.68
D	42.74	− .0240	4.68
E	34.35	− .0167	7.84
F	34.71	− .0184	5.56

[1]All material was finish rolled from 75mm square to 12.5mm round starting at 30°C below its beta transus then solution treated 15 to 30°C below its transus and aged at various temperature/time combinations.

[2]Based on Equation (1) in text.

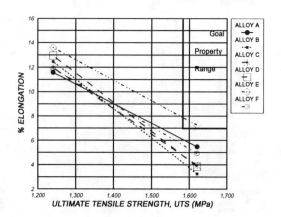

Figure 1 - Plot of Regression Curves for Data in Tables II and III.

Having established the optimum beta stabilizer content and successfully meeting the goal ductility at 1586 MPa UTS, it was then decided to check the effects of the alpha stabilizers (Al and oxygen). Table IV lists the alloys melted for this portion of the study and Table V provides the calculated ductility at the 1586 MPa strength level. From these tables, it is seen that the aluminum equivalence effect is very strong, and as shown in Figure 2, there is a critical (high) level which must be maintained in order to reach the goal ductility. It should also be noted that although oxygen is shown to substitute effectively for aluminum (compare Alloys G and J, Tables IV and V) low aluminum levels are impractical because of the necessity to use V-Al and Mo-Al master alloys in formulating this chemistry. Figure 2 also suggests than an aluminum equivalence of about 4.5 might be optimum.

Table IV   Ti-V-Mo-Fe-Al Series Melted with Varying Aluminum and Oxygen Levels[1]

Alloy	V	Mo	Fe	Al	O_2	Ti	Al Eq[2]
G	6.1	6.2	5.7	3.2	.13	Bal	4.5
H	5.2	5.5	5.2	2.7	.13	Bal	4.0
I	5.0	5.1	5.0	1.5	.14	Bal	2.9
J	5.2	5.2	5.1	1.6	.31	Bal	4.7

[1]All chemistries cited in Wt. %.
[2]Al Eq = Aluminum Equivalence = % Al + 10 x (% O_2).

Table V   Results of Regression Analysis of Tensile Data from Table IV Alloys[1]

	Regression Coefficients[2]		Calculated % El
Alloy	A	B	at 1586 MPa UTS[2]
G	48.45	- .0261	6.92
H	49.64	- .0277	5.65
I	85.27	- .0534	0.60
J	37.79	- .0196	6.73

[1]Material melted and processed similar to Table IV material.
[2]Based on Eqn (1) in text.

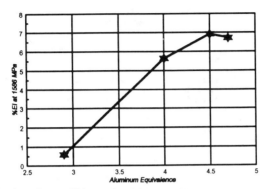

Figure 2 - Calculated Ductility of Table IV Alloys at 1586 MPa UTS as a
Function of Aluminum Equivalence.

In light of the above results, the selected chemistry for scale-up evaluation
was: Ti-6.0V-6.2Mo-5.7Fe-2.7Al-.18O_2.

An 820 kg (1800-lb) x 457mm (18-inch) diameter ingot of the above chemistry
was melted. There were two primary reasons for immediately scaling to this
ingot size:

a)   It provided enough material to enable hot rolling of coil stock on
      a production mill. This, it was felt, would allow processing to a
      finer grain size than with the smaller heats.

b)   It would provide a large enough ingot to evaluate segregation
      tendencies.

The ingot was forged to 89mm (3.5-inch) diameter as input stock for rolling
to small diameter coil. Since several pieces were cut at this size, slices
were taken at positions representing various positions along the ingot length
and chemistry samples were taken at center and outside edge locations of each
slice. These samples were analyzed for major chemistry variations in order
to assess the macrosegregation tendency for this formulation. As the results
of Table VI indicate, the chemistry variation was very small, thus indicating
that scale-up to a full sized ingot is feasible. Of course, the degree of
segregation is a function of the melt practice employed. Also, microsegregation tendency is not assessed by this technique.

Table VI Chemical Analysis of 820 kg Heat Chemistry, Wt. %

Location[1]	V	Mo	Fe	Al	O_2
2%/O	5.47	6.02	5.25	2.53	.186
2%/C	5.42	5.95	5.27	2.52	.197
25%/O	5.78	6.43	5.27	2.60	.200
25%/C	5.58	6.19	5.34	2.56	.179
75%/O	5.69	6.30	5.51	2.58	.190
75%/C	5.61	6.29	5.43	2.57	.220
90%/O	5.72	6.34	5.32	2.59	.196
90%/C	5.79	6.42	5.52	2.61	.178
Overall Avg	5.63	6.24	5.36	2.57	.193
Outside Avg	5.67	6.27	5.34	2.57	.193
Center Avg	5.60	6.21	5.39	2.57	.194

[1]% values correspond to approximate distance from ingot top; "O"
   refers to outside edge of sample slice; "C" refers to center.

The beta transus was 749°C (1380°F) for this heat.

43

The ingot was forged and hot rolled to 9.5mm diameter rod. A series of solution and age heat-treatments were performed on the material. The solution temperature was 710°C (1310°F) and the age temperature ranged from 482°C to 593°C (900°F to 1100°F) for 2 to 24 hours. These heat treatments resulted in ultimate tensile strengths from 1004 to 1373 MPa - below the desired level of 1586 MPa.

In order to increase the strength level, another series of heat treatments were performed utilizing a solution and duplex aging (STDA) cycle. The material was solution treated at 710°C, pre-aged at 427°C 4 to 8 hours, and final aged at 468°C or 482°C 16 hours. These heat treatments generally resulted in higher strengths. The over 1586 MPa ultimate tensile strength specimens had yield strengths ranging from 1475 to 1493 MPa and 6.0 to 8.0% elongation. The highest strengths were produced by a STDA heat treatment - 710°C/2hr/fan air cool plus 427°C/8hr/air cool plus 468°C/16hr/air cool. Table VII contains the results of the six highest tensile strengths.

Table VII Tensile Test Results from STDA[1] Material

Sample	UTS (MPa)	.2% YS (MPa)	% RA	% Elong
1	1617	1493	11.8	6.0
2	1616	1492	12.1	7.5
3	1609	1491	10.0	7.0
4	1599	1473	12.5	7.0
5	1594	1473	14.4	8.0
6	1587	1475	13.1	8.0

[1]710°C/2hr/fan air cool plus 427°C/8hr/air cool plus 468°C/16hr/air cool.

The balance of the approximately 160 tensile tests are not listed and show generally lower strengths and concomitant increasing ductility at the 482°C age temperature. However, some of the elongation values were unexpectedly low.

The double shear results from 46 tests from the 9.5mm rod heat treated at 710°C/2hr/fan air cool plus 427°C/8hr/air cool plus 468°C/16hr/air cool are 837 MPa on average and 868 MPa at the maximum. The average shear values are less than the goal of 862 MPa on average but it was obtained in some cases. It is felt that a more optimum processing schedule to produce a finer grain sized product will provide the 862 MPa shear strength and ≥ 7% elong. goals.

The microstructure of the STDA (710°C/2hr/fan air cool plus 427°C/8hr/air cool plus 468°C/16hr/air cool) material is typically a very fine-grained alpha precipitate in a beta matrix. (See Figure 3).

Three studs with a thread size of .250" diameter with 28 threads per inch were manufactured by Valley-Todeco from 6.35mm diameter rod from the first 820 kg ingot. The rod had been cold drawn 34% reduction in area during its manufacture. A sample was heat treated at 710°C/2hrs/fan air cool + 427°C/8hrs/air cool + 468°C/16hrs/air cool. Valley-Todeco rolled threads at room temperature and reported tensile strengths of 1420, 1460, and 1500 MPa based on 26.0mm^2 area. The microstructure in the threaded area shows heavily cold worked material at the bottom of the threads and flow lines following the outline of the threads. (See Figure 4). The internal microstructure shows a fine grained aged structure with some lighter etching steaks in the longitudinal direction.

Figure 3 - Microstructure of aged rod. 200X, L.

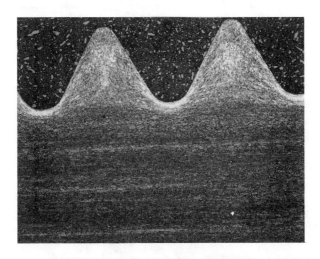

Figure 4 - Cross-section of STDA and cold threaded .250" diameter stud from
6.35mm rod produced from first 820 Kg ingot. 50X.

Preliminary tensile properties from the second 820 kg ingot were determined
from a portion of the ingot which had been forged, rolled, and cold drawn 30%
to finish at 9.5mm.   After a treatment of 746°C (beta transus minus
17°C)/2hrs/air cool + 427°C/4hrs/air cool + 510°C/8hrs/air cool, tensile
elongation and double shear goals of 7% and 860 MPa were met.   (See Table
VIII.)

Table VIII   Tensile and Double Shear Results on STDA[1] From
Second 820 Kg Heat in 9.5mm Bar

Sample	UTS (MPa)	.2% YS (MPa)	% RA	% Elong.	Double Shear (MPa)
Goal	1586			>7	860
1	1573	1456	11.9	8.0	
2	1551	1434	12.4	8.5	
3 & 4					860 & 876
5 & 6					876 & 881
7 & 8					881 & 892

[1]746°C/2hrs/air cool + 427°C/4hrs/air cool + 510°C/8hrs/air cool.

A solution time and temperature study was done in order to determine if the
solution cycle could be further optimized.   Rod measuring 9.5mm which had
been cold drawn from the second 820 Kg heat was solution treated at 677, 718,
or 746°C at 5, 15, 30, or 120 minutes and aged at 427°C/4hrs/air cool +
510°C/8hrs/air cool.   Two tensile tests were performed from each condition
and the percentage elongation was extrapolated at the 1552 MPa UTS level
using the slope of the line from Figure 1 for alloy E.   (See Figure 5.)   The
solution treatments from 5 to 30 minutes resulted in both a large variation
and low elongation.   Only the 677°C/120 minute solution treatment followed
by the age cycle showed a consistent elongation exceeding 8.5%.

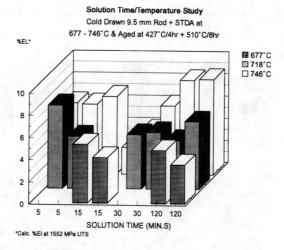

Solution Time/Temperature Study
Cold Drawn 9.5 mm Rod + STDA at
677 - 746°C & Aged at 427°C/4hr + 510°C/8hr

%EL*

- 677°C
- 718°C
- 746°C

SOLUTION TIME (MIN.S)

*Calc. %El at 1552 MPa UTS

Figure 5 - Effect of solution time and temperature on % elongation in cold
drawn and treated rod.   The % elongation was extrapolated from
the slope of the line in Figure 1 for alloy E at the 1552 MPa
strength level.

Since corrosion is an important aspect of bolt performance, several general
corrosion tests were performed on beta rolled sheet[1].   *TIMETAL*®125 is equal
to or better than CP grade 2 and Ti-6Al-4V in general corrosion rates in
boiling acid solutions as shown in Table IX.

Table IX   General Corrosion Rates (MM/Year)

Environment	TIMETAL®125	CP Grade 2	Ti-6Al-4V
H₂SO₄, 1%, Boiling	.094	17.4	2.5
HCl, 1%, Boiling	.475	1.8	.845
HNO₃, 40%, Boiling	.470	1.5	

Summary

Since no existing class of titanium alloys showed the potential of meeting the requirements of an ultra-high strength fastener alloy, an alloy development program was initiated. After evaluating Ti-6/2/2, Ti-10/2/3, Ti-15/3, Ti-3/8/6/4/4, and Ti-6/2/4/6 and modifications thereof, the alloy development program resulted in the alloy, *TIMETAL*®125 with the composition of Ti-6.0V-6.2Mo-5.7Fe-2.7Al (wt. %). *TIMETAL*®125 shows potential for fasteners requiring 860 MPa shear strength, 1520 MPa yield strength, and 1586 MPa tensile strength with elongation greater than 7%. These properties were met with the exception of occasional values below the 1586 MPa tensile strength target in heat treated hot-rolled bar without the benefit of cold working before aging.

The heat treatment was optimized at 710°C/2hr/fan air cool plus 427°C/8hr/air cool plus 468°C/16hr/air cool or 746°C/2hr/air cool plus 427°C/4hr/air cool plus 510°C/8hr/air cool.

General corrosion rates of *TIMETAL*®125 are equal to or better in severe boiling acid environments than Ti-6Al-4V or CP grade 2.

Reference

1. Bergman, D. D., "HTL Monthly Progress Report for April 1992", TIMET Henderson Technical Lab.

THE CHARACTERIZATION OF Ti-12Mo-6Zr-2Fe - A NEW BIOCOMPATIBLE

TITANIUM ALLOY DEVELOPED FOR SURGICAL IMPLANTS

K. Wang, L. Gustavson and J. Dumbleton

Howmedica, Pfizer Hospital Products Group Inc.
359 Veterans Boulevard
Rutherford, New Jersey 07070

Abstract

A new beta titanium alloy, Ti-12Mo-6Zr-2Fe (TMZF) was developed for orthopaedic use. This alloy has a unique combination of properties, i.e. low modulus of elasticity, excellent mechanical strength, corrosion resistance, and formability, coupled with good wear and notch fatigue resistance. Also, no vanadium and aluminum in this alloy offer a biocompatibility advantage over the Ti-6Al-4V alloy. The processing development, metallurgical characteristics, physical and mechanical properties, and wear resistance of this new alloy are reported here.

Beta Titanium Alloys in the 1990's
Edited by D. Eylon, R.R. Boyer and D.A. Koss
The Minerals, Metals & Materials Society, 1993

Introduction

The most widely used orthopaedic titanium alloy, Ti-6Al-4V, is known for its excellent corrosion resistance and good mechanical properties. It is also known for its notch fatigue sensitivity and relatively poor wear resistance. High levels of titanium, vanadium, and aluminum debris have been found in surrounding tissues under conditions of high wear. Although no toxic effect has been connected to these debris, safety concerns on vanadium and aluminum have been reported since 1980.

Due to the perceived safety concerns, two vanadium free titanium alloys, Ti-5Al-2.5Fe and Ti-6Al-7Nb were developed in Europe in the 1980's.[1,2] Both of these alloys offer a potential biocompatibility advantage over Ti-6Al-4V due to the absence of vanadium. But in terms of mechanical properties they are quite similar to Ti-6Al-4V since they are still in the α-β alloy family.

Recent finite element studies suggest that a lower modulus (more flexible) hip prosthesis may better simulate the natural femur in distributing stress to the adjacent bone tissue.[3,4] Animal studies also suggest that the bone resorption problem commonly experienced by hip prosthesis patients may be alleviated by a prosthesis having a lower modulus.[5,6] These studies aroused significant interest in producing materials with a lower modulus of elasticity.

The preferred orthopedic titanium alloy should have a lower elastic modulus, a lower notch sensitivity, and a better biocompatibility than Ti-6Al-4V. In our new alloy design, beta alloys were chosen for their advantages in low modulus and lower notch sensitivity. Vanadium and aluminum were purposely avoided while more biocompatible elements, i.e. Mo, Zr, and Fe were introduced. Our recent achievement, Ti-12Mo-6Zr-2Fe (TMZF), is unique in having a modulus of elasticity as low as 74 GPa, excellent mechanical strength and corrosion resistance coupled with good wear and notch fatigue resistance. As will be shown in this paper the Ti-12Mo-6Zr-2Fe alloy is successful in meeting the alloy development goals.

Alloy Processing

It is very difficult to produce a homogeneous Ti-12Mo-6Zr-2Fe (TMZF) ingot using the conventional vacuum arc remelting method, because Mo and Zr have a much higher melting point and density than titanium. Thus, a triple melting technique, which included two vacuum arc remelts and one electron beam or two plasma arc remelts and one vacuum arc remelt was developed for this new alloy. One 45 kg pilot lot and one 4545 kg production lot of 25.4-28.6mm diameter bars were successfully produced through the conventional forging and rolling process.

Workability of the new alloy was excellent. The solution annealed bars were cold drawn into rods of 14mm in diameter and close-die forged into hip prosthesis. In all trials, Ti-12Mo-6Zr-2Fe alloy showed excellent hot and cold formabilities.

General Alloy Microstructure

Ti-12Mo-6Zr-2Fe (TMZF) is a metastable beta alloy. It retains an all-beta structure after rapid cooling from its beta transus temperature of 754°C or higher. The all-beta structure will precipitate fine alpha phases upon subsequent aging.

The solution-annealed single beta phase Ti-12Mo-6Zr-2Fe was chosen for orthopaedic implants for its low modulus and excellent ductility (Figure 1).

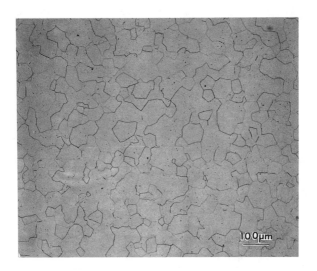

Figure 1 - Microstructure of 788°C (1450°F) solution annealed TMZF.

<u>**Physical Properties**</u>

The physical properties of Ti-12Mo-6Zr-2Fe (TMZF) alloy are listed in Table I.

The thermal expansion of Ti-12Mo-6Zr-2Fe alloy was measured from 25 to 900°C in accordance with ASTM E-228 (Figure 2). The thermal expansion of Ti-12Mo-6Zr-2Fe alloy is 0.884% between 25 and 900°C. The thermal expansion coefficient of Ti-12Mo-6Zr-2Fe is 8.8 x 10^{-6}/°C between 25 and 250°C; and 11.5 x 10^{-6}/°C between 525 and 900°C.

A departure from a linear response becomes evident in the range 250 to 525°C and indicates crystalline transitions of beta and alpha phases in this alloy. At temperatures above 550°C, the alpha precipitates were re-dissolved into the beta matrix.

The modulus of Ti-12Mo-6Zr-2Fe is 74-85 GPa, which is 25% lower than that of Ti-6Al-4V. The density of this alloy is 12% higher than that of Ti-6Al-4V. The hardness of this alloy (Rc 34-35) is slightly higher than the Rc 31 of Ti-6Al-4V.

Table I - Typical Physical Properties of Ti-12Mo-6Zr-2Fe Alloy

Thermal Expansion Coefficient*	x 10^{-6}/°C x 10^{-6}/°C	8.8 at 25-250°C 11.5 at 525-900°C
Modulus*	x 10^6 psi (GPa)	10.7 - 12.3 (74-85)
Beta Transus	°F °C	1350 - 1390 732 - 754
Density*	lb/cu-in gm/cu-cm	0.18 5.0
Hardness*	Rc	34-35

*Solution treated condition

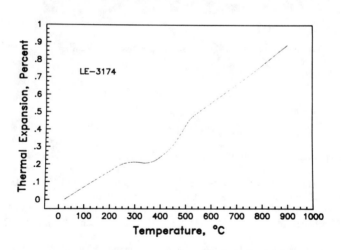

Figure 2 - Thermal expansion of TMZF alloy from 25° to 900°C

Tensile Properties

Tensile tests were conducted on solution annealed Ti-12Mo-6Zr-2Fe (TMZF) bars at room temperature according to ASTM E-8. The tensile properties of Ti-12Mo-6Zr-2Fe and mill annealed Ti-6Al-4V are listed in Table II. Results show that Ti-12Mo-6Zr-2Fe has a much higher yield strength and better elongation than Ti-6Al-4V. As noted in the physical property section of this report, the elastic modulus of Ti-12Mo-6Zr-2Fe alloy is lower than that of Ti-6Al-4V.

Table II - Tensile Properties of Solution Annealed Ti-12Mo-6Zr-2Fe

Alloy	YS (MPa)	UTS (MPa)	EL (%)	RA (%)	E (GPa)
TMZF	1000-1060	1060-1100	18 - 22	64 - 73	74 - 85
Ti-6Al-4V	850 - 900	960 - 970	10 - 15	35 - 47	110

Rotating Beam Fatigue Properties

The solution annealed Ti-12Mo-6Zr-2Fe (TMZF) bars were subjected to smooth and notched Krouse fatigue tests. The testing modes include high cycle fatigue, rotating beam, and cantilever bending. The test loading was constant force sinusoidal at approximately 167 Hz. The test stress ratio (R) was -1, i.e. fully reversed. All testing was conducted at room temperature in air.

Tests normally stopped when 10 million cycles were achieved. The notched samples had a stress concentration factor of $K_t = 1.6$. For comparison purposes mill annealed Ti-6Al-4V alloy bars were also tested under the same conditions.

The smooth rotating fatigue properties of Ti-12Mo-6Zr-2Fe and Ti-6Al-4V are plotted in Figure 3. The smooth fatigue strength of Ti-12Mo-6Zr-2Fe is comparable to that of Ti-6Al-4V. At 10^7 cycles, the maximum stress of Ti-12Mo-6Zr-2Fe alloy is 585 MPa (85ksi).

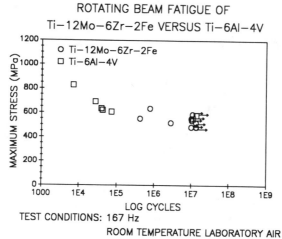

Figure 3 - Smooth rotating beam fatigue strength of TMZF and Ti-6Al-4V

The notched rotating beam fatigue properties of Ti-12Mo-6Zr-2Fe and Ti-6Al-4V are shown in Figure 4. At 10^7 cycles, the Ti-12Mo-6Zr-2Fe alloy showed a notched strength of 410 MPa (60 ksi) at 10 million cycles (70% of its smooth fatigue strength). Under the same condition, Ti-6Al-4V showed a notched strength of 280 MPa (45 ksi), or 53% of its smooth fatigue strength. The notched fatigue results are in keeping with the characteristic property of beta titanium alloys being less fatigue notch sensitive than alpha beta alloys such as Ti-6A1-4V alloy.

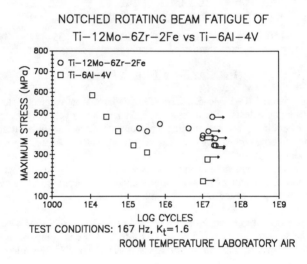

Figure 4 - Notched rotating beam fatigue strength of TMZF and Ti-6Al-4V

Fracture Toughness

The solution annealed Ti-12Mo-6Zr-2Fe bar samples were tested for precracking and fracture
toughness at 33°C according to ASTM E399. The crack area was saturated with saline solution (0.9% NaCl) before testing. Mill annealed Ti-6Al-4V bar samples were also tested for comparison.

The mean toughness of Ti-12Mo-6Zr-2Fe was 90 MPa $\sqrt{m}$ (81.7 ksi $\sqrt{in}$). That of Ti-6Al-4V was 52 MPa $\sqrt{m}$ (47.2 ksi $\sqrt{in}$) (Table III). These results show that the fracture resistance of the solution annealed Ti-12Mo-6Zr-2Fe alloy is much better than that of Ti-6Al-4V.

Table III. Fracture Toughness of TMZF and Ti-6Al-4V Bar Stock (28.6 mm dia.)

Alloy	Specimen Number	Toughness MPa√m	Kq (ksi√ in)
TMZF	1 - 1	90	(81.7)
TMZF	1 - 2	88	(79.7)
TMZF	1 - 3	92	(83.7)
Ti-6Al-4V	2 - 1	54	(48.8)
Ti-6Al-4V	2 - 2	51	(46.4)
Ti-6Al-4V	2 - 3	51	(46.4)

Corrosion Property

Anodic polarization testing was performed on the solution annealed Ti-12Mo-6Zr-2Fe and mill annealed Ti-6Al-4V specimens. Specimens were in the form of 16 mm diameter x 3 mm thick disks. Tests were run in deaerated saline (0.9% Nacl) at 37°C. Each disk specimen was finished to a 600 grit silicon carbide immediately before testing.

ASTM G5, Standard Reference Test Method for Making Potentiostatic and Potentiodynamic Anodic Polarization Measurements was followed for the testing. The anodic polarization curves of Ti-12Mo-6Zr-2Fe and Ti-6Al-4V are shown in Figure 5. Data show excellent corrosion resistance for both alloys. As the potential increases each alloy reaches a stable passive current density. The difference in the passive current density between the two alloys is not directly related to their relative free corrosion rates in the body environment. Further increases in potential up to +1000 mv v.s. S.C.E. cause no further increase in current density for either alloy. This indicates that the protective oxide on either alloy resists breakdown equally well at this excessive polarization. Resistance to breakdown is the key result of this test.

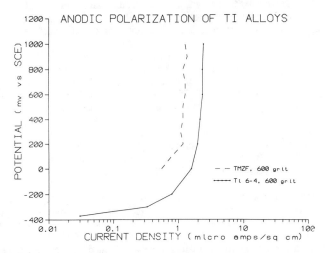

Figure 5. Anodic polarization of Ti-12Mo-6Zr-2Fe and Ti-6Al-4V

Wear Properties

The friction and wear properties of annealed Ti-12Mo-6Zr-2Fe (TMZF) and Ti-6Al-4V versus Surgical Simplex® P bone cement were determined using an ISC-200 tribometer. The ISC-200 is a pin-on-disk type wear and friction measurement system. The bone cement sample was held stationary by a cantilever beam while the metal disc counterface rotated against it. The surface roughness of bone cement ball was 1μm.

Abrasive wear of loose cemented femoral stems has been shown to be a source of debris in vivo, particularly for Ti-6Al-4V systems.[7] For this reason PMMA bone cement was chosen as the counterface material in this study of the abrasive properties of TMZF titanium alloy versus Ti-6Al-4V alloy.

Five polished disc samples of each alloy were individually tested against a 12.7 mm diameter bone cement ball at a 100 gm load, in deionized water. Tests were run at 80 cycles/min. for 1.0 x 10^5 cycles. The sliding speed for each test was 7.4 cm/sec. Wear was determined by the weight loss and surface roughness of the disc samples. The mass worn away from a ball was calculated by the radius of the worn flat spot and the density of the bone cement sample.

Coefficient of Friction

Typical coefficient of friction measurements for TMZF and Ti-6Al-4V alloy against PMMA bone cement are shown in Figures 6. The friction coefficient for TMZF was in the range of 0.30 to 0.44 while that for Ti-6Al-4V was at least twice as high. The measurement band for the Ti-6Al-4V samples was considerably broader than that for the TMZF samples, possibly indicative of a sticking and sliding phenomena associated with severe surface wear.

Metallic Disk Wear

The wear data, (metallic disk surface roughness changes and PMMA pin weight loss) for TMZF alloy and Ti-6Al-4V alloy against PMMA bone cement are shown in Table IV.

After 1.0 x 10^5 cycles, the TMZF disk samples exhibited only minimal change in surface roughness (0.01 to 0.06 μm) and no weight loss. Three out of five samples exhibited only a contact track from the PMMA pin with no associated surface scratching (Figure 7). The remaining two samples exhibited some scratching but did not initiate self-perpetuating abrasive wear and only bone cement debris was noted on the disk surface (Figure 8).

Severe abrasive wear of the Ti-6Al-4V disks was observed after only a few hundred cycles. Surface roughness increased on average over 5.0 μm with noticeable disk surface scratching. Significant black wear debris was generated (Figure 9), resulting in sample weight losses between 0.0129 and 0.0176 grams. The black debris when analyzed using SEM/EDS was found to be rich in titanium, aluminum, vanadium, oxygen and carbon.

PMMA Pin Wear

The surfaces of all bone cement pins run against Ti-6Al-4V alloy were covered with fine black particles (Figure 8). No black particles were seen on the surfaces of the bone cement pins that were tested against the TMZF disc samples.

The weight loss of bone cement pins abraded against Ti-6Al-4V discs was calculated to be 1.04 - 1.33 mg which was significantly higher than that of the bone cement pins against the TMZF discs (0.05 - 0.22 mg). The increased bone cement weight loss further illustrates the nature of the wear which occurred between Ti-6Al-4V and PMMA bone cement, namely a form of self-perpetuating wear accelerated by the presence of metallic and bone cement third body particles.

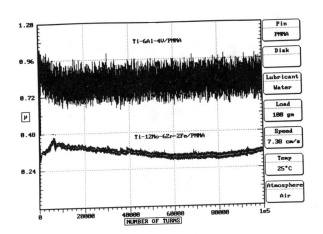

Figure 6 - Friction coefficient of TMZF and Ti-6Al-4V against bone cement.

Table IV - <u>Results of Wear Test on TMZF and Ti-6Al-4V Against bone cement.</u>

Sample No.	Disc Sample Surface Roughness, μm		Disc Sample Wt.loss, gm	Bone Cement Sample Wt. Loss, mg
	Initial	Final	(gain)	
TMZF - 1	0.02	0.07	(0.0008)	0.09
TMZF - 2	0.02	0.08	(0.0006)	0.09
TMZF - 3	0.02	0.03	(0.0000)	0.05
TMZF - 4	0.05	0.09	(0.0009)	0.22
TMZF - 5	0.07	0.09	(0.0000)	0.09
Ti-6-4 - 1	0.01	4.85	0.0129	1.04
Ti-6-4 - 2	0.02	5.73	0.0135	1.33
Ti-6-4 - 3	0.02	4.47	0.0176	1.04
Ti-6-4 - 4	0.06	4.13	0.0149	1.33
Ti-6-4 - 5	0.02	6.23	0.0134	1.33

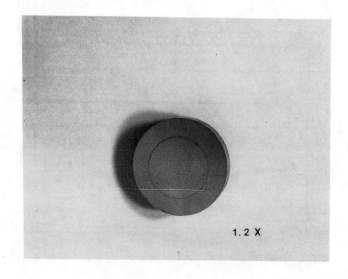

1. 2 X

Figure 7. A TMZF wear disc (Sample No. 3) abraded against bone cement for 1.0 x 10⁵ cycles showing contact track and no surface scratching.

Ti-Mo-Zr-Fe
Distance 5375m

1.0 X

Figure 8. A TMZF disc (Sample No. 4) abraded against bone cement for 1.0 x 10⁵ cycles showing bone cement debris on the disc.

Figure 9.   Typical Ti-6A1-4V wear disk abraded against PMMA bone cement for 1.0 x 10⁵ cycles showing black debris on the disk.

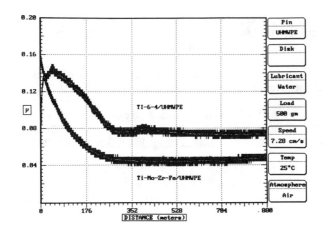

Figure 10 - Friction Coefficient of TMZF against UHMWPE

The friction coefficients of annealed Ti-12Mo-6Zr-2Fe and Ti-6Al-4V versus ultra high molecular weight polyethylene (UHMWPE) pins were also measured. Two tests for each alloy were conducted in deionized water at a 500gm load and at 80 cycles/min. The sliding speed for each test was 7.3 cm/sec. The disk samples had a surface roughness of $0.02\mu m$. The surface roughness of the UHMWPE pins was $2.5\mu m$. The friction coefficients of Ti-12Mo-6Zr-2Fe and Ti-6Al-4V against UHMWPE at various distances are shown in Figure 10. The steady state friction coefficient of Ti-12Mo-6Zr-2Fe is 0.04 which is much lower than that of Ti-6Al-4V.

SUMMARY

A new biocompatible titanium alloy, Ti-12Mo-6Zr-2Fe has been developed. The processing and formability advantages of this alloy have been briefly reviewed. This beta titanium alloy exhibits higher yield strength, lower modulus of elasticity, higher fracture toughness and better wear resistance than mill annealed Ti-6Al-4V. It also has an excellent corrosion resistance and a smooth fatigue strength similar to that of Ti-6Al-4V. With the characteristics of high strength, low modulus, biocompatibility and excellent corrosion resistance, the alloy is well suited for orthopaedic applications.

REFERENCES

1. R. Zwicker, K. Buehler, R. Mueller et al., "Mechanical Properties and Tissue Reactions of a Titanium Alloy for Implant Material," TITANIUM '80, Science and Technology Proc. 4th Int. Conf. on Titanium (Kyoto, May 1980). The Met. Soc. AIME (ISBM-No. 89520-370-7); 505-514.

2. M. Semlitsch, F. Staub and H. Webber, "Titanium-Aluminum-Niobium Alloy, Development for Biocompatible, High Strength Surgical Implants," Biomed. Technik 30 (1985), 334-339.

3. E. Cheal, M. Spector and W. Hayes, "Role of Loads and Prosthesis Material Properties on the Mechanics of the Proximal Femur After Total Hip Arthroplasty," J. Orthop Res, (1992) 10:405-422

4. P. Prendergast and D. Taylor, "Stress Analysis of the Proximo-Medial Femur After Total Hip Replacement," J. Biomed.Eng. (1990), Vol. 12, No. 5, 379-382.

5. J.D. Bobyn, A.H. Classman, H. Goto, J. Krygier, J. Miller, and C. Brooks, "The Effect of Stem Stiffness on Femoral Bone Resorption After Canine Porous-Coated Total Hip Arthroplasty." Clin. Orthop. Relat. Res. (1990), 261:196-213.

6. J.D. Bobyn, E.S. Mortimer, A.H Glassman, C.A. Engh, J. Miller, C. Brooks, "Producing and Avoiding Stress Shielding: Laboratory and Clinical Observation of Noncemented Total Hip Arthroplasty." Clin. Orthop. Relat. Res. (1992), 274:79-96.

7. J.D. Witt, M. Swann, "Metal Wear and Tissue Response in Failed Titanium Alloy Total Hip Replacements." J. of Bone and Joint Surgery (1991),73B:559-563.

Ti-13Nb-13Zr: A NEW LOW MODULUS, HIGH STRENGTH, CORROSION RESISTANT NEAR-BETA ALLOY FOR ORTHOPAEDIC IMPLANTS

Ajit K. Mishra, James A. Davidson, Paul Kovacs, and Robert A. Poggie

Smith and Nephew Richards Inc.
1450 Brooks Road
Memphis, TN 38116, USA

Abstract

Ti-13Nb-13Zr is a new, near-beta alloy which has been formulated to achieve the objectives of low modulus, optimal biocompatibility, and superior wear resistance, with high strength, toughness, fatigue endurance and corrosion resistance. These properties make this alloy extremely promising for implant applications such as total hip replacement. Compared to mill-annealed Ti-6Al-4V, Ti-13Nb-13Zr has higher strength (UTS = 1,030 MPa, YS = 900 MPa) and ductility (Δl = 15%, RA = 45%), 30% lower elastic modulus (79 GPa), equivalent 10 million cycle unnotched axial fatigue endurance limit (500 MPa), significantly higher notched fatigue endurance limit at Kt=3.0 (215 vs 170 MPa), 20% higher plane strain fracture toughness (65 vs 53 MPa$\sqrt{m}$), 20% higher Charpy impact energy (30 vs 25 J), 30-40% lower flexural and shear moduli, 40% lower corrosion rate in simulated body environments, and equivalent wear resistance. The microstructure of Ti-13Nb-13Zr consists of platelets of aged hcp martensite (α') with submicroscopic precipitates of bcc beta which strengthen and harden the material by dispersion strengthening. Hot working, cold working and heat treatment procedure variations can lower its modulus to below 50 GPa while retaining the high strength. The UTS can also be increased to 1,115 MPa and the YS to 980 MPa. Ti-13Nb-13Zr does not contain any elements such as Al, V, Co, Cr or Mo which have been associated with adverse tissue reactions. This alloy has a low beta-transus temperature (735°C) and is hot workable at lower temperatures and cold workable to greater extents than Ti-6Al-4V. When Ti-13Nb-13Zr is subjected to a proprietary diffusion hardening (DH) process, a 0.2 μm thick blue, wear-resistant, ceramic surface is produced with a 2-3 μm thick solid solution hardened region underneath it. In an accelerated wear test, DH Ti-13Nb-13Zr had a wear track depth of only 0.15 μm, compared to 21 μm for Ti-6Al-4V and 7.8 μm for TiN-coated Ti-6Al-4V. When the blue, ceramic surface was removed prior to the test, the solid solution hardened region underneath it was still found to be extremely wear resistant, producing a wear track depth of 1.5 μm. Hence, DH Ti-13Nb-13Zr is an extremely promising material for THR and other orthopaedic applications.

Beta Titanium Alloys in the 1990's
Edited by D. Eylon, R.R. Boyer and D.A. Koss
The Minerals, Metals & Materials Society, 1993

Introduction

Over 200,000 total hip replacement (THR) and 200,000 total knee replacement (TKR) procedures are performed in the U.S. alone each year, primarily for arthritis. The number of such procedures continues to increase due to: (i) their exceptional success in restoring mobility and providing pain relief, and (ii) the aging population.

Many of these devices are made of Co-Cr-Mo alloy or Ti-6Al-4V. The advantages of Ti-6Al-4V are its high strength, low modulus and good biocompatibility. Hence it has been successfully used for many years in THR and TKR. However, the very success of total joint replacement procedures has resulted in their increasing use in younger patients, which requires more demanding performance criteria of these implants, driving the search for the next generation of materials in this field.

Even Ti-6Al-4V has a significantly higher modulus than bone. THR devices have been associated with stress shielding, i.e. insufficient transfer of stress to the bone due to the high modulus of the prosthesis. This can potentially lead to bone resorption in the long term and eventual loosening of the device, requiring revision surgery.

Both finite element analyses [1,2] and strain gage analyses [2-6] have demonstrated that lower modulus hips produce stresses and strains that are closer to those of the intact femur. Canine and sheep implantation studies have shown significantly reduced bone resorption in animals with low modulus hips [7-11]. A number of prominent clinicians have reported reduced incidence of thigh pain [12-14], hip pain [15] and bone resorption [10] when relatively lower stiffness hips are used in cementless applications. Sarmiento and Gruen [16] have reported a lower incidence of calcar resorption, cortical hypertrophy and prosthesis loosening when lower modulus hips are used in cemented applications as well.

The potential adverse long term effects of certain metal ions released by prosthetic materials has also gained more attention lately. V, contained in Ti-6Al-4V, has been associated with potential cytotoxic effects [17] and adverse tissue reaction [18]. Al has been associated with potential neurological disorders [19,20]. Co, contained in Co-Cr-Mo alloy, has been shown to be carcinogenic in animals [21]. Cr and Mo have also been shown to cause severe tissue reactions in animals [18].

Finally, wear debris generated by the prosthesis due to fretting with and abrasion by bone and/or bone cement, and the resultant osteolytic response, is also a potential concern. Retrieved Ti-6Al-4V hips have shown evidence of burnishing at the distal tip, possibly as a result of such abrasion. Black wear debris has at times been observed in the tissue around Ti-6Al-4V devices at retrieval.

Hence, the ideal material for orthopaedic applications such as THR should have a low modulus to minimize bone resorption and thigh pain, should not contain any of the elements described above with potential adverse tissue reactions, and should be resistant to wear and debris generation.

In an attempt to achieve these objectives, two Vanadium free alloys, Ti-5Al-2.5Fe and Protasul-100 (Ti-6Al-7Nb) were introduced in the 1980's. However, both alloys still contain Al with its potential adverse effects, and do not have a significantly lower modulus (Protasul 100: 105 GPa, Ti-5Al-2.5Fe: 116 GPa) than Ti-6Al-4V (115 GPa). Recently an alloy called TMZF (Ti-11.5Mo-6Zr-2Fe) has also been proposed. However, this alloy contains a very high percentage of Mo which has been associated with severe tissue reactions in animal studies [18].

The present paper describes the results of extensive evaluation of a new Titanium alloy, Ti-13Nb-13Zr, which has been developed to meet these criteria: low modulus, optimal biocompatibility [22] and superior wear resistance. The only constituents of this alloy are Ti, Nb and Zr, three of the only five elements (the others being Ta and Pt) which have been identified as producing no adverse tissue reaction [13]. The paper also describes the results of testing and analysis performed to characterize the diffusion hardening (DH) process, a proprietary surface hardening treatment developed for this alloy.

Materials and Methods

Materials

A triple vacuum-arc melted ingot of Ti-13Nb-13Zr, produced by Teledyne Wah Chang Albany of Albany, Oregon, was hot-rolled into plates and bars and heat treated. The heat treatment performed consisted of water quenching and aging. Most of the testing described below was performed on a 1 inch thick plate. The Ti-6Al-4V plate used for comparative testing was a 1 inch thick, standard grade, mill-annealed plate, supplied by TiMet.

Test Methods

Mechanical Testing: Mechanical testing was performed at Sherry Labs., Muncie, IN, as per ASTM E8 to determine the ultimate tensile strength, 0.2% off-set yield strength, % elongation, and % reduction of area, and as per ASTM E111 to determine the Young's modulus.

The 10 million cycle fatigue endurance limit, for both smooth and notched specimens, was determined by axial fatigue testing per ASTM E466 at 60 Hz and R=0.1 at Cincinnati Testing Labs. Two sets of notched specimens were tested, one set with stress concentration factor, $K_t = 1.6$ and the other set with $K_t = 3.0$.

The fracture toughness, K_Q, was determined per ASTM E399, also at Cincinnati Labs. The appropriate calculations were made as per ASTM E399 to determine whether a plane strain condition had been attained which would make $K_Q=K_{IC}$, the plane strain fracture toughness. The Charpy V-notch impact energy was determined at Sherry Labs. per ASTM E23. The flexural modulus was determined by 3-point bending, per ASTM D790, at Cincinnati Labs. The shear modulus was determined at the University of Dayton Research Institute by torsional shear testing. The specimens used for these tests are shown in Figure 1.

Microstructure and Phase Analysis: As-quenched and aged Ti-13Nb-13Zr specimens were etched, after polishing, by swabbing with a modified Kroll's etchant (3 ml HF + 8 ml HNO_3 + 100 ml H_2O). Microstructural analysis was performed at Vanderbilt University and at S&NR using optical microscopy, scanning electron microscopy (SEM), and transmission electron microscopy

(TEM). SEM and energy dispersive spectroscopy (EDS) at Vanderbilt was performed with a Hitachi X-650 hybrid scanning electron/microprobe microscope. TEM was performed using a CM20 Philips TEM/STEM operated at 200 KV.

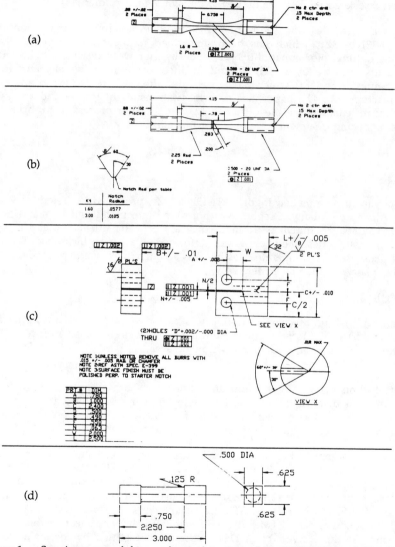

Figure 1 - Specimens used for mechanical testing of Ti-13Nb-13Zr and Ti-6Al-4V at Cincinnati Labs. and University of Dayton Research Institute. (a) Smooth fatigue specimen, (b) Notched fatigue specimen, (c) Compact tension fracture toughness specimen, (d) Torsional shear specimen.

X-ray diffraction (XRD) was performed at Vanderbilt University using a Philips XRG-3000 powder diffractometer with Cu-Kα radiation. XRD and selected area electron diffraction (SAD) were used to determine the crystallographic phases present in the as-quenched and aged materials.

Corrosion and Wear: The reciprocal polarization resistance (1/Rp), i.e. corrosion rate + electron exchange rate, in a physiological saline (Ringer's solution) at 37°C, was determined by Electrochemical Impedance Spectroscopy (EIS) using an EG&G Princeton Applied Research Model 273 potentiostat. Accelerated pin-on-disc wear testing was performed on the material, both with and without diffusion hardening, using bone cement (PMMA) pins in Ringer's solution in a reciprocating, sliding mode. Testing was performed at 2.5 Hz for 10^6 cycles at an initial stress of 107 MPa. The amplitude of motion was 15 mm, hence each cycle corresponded to a total motion of 30 mm. The depth of the wear track was determined by profilometry at the conclusion of the test using a TSA Surfcom 570 profilometer with a diamond stylus.

Surface Analysis: The hardness of the diffusion hardened surface was determined at S&NR using a Leco M-400-G-2 microhardness tester at 25 gm load. The specimens were also analyzed by Secondary Ion Mass Spectrometry (SIMS), X-ray photoelectron spectroscopy (XPS) and Auger Electron Spectroscopy (AES) at Evans East Lab., and X-ray diffraction (XRD) at Vanderbilt University.

Results And Discussion

Mechanical Properties

The tensile strength, ductility and elastic modulus of Ti-13Nb-13Zr are shown in Figure 2 along with typical values for mill-annealed Ti-6Al-4V. Clearly, the tensile strength, yield strength, % elongation and % reduction of area of Ti-13Nb-13Zr are comparable to or higher than that of Ti-6Al-4V. In addition, the elastic modulus of Ti-13Nb-13Zr is more than 30% lower than that of Ti-6Al-4V (79 GPa vs 115 GPa).

The results of the axial fatigue testing, in the unnotched and notched conditions, are given in Figure 3. A logarithmic equation was used to "curve-fit" each set of data using the least-squares technique to generate the S-N (stress vs no. of cycles) curve in each case. The equation which correlated best with the data was used to determine the 10 million cycle fatigue endurance limit.

The 10 million cycle smooth fatigue endurance limit of both Ti-13Nb-13Zr and mill-annealed Ti-6Al-4V were found to be 500 MPa. The fatigue specimens described above had an average surface roughness (Ra) of 0.12 to 0.2 μm. Specimens with a mirror-finished surface (Ra = 0.05 to 0.07 μm) may be expected to exhibit a higher fatigue endurance limit.

As shown in Figure 3, at Kt=1.6, the 10 million cycle fatigue endurance limit of Ti-6Al-4V was determined to be 320 MPa while that of Ti-13Nb-13Zr was 335 MPa. At Kt=3.0, the 10 million cycle fatigue endurance limit of Ti-6Al-4V was determined to be 170 MPa while that of Ti-13Nb-13Zr was 215 MPa. These results indicate that Ti-13Nb-13Zr is significantly less notch sensitive than Ti-6Al-4V.

The fracture toughness test results are given in Table I. The average plane strain fracture toughness (KIC) of Ti-13Nb-13Zr was 65 MPa√m, while it was 53 MPa√m for mill-annealed Ti-6Al-4V. Hence, the plane strain fracture toughness of Ti-13Nb-13Zr is 20-25% greater than that of Ti-6Al-4V.

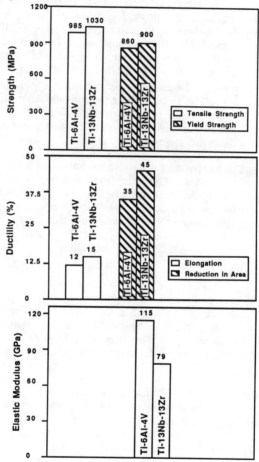

Figure 2 - Mechanical Properties of heat treated Ti-13Nb-13Zr and mill-annealed Ti-6Al-4V. Note that Ti-13Nb-13Zr has a 30% lower modulus than Ti-6Al-4V.

This is consistent with information in the literature which indicates that beta and near-beta alloys have greater fracture toughness than alpha+beta alloys such as Ti-6Al-4V [23]. Furthermore, fatigue crack propagation behavior in Titanium alloys parallels fracture toughness [24], which explains the superior notched fatigue strength of Ti-13Nb-13Zr.

As shown in Table II, the Charpy V-notch impact energy of Ti-13Nb-13Zr was 30 Joules while the corresponding value for mill-annealed Ti-6Al-4V was 25 Joules. This is consistent with the greater plane strain fracture toughness of Ti-13Nb-13Zr as described in the previous paragraph.

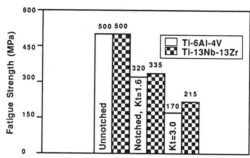

Figure 3 - Results of axial fatigue testing on unnotched and notched specimens, per ASTM E466, from a heat treated plate of Ti-13Nb-13Zr and a mill-annealed plate of Ti-6Al-4V. Frequency = 60Hz, R = 0.1. Results obtained by extrapolation of the logarithmic equation obtained by least squares curve fitting of the stress vs no. of cycles data.

Table I Plane strain fracture toughness of a heat treated plate of Ti-13Nb-13Zr and a mill-annealed plate of Ti-6Al-4V, determined per ASTM E399.

Ti-6Al-4V		Ti-13Nb-13Zr	
Specimen No.	K_{IC} (MPa√m)	Specimen No.	K_{IC} (MPa√m)
1	53.7	1	65.3
2	52.0	2	65.2
3	51.6	3	63.3
4	52.4		
5	54.0		
Average	52.7±0.9		64.6±0.9

Table II Charpy V-notch impact energy of Ti-13Nb-13Zr and Ti-6Al-4V determined per ASTM E23.

Ti-6Al-4V		Ti-13Nb-13Zr	
Specimen No.	Toughness	Specimen No.	Toughness
1	24.4	1	28.5
2	23.0	2	31.2
3	24.4	3	29.8
4	27.1		
Average	24.7±1.5 Joules		29.8±1.1 Joules

The flexural and shear moduli are given in Table III. The flexural modulus of Ti-13Nb-13Zr was determined to be 87 GPa while that of Ti-6Al-4V was 123 GPa.

Similarly, the shear modulus of of Ti-13Nb-13Zr was 30 GPa compared to 48 GPa for Ti-6Al-4V. Hence, by both these measures as well, the modulus of Ti-13Nb-13Zr is 30% lower than that of Ti-6Al-4V.

Table III Flexural and shear moduli of a heat treated plate of Ti-13Nb-13Zr and a mill-annealed plate of Ti-6Al-4V. Flexural modulus determined by three-point bend testing per ASTM D790 and shear modulus determined by torsional shear testing.

Ti-6Al-4V		Ti-13Nb-13Zr	
Specimen No.	Flexural Modulus	Specimen No.	Flexural Modulus
1	120.5	1	87.8
2	123.3	2	87.8
3	123.9	3	86.9
Average	122.6±1.5 GPa		87.5±0.4 GPa
Specimen No.	Shear Modulus	Specimen No.	Shear Modulus
1	48.0	1	30.5
2	48.3	2	30.1
		3	29.9
Average	48.2±0.2 GPa		30.2±0.2 GPa

Microstructure and Phase Analysis

The microstructure of the water quenched plates and bars of Ti-13Nb-13Zr was observed to consist of hcp martensite (α'). Figure 4 is a typical optical micrograph of the unaged microstructure in plates and bars. XRD and TEM analysis showed that only the hcp phase is present.

The microstructure of the heat treated plates and bars of Ti-13Nb-13Zr was observed to consist of aged hcp martensite (α'). Figure 5 is a typical optical micrograph of the aged material. XRD and SAD confirmed the presence of the hcp crystal structure and revealed the presence of submicroscopic bcc beta precipitates as well. The beta precipitates strengthen and harden the material by dispersion strengthening.

The unaged and aged microstructures of Ti-13Nb-13Zr are both acicular in nature. The aging treatment increases the size of the acicular needles but does not alter the prior beta grain size.

Figure 6 is a transmission electron micrograph showing the twinned α' microstructure which is typical of both the unaged and aged materials. The accommodation of mechanical stress via twinning is a common phenomenon in hcp metals due to the lack of five independent slip systems. The martensitic transformation from β to α' results in thermally induced internal stresses which are in part accommodated by the twinning process.

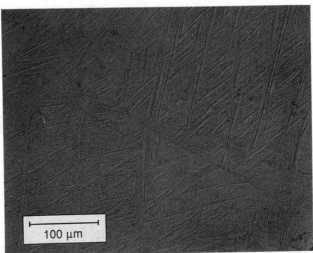

Figure 4 - Optical micrograph of a water quenched Ti-13Nb-13Zr bar showing an unaged, acicular martensitic (α') microstructure.

Figure 5 - Optical micrograph of a heat treated Ti-13Nb-13Zr bar. The micrograph shows an aged, acicular martensitic (α') microstructure. XRD and electron diffraction revealed the presence of the submicroscopic β phase.

Corrosion and Wear

The reciprocal polarization resistance (corrosion rate + electron exchange rate), as determined by EIS, was 40% lower for Ti-13Nb-13Zr (0.41 MΩ^{-1}cm^{-2}) than for Ti-6Al-4V (0.68 MΩ^{-1}cm^{-2}). This is attributable to the more protective passive oxide layer formed on Ti-13Nb-13Zr, which would result in a reduction in metal ion release as well.

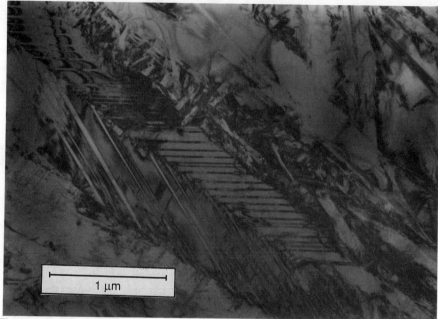

Figure 6: Transmission electron micrograph of aged Ti-13Nb-13Zr showing twinned hcp α' martensite.

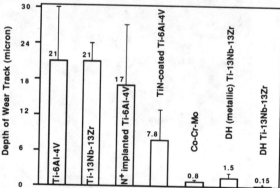

Figure 7 -   Abrasive wear resistance of Ti-13Nb-13Zr, with and without diffusion hardening, in comparison to other orthopaedic materials. Results obtained by pin-on-disc testing in Ringer's solution at an initial stress of 107 MPa.

Figure 7 shows that the abrasive wear resistance of untreated Ti-13Nb-13Zr is the same as that of Ti-6Al-4V. However, after diffusion hardening, the wear resistance of Ti-13Nb-13Zr is significantly better than that of Ti-6Al-4V and even TiN-coated Ti-6Al-4V. Even when the blue, ceramic surface was removed prior to the test, the solid solution hardened region underneath it, referred to as DH (metallic), was still found to be extremely wear resistant, producing a wear track depth of 1.5 µm. Furthermore, DH Ti-13Nb-13Zr and DH (metallic) Ti-13Nb-13Zr did not generate any black debris even in this severe test while Ti-6Al-4V generated considerable amounts of black wear debris.

Surface Analysis

The average surface microhardness of DH Ti-13Nb-13Zr was 930±74 VHN which is significantly greater than that of both Ti-6Al-4V (330 VHN) and Co-Cr-Mo (400 VHN). XPS analysis revealed that the blue, ceramic surface of DH Ti-13Nb-13Zr consists primarily of TiO_2, TiO and ZrO_2. Profilometry and SIMS revealed that the depth of this blue ceramic surface was about 0.2 µm, but the depth of the solid solution hardened layer beneath it was 2-3 µm, more than an order of magnitude greater than the thickness of the hardened layer produced by Nitrogen ion implantation (0.1 µm). This is one of the reasons behind the excellent performance of the DH surface.

Furthermore, the diffusion hardening process produces a hardened surface by diffusion of Oxygen into the substrate, not deposition of an overlay coating on the substrate. The hardened surface produced is an integral part of the substrate, thus eliminating issues such as delamination or spalling which would need to be addressed in the case of an overlay coating. XRD revealed that the presence of Oxygen in interstitial solid solution form had caused lattice expansion, which would result in the presence of compressive stresses, in the diffusion hardened region. This would be expected to have a beneficial effect on the fatigue endurance of the material and tend to counteract the effect of Oxygen content.

Conclusions

- Heat treated mill-products of Ti-13Nb-13Zr attain higher strength and ductility, with a 30% lower elastic tensile modulus, than mill-annealed Ti-6Al-4V.
- The smooth fatigue strength of Ti-13Nb-13Zr is equivalent to that of Ti-6Al-4V.
- Ti-13Nb-13Zr is less notch sensitive than Ti-6Al-4V. The 10 million cycle notched (Kt =3.0) fatigue strength of Ti-13Nb-13Zr is greater than that of Ti-6Al-4V.
- The fracture toughness and impact energy of Ti-13Nb-13Zr are 20-25% greater than the corresponding values of mill-annealed Ti-6Al-4V.
- The corrosion resistance of Ti-13Nb-13Zr is 40% greater than that of Ti-6Al-4V.
- Diffusion hardened Ti-13Nb-13Zr has significantly superior wear resistance than Ti-6Al-4V and even TiN-coated Ti-6Al-4V.

References

1. J. B. Koeneman, T. M. Hansen, and T. R. Toal, Proc. Biomechanics Symposium, ASME, 120 (1991), 117.

2. J. L. Lewis et al., J. Bone and Joint Surgery, 66-A (1984), 280.

3. J. A. Engelhardt and S. Saha, Medical and Biological Eng. and Computing, 26 (1988), 38.

4. M. T. Manley et al., <u>Orthopaedic Transactions</u>, 7 (1983), 344.

5. M. T. Manley et al., <u>Orthopaedic Transactions</u>, 6 (1982), 304.

6. D. O'Connor and W. H. Harris, <u>Proc. Orthopaedic Research Soc.</u> (1992), 295.

7. D. R. Sumner et al., <u>Proc. Orthopaedic Research Soc.</u> (1992), 302.

8. G. L. Maistrelli et al., <u>J. Bone and Joint Surgery</u>, 73-B (1991), 43.

9. V. G. Langkamer, D. M. O'Doherty, and A. E. Goodship, <u>Proc. Orthopaedic Research Soc.</u> (1992), 240.

10. J. D. Bobyn, E. S. Mortimer, A. H. Glassman, C. A. Engh, J. E. Miller, and C. E. Brooks, <u>Clinical Orthopaedics and Related Research</u>, 274 (1992), 79.

11. J. D. Bobyn et al., <u>Clinical Orthopaedics and Related Research</u>, 261 (1990), 196.

12. E. J. Vresilovic, Jr. et al., <u>Proc. Orthopaedic Research Soc.</u> (1992), 337.

13. E. Franks et al., <u>Proc. Orthopaedic Research Soc.</u> (1992), 296.

14. B. C. Burkart et al., <u>Proc. American Academy of Orthopaedic Surgeons</u> (1992), 120.

15. H. B. Skinner and F. J. Curlin, <u>Orthopaedics</u>, 13(11) (1990), 1223.

16. A. Sarmiento and T. A. Gruen, <u>J. Bone and Joint Surgery</u>, 67-A (1985), 48.

17. S. G. Steinemann, "Corrosion of Surgical Implants - in vivo and in-vitro Tests", in <u>Evaluation of Biomaterials</u>, G. D. Winter, et al., Eds., John Wiley, N. Y. (1980), 1.

18. P. G. Laing, A. B. Ferguson, Jr., and E. S. Hodge, <u>J. Biomedical Materials Research</u>, 1 (1967), 135.

19. D.R.C. McLachlan et al., "Aluminum in Human Brain Disease", in <u>Biological Aspects of Metals and Metal-Related Diseases</u>, B. Sarkar, Ed., Raven Press, N.Y. (1983), 209.

20. D.P. Perl, A.R. Brody, <u>Science</u>, 208 (1980), 297.

21. J. C. Heath, <u>British J. Cancer</u>, 10(4) (1956), 668.

22. A. K. Mishra et al., <u>Proc. 12th. Southern Biomedical Engineering Conf.</u>, New Orleans, LA, April 2-4, 1993).

23. F. H. Froes and H. B. Bomberger, in <u>Titanium Technology: Present Status and Future Trends</u>, F.H. Froes, D. Eylon, and H. B. Bomberger, Eds., Titanium Development Association, Dayton, OH (1985), 103.

24. M. J. Donachie, Jr., <u>Titanium - A Technical Guide</u>, ASM International, Materials Park, OH (1988).

ENVIRONMENTAL ASPECTS OF BETA TITANIUM ALLOYS

AN OVERVIEW OF BETA TITANIUM ALLOY ENVIRONMENTAL BEHAVIOR

--Keynote Lecture--

R. W. Schutz

RMI Titanium Company, 1000 Warren Avenue, Niles, Ohio 44446 USA

Abstract

Stemming from their unique combination of elevated strength, low density, and good overall corrosion resistance, beta titanium alloys have become attractive candidate materials for critical, high-stress, components in corrosive services. An overview of the comparative corrosion resistance of beta alloys to conventional alpha and alpha/beta titanium alloys in common industrial and aerospace service environments generally reveals attractive behavior depending on the environment and alloy composition and, in some cases, alloy condition. Expanded performance windows are especially noted for the molybdenum-rich beta alloys, particularly in regard to resisting reducing acids, stress corrosion, and high temperature localized chloride attack, along with hydrogen and oxidation resistance. Where applicable, implications of this enhanced corrosion performance on current and perspective beta alloy applications are also noted.

Introduction

Beta titanium alloys are rapidly becoming choice, attractive materials for critical high-stress, corrosive service based on their high strength, low density and superior corrosion resistance. By virtue of the rather large alloying additions possible and the unique intrinsic characteristics of the beta phase and its crystal structure, the useful windows of alloy application can be expanded beyond the scope of conventional titanium alloys in many aggressive environments. Current prominent and evolving applications of these beta alloys include deep oil/gas and geothermal brine well tubulars and components (1,2,3), offshore production and Naval ship marine fasteners (4), structural components for hydrogen-cooled hypersonic aircraft (i.e., NASP), and hot, critical aircraft components exposed to Skydrol hydraulic fluid.

This paper presents an overview of the environmental behavior of beta titanium alloys, with direct comparisons in corrosion performance to conventional alpha (α) and alpha-beta (α/β) alloys. Resistance to uniform, localized, and stress corrosion, as well as hydrogen, oxidation, and ignition resistance is surveyed with respect to beta alloy composition and metallurgical condition. Although corrosion mechanisms are not addressed for brevity sake, implications of specific beta alloy corrosion behavior to relevant applications are noted.

Beta Titanium Alloys in the 1990's
Edited by D. Eylon, R.R. Boyer and D.A. Koss
The Minerals, Metals & Materials Society, 1993

Beta Alloys Surveyed

The beta titanium alloys surveyed in this review include the well-characterized, commercial metastable beta alloys listed in Table 1. These alloys range from near-beta, solute lean beta alloys (i.e., Ti-8Mn, Ti-10-2-3) to the solute-rich beta alloys (i.e., Beta-C, Ti-13-11-3), exhibiting a wide variation in V, Mo, and Cr alloy content. Where possible, alloys are assessed in both the low strength solution-treated (ST) and high strength aged (STA) conditions. Corrosion performance of these alloys is directly compared to that of the α alloy, Gr. 2 (unalloyed) titanium, and the annealed α/β alloy, Ti-6-4.

Table 1 Beta Titanium Alloys Surveyed

ALLOY DESIGNATION	NOMINAL COMPOSITION (wt.%)
Ti-8Mn	Ti-8Mn
Ti-45Nb	Ti-45Nb
Ti-10-2-3	Ti-10V-2Fe-3Al
Ti-15-3-3-3	Ti-15V-3Al-3Sn-3Cr
Transage 207	Ti-8Mo-2.5Al-9Zr-2Sn
Ti-8-8-2-3	Ti-8Mo-8V-2Fe-3Al
Ti-13-11-3	Ti-13V-11Cr-3Al
Ti-15-5	Ti-15Mo-5Zr
Beta-C	Ti-3Al-8V-6Cr-4Zr-4Mo
Beta-C/Pd	Ti-3Al-8V-6Cr-4Zr-4Mo-0.05Pd
Beta III	Ti-11.5Mo-6Zr-4.5Sn
Beta-21S	Ti-15Mo-2.7Nb-3Al-0.2Si

General Corrosion Behavior

As with α and α/β titanium alloys, the beta alloys rely on the formation of an extremely thin (typically 50-200Å thick) adherent, protective titanium oxide (primarily TiO_2) film for passivity and corrosion resistance. The extent to which the alloying elements in these beta alloys either interfere with or enhance oxide film formation and influence film morphology will determine whether the alloys exhibit expanded performance windows in aqueous media over leaner titanium alloys.

Natural Waters and Salt Solutions

Like other titanium alloys (5), beta titanium alloys are considered to be fully resistant to uniform corrosion in all natural waters to temperatures in excess of 200°C. This includes exposures to distilled water, fresh waters, seawater (aerated, deaerated, or contaminated), and natural brines. This general passivity also extends into practically all common salt solutions, including those of chloride (whether sweet or sour), sulfates, carbonates, and phosphates over the pH range of 3-12 and to temperatures as high as 200°C. Practically speaking, this general resistance to salt solutions means that beta alloys do not require special or additional corrosion protection (i.e., cathodic or anodic protection, or coatings) in any atmospheric, marine, or downhole brine service exposures, and will not "rust" like steels.

Oxidizing Acids

The resistance of beta alloys to oxidizing acid media is highly dependent on the oxidizing potential and temperature of the solution. Table 2 (6,7) reveals that beta alloys exhibit excellent resistance to mildly-reducing aerated HCl (pH 1) and moderately-oxidizing $FeCl_3$ solutions up to at least the boiling point. However, significantly enhanced uniform corrosion can be expected in hot, severely-oxidizing acids such as boiling HNO_3 compared to unalloyed titanium. Except for niobium, addition of most common beta alloying elements diminishes alloy resistance to hot, severely oxidizing acid media due to their inherent reactivity with oxidizing acids. On the other hand, these results do imply that the useful passive range of beta alloys in reducing acids (i.e., HCl, H_2SO_4) may be expanded by the presence or addition of oxidizing, inhibitive species, such as oxidizing metal ions (i.e., Fe^{+3}, Cu^{+2}, Pt^{+2}), nitrates, chromates, and or chlorine. This suggests application possibilities in metallic ore a c i d - l e a c h i n g (hydrometallurgical) process equipment or mild-moderate oxidizing chemical plant service where higher strength components are required.

Table 2 Resistance of Beta Titanium Alloys to Boiling Oxidizing Acid Media

	Corrosion Rate (mm/yr)		
ALLOY	**AERATED HCl (pH = 1)**	**10% FeCl₃ (pH ~0.3)**	**10% HNO₃**
Ti-Gr. 2	0.64	0.01	0.08
Ti-6-4	0.60	0.00	0.33
Ti-8-8-2-3	0.00	0.01	--
Beta-C*	0.00	0.01	0.18
Beta-C/Pd*	0.00	0.00	0.15
Beta 21S*	0.00	0.01	--

* ST and STA conditions

Reducing Acids

Due to predicted oxide film instability and breakdown at negative potentials and low pH (5), titanium alloys, in general, exhibit their most serious performance limitations in pure reducing acid media. Alloy resistance is highly dependent on acid concentration and temperature, which diminishes as either of these two factors increases. Common industrial acids encountered include HCl, HBr, H_2SO_4, and H_3PO_4 inorganic acids, and acetic, formic, and citric organic acids.

As a common, highly corrosive acid, HCl solutions are frequently utilized as a discriminating alloy screening media. Figure 1 presents corrosion rate profiles of aged beta alloys in boiling naturally-aerated HCl media as a function of acid concentration (6-7). Each alloy curve exhibits a passive-to-active corrosion transition acid concentration

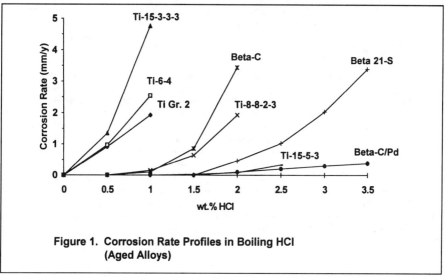

Figure 1. Corrosion Rate Profiles in Boiling HCl (Aged Alloys)

Table 3. Active-to-Passive Transition Acid Concentrations for Beta Titanium Alloys in Boiling HCl

(0.13 mm/yr transitions)

ALLOY	wt. % HCl	
	ANNEALED CONDITION	AGED CONDITION
TI-10-2-3	–	0.08
TI-13-11-3	0.10	–
TI-15-3-3-3	0.12	0.08
TI-6-4	0.12	0.13
TI-Gr. 2	0.16	–
Beta-C	1.1	0.87
TI-8-8-2-3	1.1	1.0
TI-15-5-3	2.1	2.1
Beta-21S	5.0	1.5
Beta-C/Pd	2.1	2.1

Increasing Acid Resistance ↓

assumed to correspond to 0.13 mm/yr corrosion rate, below which useful resistance can be expected. These transition acid concentrations, listed in Table 3, permit direct ranking of beta alloy corrosion resistance and reveal that exceptional increases in HCl resistance relative to titanium grades 2 and 5 are exhibited by the Mo-containing beta alloys. The non-Mo beta alloys, on the other hand, are substantially less resistant, and comparable to the Gr. 5 titanium alloy.

Regression analyses of a wide range of commercial titanium alloys in boiling HCl (6) indicate that an alloy content of >3% Mo and/or >8% Zr is exceptionally beneficial, vanadium is of minor benefit, and Al levels above 3% are increasingly detrimental to reducing acid resistance. Table 3 and Figure 1 results also show the potent beneficial influence of minor (~0.05%) Pd (or other platinum group metals [7]) additions, which is evident when comparing the Beta-C and Beta-C/Pd alloys. Significant positive synergism between "Mo" and "Pd" alloy content in beta alloys is also indicated (8,9). It is also noteworthy that high (>5%) alloy Nb content greatly enhances acid resistance, whereas silicon levels above ~0.1% detract from alloy performance. Except for Beta-21S, the beta alloys generally display comparable reducing acid performance in either ST or STA condition (7) at low-medium strength levels (see Table 3). At elevated strength levels and/or where highly-active, severely corroding conditions prevail, the STA condition is generally least resistant.

Practically speaking, these results indicate that Mo-rich and/or Pd-enhanced beta alloys offer expanded use over most common industrial titanium alloys in hotter and/or stronger reducing acid streams found in various chemical processes, hydrometallurgical ore leaching, and downhole (well) acidizing treatments (1).

Skydrol Hydraulic Fluid

Skydrol hydraulic fluids used on commercial and military aircraft consist of phosphate esters that are known to thermally decompose at temperatures above ~140°C. Decomposition by-products include a thick syrupy organic-phosphate acid (i.e., H_3PO_4) which is highly corrosive to all commercial titanium alloys when hot, and has led to localized attack and perforation, and excessive hydrogen absorption of critical aircraft components such as hot air ducting and engine nacelle structures. Recent RMI Titanium lab data generated from the Boeing Skydrol fluid drip test at ~230°C (10) reveals localized attack on the order of 0.10 - 0.15 mm deep for Ti-6Al-2Sn-4Zr-2Mo and Gr. 7 titanium alloys after 1000 hours exposure. In contrast, the Beta-21S alloy in this test experienced only ~20% of this attack depth with minimal hydrogen absorption. These and other findings strongly suggest that Mo-rich (>10-15% Mo) beta alloys may be uniquely used or further developed to satisfactorily resist thermal attack by Skydrol fluid decomposition in hot critical aircraft components.

Titanium alloys, in general, are recognized for their elevated resistance to pitting in most aqueous media including halide brines. Alloy pitting resistance is readily determined electrochemically by anodic breakdown and repassivation pitting potential

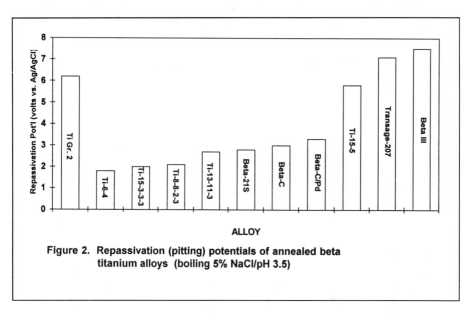

Figure 2. Repassivation (pitting) potentials of annealed beta titanium alloys (boiling 5% NaCl/pH 3.5)

measurements, and is usually conducted in aqueous chloride solutions where conservative pitting potentials are obtained. Repassivation potential, which is defined as the lowest potential at which anodic pitting is possible, has been measured by the author (11) in boiling 5% NaCl (pH 3.5) using the galvanostatic method at a constant current density of 200 ma/cm². The repassivation potential values for annealed beta alloys surveyed are plotted in Figure 2, revealing a wide range of values highly dependent on alloy composition. Regression analysis of this data (11) suggests that Al above 3%, V, and Si all significantly diminish pitting potentials, with very minor negative effects of Cr, Fe and Sn. Although Mo may be slightly beneficial at lower levels, a neutral influence is indicated at high (>4%) alloy levels. Low Al (<3%) and high zirconium alloy additions elevate alloy pitting potential to levels as high as or even greater than those of unalloyed titanium.

Since all potential values were well above +1 volt, these results imply that all beta titanium alloys can be expected to fully resist to spontaneous chloride pitting in hot brine service. If elevated anodic pitting potentials are required for anode service, impressed current situations, or high temperature bromide process streams, then beta alloys with high Zr and Mo, and low Al alloy content can be selected or formulated to optimize pitting potential. This has practical application in high temperatures (HBr-catalyzed) organic synthesis process components and impressed current anodes for halide electrolytes.

Crevice Corrosion Resistance

Crevice corrosion is often the critical mode of attack in halide (i.e., Cl⁻, Br⁻) brines and sulfate solutions that can limit use of many titanium alloys when temperatures exceed ~75-90°C. Crevice attack requires the presence of a tight metal-to-metal gasket-to-metal, or salt deposit-to-metal crevices, and is highly dependent on and aggravated by increasing solution temperature and decreasing pH (5).

Approximate temperature limits for the beta titanium alloys surveyed are depicted in Figure 3 in various common industrial brine environments. Since local crevice environments consist of hot deaerated reducing acids (5), it is not surprising that the crevice resistance of beta titanium alloys reflects and follows the alloy composition trends discussed in the reducing acid (boiling HCl) corrosion section. As before, the Mo-rich and Pd-enhanced alloys such as Beta-21S (5), Ti-15-5 (12), and Beta-C/Pd (7,13), and Beta III (14) offer optimal crevice resistance in sweet or sour brines to as high as 250°C when pH's equal or exceed 3. This resistance extends down to pH 1 in NaCl brines to temperatures as high as 100°C (boiling point). Resistance to localized attack in concentrated hydrolyzable brines, such as $MgCl_2$, $CaCl_2$, and CaBr up to 232°C has been demonstrated as well (1). The leaner-Mo alloys (Group "B") exhibit a somewhat reduced upper shelf, but offer measurably expanded temperature limits (1,6,7) over the non-Mo or -Pd containing (Group "A") alloys. Very little influence of alloy strength level or metallurgical condition on brine crevice corrosion resistance is indicated (7,12).

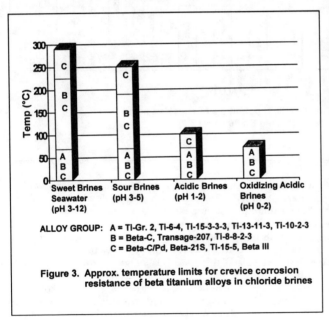

ALLOY GROUP: A = Ti-Gr. 2, Ti-6-4, Ti-15-3-3-3, Ti-13-11-3, Ti-10-2-3
B = Beta-C, Transage-207, Ti-8-8-2-3
C = Beta-C/Pd, Beta-21S, Ti-15-5, Beta III

Figure 3. Approx. temperature limits for crevice corrosion resistance of beta titanium alloys in chloride brines

Practically speaking, it is important to note that all titanium alloys, regardless of composition or solution pH or chemistry, are not susceptible to crevice attack below the threshold temperature of approximately 75°C. Thus, near-ambient brine service environments, such as seawater exposures in offshore or Navy ship applications, pose no threat of localized chloride attack to any beta titanium alloys. For elevated temperature geothermal brine and deep sour gas well service (1-3) or hot halide chemical process streams, the Mo-rich and/or Pd-enhanced beta alloys are prime high strength materials that fully resist localized attack.

Stress Corrosion Cracking (SCC) Behavior

Ambient Salt Solutions

Like most commercial titanium alloys, beta titanium alloys are generally resistant to SCC in aqueous salt solutions in stressed smooth and notched component configurations (15,16). One of the few exceptions is the Ti-13-11-3 alloy, which may stress crack in distilled water or 3.5% NaCl solution in sustained bend specimens in certain metallurgical conditions (16). Most other beta alloys require a highly loaded crack or sharp flaw to reveal possible indications of environmental degradation, manifesting itself as accelerated crack growth rate and K_{Ic} reduction (e.g., K_{Iscc}).

A thorough review of beta alloy SCC behavior found in Reference Nos. 15 and 16 reveal that alloy SCC resistance is highly dependent on both alloy composition and metallurgical condition. Beneficial alloying elements include the beta-isomorphous elements: Mo, V, Nb and Ta; whereas, $\geq$5% Al, Sn, and the beta eutectoid-and compound-forming elements: Cr, Mn, and Si, are clearly detrimental. Thus, the Mo-, Nb-, and/or V-rich beta alloys generally tend to exhibit little or no SCC susceptibility in ambient salt solutions in most product forms. However, one must factor in the influence of metallurgical structure, which can impart some SCC susceptibility even on these more resistant alloys. More specifically, it has been found that increasing SCC tendency stems from: 1) increasing beta grain size (15,17); 2) increasing volume fraction of alpha phase (via aging)(8,15); 3) increasing degree of grain boundary alpha (8,15,18,19); and 4) nonuniform aging (8). Therefore, optimal saltwater K_{Iscc} values for many beta alloys can be achieved at almost any strength level if good age uniformity with no or minimal grain boundary alpha is achieved.

Table 4 Other Aqueous SCC Studies on Mo-Rich Beta Alloys

Alloy/Cond	YS (MPa)	Environ-ment	SCC Test	Cathodic Charging Level	Result
Beta-C (ST & STA)	800 - 1620	NACE Sour Brine (25°C)	smooth/precracked tensiles	0 and 0.5 mA/cm^2	No SCC, except for charged STA condition
Beta-C (STA)	1262	Seawater (25°C)	SSRT	None, -1.0, -1.5V	No SCC
Beta-C Beta-21S (ST)	827 827	Seawater (25°C)	SSRT	None, -0.85, -1.0V	No SCC
Beta-21S (STA)	1241	3.5% NaCl (25°C)	SSRT	None, -1.5 to +1.5V	No SCC

The excellent SCC resistance of the Mo-rich beta alloys, Beta-C and Beta-21S, in ambient seawater and sweet and sour brines has been demonstrated as shown in Table 4. Discriminating slow strain rate (SSRT) tests by various investigators (20-23) reveal immunity to chloride SCC (in seawater) over a wide range of strength levels, even when cathodically charged up to -1.5 volts (vs. Ag/AgCl) to stimulate hydrogen embrittlement. Sustained load tensiles of ST and STA Beta-C titanium in ambient sour (H_2S-sat.) brines, with and without cathodic charging (24), also failed to show susceptibility to SCC. The one exception was the charged sample in the elevated strength (1622 MPa YS) STA condition, where high alpha phase volume fraction reduced alloy hydrogen tolerance. Excellent SCC resistance implies that most beta alloys are ideal for high strength components (i.e., fasteners, pumps, valves, risers) in marine, offshore, and Naval ship service.

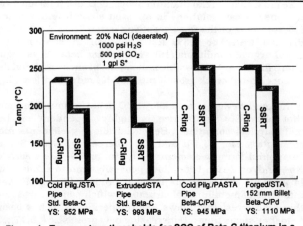

Figure 4. Temperature thresholds for SCC of Beta-C titanium in a worst-case deep gas well sour brine.

Hot Brines

The same alloy composition and metallurgical condition trends and guidelines for beta alloys in ambient salt solutions also apply to hot sweet and sour brines. For this reason, the Mo-rich beta alloys, such as the Beta-C and Beta III alloys (1,13,14,25) are naturals for geothermal brine and deep sour gas well service. Constant deflection test exposures of various beta alloys (14,25,27) including Beta-C, Transage 207, Ti-8-8-2-3, and Ti-15-5 in sour NaCl brines up to 232°C fail to induce alloy SCC in either ST or STA conditions. As indicated in Figure 4, temperature limits for Beta-C alloy SCC resistance in a worst-case deep sour gas well fluid (13) are a function of SCC test method, alloy product form (metallurgical condition) and alloy Pd content. Minor Pd additions to certain α/β and beta alloys have been shown to significantly elevate alloy SCC resistance in high temperature sweet and sour brines (1,13,14,25,27), allowing beta alloys to outperform Ni-Cr-Mo alloys in worst-case deep sour gas well environments.

Hot Salt Residues

Many titanium alloys may experience brittle cracking above ~240°C when moist halide salt residues exist on stressed metal surfaces. This phenomenon, known as "hot-salt SCC" and thoroughly reviewed in Reference 15, is dependent on many factors including temperature, stress level, exposure time, salt residue type, and alloy composition on microstructure. As with saltwater SCC, beneficial alloying elements are Mo, Nb, V, and Zr; whereas >3% Al and higher

Table 5 Relative Susceptibility of Beta Alloys to Hot Salt SCC

ALLOY	RELATIVE SUSCEPTIBILITY
Ti-8Mn	high
Ti-13-11-3	intermediate
Ti-6-4	intermediate
Ti-15-3-3-3	intermediate
Beta-C	more resistant
Ti-8-8-2-3	more resistant
Beta III	more resistant
C.P. Ti	immune

interstitial levels are clearly detrimental. Grain boundary alpha and increasing alpha phase volume fracture exacerbate alloy susceptibility. As the titanium alloy ranking in Table 5 suggests, the Mo-rich (and probably Pd-containing) beta alloys represent highly resistant materials to hot salt residues compared to most α/β and beta alloys. This means that these more resistant beta alloys can tolerate greater stresses, higher temperatures, and/or longer exposures without experiencing cracking problems in halide salt environments above 250°C.

Methanol

Like all other titanium alloys, beta alloys are highly susceptible to stress cracking in anhydrous or relatively dry methanol (15,28,29). With the exception of Beta-C and Beta-C/Pd alloys, little data is available addressing practical SCC thresholds for beta alloys as a function of methanol water content. Recent studies on the Beta-C alloy (29) reveal that methanolic SCC can be fully inhibited by water additions, is aggravated by chlorides, exhibits decreasing susceptibility with increasing temperature, and depends strongly on SCC test method. Furthermore, SCC susceptibility is somewhat enhanced by minor Pd alloy additions, and increasing strength level, and in products exhibiting coarse grain structures (i.e., billet). These studies indicate a wide range of minimum water additions of 1.5 - 10 wt.% to fully inhibit methanolic SCC, depending on product form and test method. In most cases, water levels in excess of 3-5 wt.% are adequate, which are somewhat greater than the 1.5 - 2.0% water levels required for unalloyed titanium and Ti-6-4 (15). Practically speaking, these minimum water levels for preventing methanolic SCC in beta alloys are compatible with water levels (5 - 30%) in commercial methanol grades often used offshore or injected into wells to stimulate flow.

Corrosion Fatigue

Titanium alloys, such as unalloyed titanium and Ti-6-4, are known for their excellent resistance to corrosion fatigue (5), exhibiting no significant reduction in endurance limit in saltwater or seawater media. This feature appears to apply to beta alloys as well, based on fatigue testing of Beta-C, Beta-21S, and Ti-10-2-3 alloys in 3.5% NaCl brine at 25°C. Studies by Krugman and Gregory (30), and G. Yoder (31) revealed no accelerated cracking and no change in ΔK_{TH} during da/dn (R=0.1) testing of ST and STA (827 - 1220 MPa YS) Beta-C and Ti-10-2-3 alloys down to 0.1 Hertz. B. Bavarian (23) also found that smooth S-N curves (R=0.1, -1) remained unchanged for STA (1241 MPa) Beta-21S in air versus saltwater. Resistance to corrosion fatigue in beta alloys is a highly desirable feature in dynamic high-strength components for marine, offshore, downhole, or aerospace service, including fasteners, springs, valves, and risers.

Resistance to Hydrogen

Alpha and alpha-beta titanium alloys are recognized for their limited tolerance to absorbed or internal hydrogen due to formation of brittle titanium hydride phase (5,32,33). This tolerance limit is on the order of several hundred ppm hydrogen, beyond which fast and/or slow strain rate embrittlement of the alloy can be expected. By virtue of the bcc crystal structure, beta alloys do not form hydrides (32,33) and possess very high hydrogen solubilities. In the annealed (all-beta) condition, hydrogen tolerance (solubility) is on the order of 4,000 - 5,000 ppm, and may be even greater in Mo- or V-rich alloys (32,33), but lower in Fe-containing alloys (33). Aging of beta alloys to higher strengths increases volume fraction of alpha phase and may produce grain boundary alpha, both of which reduce this hydrogen tolerance window significantly as shown by Young and co-workers (18,33,34). The beta phase is also characterized by its relatively high hydrogen diffusion coefficient, which is on the order of 1,000 times greater than that of alpha phase at room temperature (i.e., ~5 x 10^{-7} vs. ~10^{-10} cm^2/sec.)(33, 35). Unlike α and α/β alloys, most beta titanium alloys are naturally resistant to sustained load cracking over a wide range of strengths and hydrogen levels (36) stemming from elevated hydrogen solubility.

Absorption in Aqueous Solutions

Beta alloys effectively resist absorption of nascent hydrogen that may be produced by minor corrosion or cathodically charged due to their protective titanium oxide surface films (33,35). The Mo-rich beta alloys, Beta-C and Beta III, resist hydrogen absorption and embrittlement under sustained tensile load in high temperature sour brines (14). Similar hydrogen uptake resistance is indicated for Beta-C/Pd exposed to dilute boiling HCl and high temperature NaOH solutions (7). Various cathodic charging studies of ST and STA Beta-C, Beta-21S, and Ti-10-2-3 alloys in ambient sweet and sour NaCl brines (24), seawater (20-23), and NaOH solutions (8) involving impressed potentials of -1.0 and -1.5 volts SCE also failed to show excessive absorption or embrittlement by hydrogen. Superimposed slow straining or sustained tensile loads on these samples during charging did not induce embrittlement either, indicating high SCC resistance in these alloys as well (20-23). Cathodic charging tests in hot acidic brines (12) indicate that the Mo-rich alloys may resist hydrogen uptake better than most α and α/β alloys. Practically speaking, this enhanced hydrogen absorption resistance suggests that beta alloys can resist degradation by cathodic protection systems in marine/offshore applications, and offer promise in aqueous service (i.e., high temp. organic acids) where excessive hydrogen absorption and embrittlement of α and α/β alloys have limited titanium use. Combined with high hydrogen solubility, beta alloys may also be reasonable candidates for cell cathodes when current densities are not excessive.

Gaseous Hydrogen

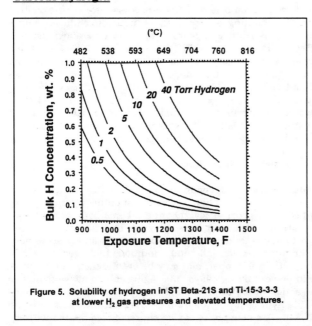

Figure 5. Solubility of hydrogen in ST Beta-21S and Ti-15-3-3-3 at lower H$_2$ gas pressures and elevated temperatures.

The oxide surface film on beta titanium alloys also acts as an effective barrier to both ingress (absorption) and egress (degassing) of hydrogen in atomic or molecular forms. This oxide barrier remains effective if oxygen or water contaminants exist in the hydrogen gas even at higher temperatures, retarding significant hydrogen absorption occurs. In pure H$_2$ gas, however, the oxide film may be dissolved and eliminated above ~600°C, and more rapid absorption (or degassing) of hydrogen occurs in the beta alloys. Extensive hydrogen gas exposure tests on oxide film-free ST Beta-21S and Ti-15-3-3-3 alloys by H. G. Nelson (37) have provided great insight into the equilibrium behavior of hydrogen in beta alloys. Figure 5 (37) reveals that hydrogen solubility increases with decreasing temperature, such that even rather low (~1 torr) H_2 pressures can result in high hydrogen solubility at temperatures above ~425°C. This hydrogen uptake is reversible at these higher temperatures consistent with gas pressure and equilibrium considerations. Although very tolerant at these higher temperatures, these hydrogen-rich beta alloys may then become brittle at lower temperatures depending on the extent the absorbed hydrogen has elevated the alloy's ductile-to-brittle transition. Figures 6 a, b (37) quantify this phenomenon, and confirm that the hydrogen tolerance of these beta alloys at 25°C is in the 4,000 - 5,000 ppm range. Similar behavior and hydrogen tolerances for these two alloys are indicated in hydrogen gas charging studies by Schwartz (38). Resistance to hydrogen embrittlement in high pressure (34.5 MPa) hydrogen gas exposures is also noted for the Beta-21S alloy below ~200°C (39).

The significance of these findings is that safe windows of hydrogen gas exposure can be identified for beta alloys where embrittlement at very high (>650°C) or low (<400°C) temperatures can be avoided at lower gas pressures. Although the natural oxide film may also prevent excessive absorption below ~550°C at low H_2 pressures, surface barrier coatings can expand exposure windows. This enhanced, reversible hydrogen solubility explains why beta alloys, such as

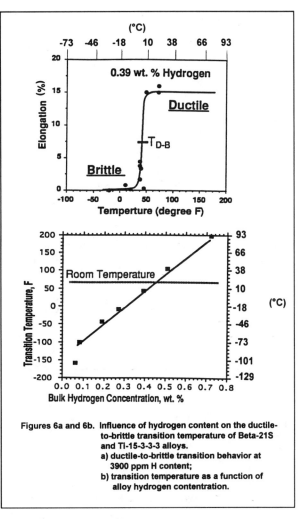

Figures 6a and 6b. Influence of hydrogen content on the ductile-to-brittle transition temperature of Beta-21S and Ti-15-3-3-3 alloys.
a) ductile-to-brittle transition behavior at 3900 ppm H content;
b) transition temperature as a function of alloy hydrogen contentration.

Beta-21S, are candidate materials for hydrogen-cooled aerospace vehicles such as NASP. Tolerance to absorbed hydrogen in hydrogen-rich chemical and organic synthesis reactions also make beta alloys attractive for chemical plant service.

Oxidation Resistance

The upper temperature limit for practical use of titanium alloys in atmospheric exposures is generally identified by the onset of excessive surface oxidation (i.e., metal wastage) and absorbed oxygen embrittlement. This limit typically occurs around 540-600°C for conventional titanium alloys in long term exposures.

The common vanadium-rich beta alloys (i.e., Ti-15-3-3-3) exhibit practical use limits slightly below these temperatures. Alloying elements known to interfere with protective titanium oxide film formation and accelerate oxidation rates in titanium include V, Sn and Zr. The detrimental influence of these elements in the beta alloy, Ti-15-3-3-3, is clearly indicated relative to unalloyed titanium in the oxidation vs. time curves

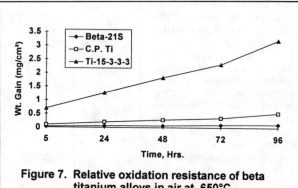

Figure 7. Relative oxidation resistance of beta titanium alloys in air at 650°C

plotted in Figure 7 (39). On the other hand, alloy additions of Mo, Cr (>10%), Nb, Ta, Al and Si all significantly enhance oxidation resistance (i.e., decrease sample weight gain) to temperatures as high as 816°C (40). Optimization of these beneficial elements in a Mo-based beta titanium alloy has led to the development of the Beta-21S alloy. The superior oxidation resistance of this alloy relative to other titanium alloys is shown in Figure 7 at 650°C, and has been demonstrated at 816°C as well (40). The Beta-21S alloy is ~100 times more resistant than Ti-15-3-3-3, and only 3-4 times less resistant that titanium aluminides at 816°C.

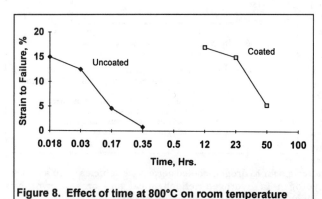

Figure 8. Effect of time at 800°C on room temperature strain to failure for 0.11 mm Beta-21S sheet

While the Beta-21S formulation does alleviate metal wastage via surface oxide formation up to ~800°C, alloy embrittlement from diffused-in surface oxygen still presents serious use limitations above 550-600°C. The diffusion coefficient for oxygen in beta titanium is substantially greater than that in alpha titanium (41). At 538°C, for example, it is approximately 15 times greater (1.5×10^{-16} vs. 10^{-17} cm^2/sec.). This means enhanced inward growth of a brittle oxygen-rich alpha case into the surface, compromising alloy ductility and fatigue resistance with time. Reduction of Beta-21S sheet ductility with time at 510-615°C has been demonstrated (42), becoming increasing detrimental with decreasing sheet thickness. This effect is evident for the uncoated thin Beta-21S foil exposed with time at 800°C plotted in Figure 8 (43,44).

The inward diffusion of oxygen can be prevented or inhibited by application of a surface coating to thwart beta alloy embrittlement. Studies by Wiedemann (45) suggest that a thin (~3 micron) coating, consisting of vacuum evaporation -deposited Al to form a Al_3Ti reaction barrier covered with a top layer of silicophosphate glass (sol-gel), can effectively retard oxygen diffusion effects. The curve in Figure 8 (45) for coated Beta-21S foil reveals that reasonable ductility is retained even after 24 hours at 800°C. Since thin, beta titanium alloy foils are often utilized in the manufacture of titanium metal matrix composites (MMC's) for hot, critical aerospace structures, thin surface barrier coatings for beta alloys will be vital for expanding temperature windows.

Ignition Resistance

Air

All titanium alloys are generally resistant to ignition or burning in ambient air exposures in most common mill product forms. It is only when extremely high surface area to volume ratios exist, such as in titanium powder, where ignition in air is possible. Ignition and burning may occur when sufficient exothermic heat from the metal oxidation reaction raises and maintains metal surface temperatures at or above the alloy's melting point. When inadequate local heat extraction exists, the normally protective titanium oxide film continues to dissolve into the molten metal surface, allowing rapid exothermic oxidation to sustain itself. The enhanced oxygen activity associated with elevated pressures and high air flow velocities in gas turbine jet engine compressors represents conditions where ignition/burning is possible for many titanium alloys (especially in thin-section blades).

Ignition tests on various titanium alloy blade samples (46,47) employing a laser beam ignitor in pressurized, flowing air have ranked the burn severity of various titanium alloys. Table 6 data (46) reveals that the only alloys that resisted burning were the two beta alloys, Ti-13-11-3 and Alloy C (Ti-35V-15Cr). In fact, the Alloy C resisted ignition at an air pressure of 2.8 MPa and 454°C (48), which no other commercial titanium alloy can survive, with the exception of Ti aluminides. These results suggest that chromium content above 11% is especially effective for enhancing ignition/burn resistance in air. Other possible beneficial alloy additions include Cu (>10%), Fe, Mo, Ni, and Si. It appears that the highly alloyed beta alloys, particularly those with high Cr, offer the most promise for extending the windows of safe, ignition-resistant use of titanium alloys in gas turbine engines.

Table 6 Relative Burn Resistance of Beta Alloys Ignited in Flowing, Pressurized Air*

ALLOY	BURN SEVERITY
Ti-6-4	37 (% sample area burned)
Ti-6-6-2	31
Ti-6-2-4-6	87
Ti-8Mn	37
Ti-13-11-3	0
Ti-35V-15Cr (Alloy C)	0 (at 2.8 MPa and 454°C)

* Test Conditions:
  0.48 MPa air pressure
  2.13 m/sec. flow rate
  427°C

Oxygen Gas

The susceptibility of titanium alloys to ignition and burning increases dramatically in pure oxygen gas relative to air. Ignition studies on C.P. titanium and Ti-6-4 (5) suggest that increasing oxygen content (above 35 vol.%), pressure, and temperature all radically stimulate ignition tendencies. Recent tensile-overload ignition tests (49) indicate that the Ti-45Nb beta alloy uniquely resists ignition in pure oxygen gas to pressures as high as ~3.1 MPa at 250°C. In comparison, unalloyed titanium exhibits susceptibility at oxygen pressures of ~1.7 MPa in this test. The availability of an ignition resistant beta titanium alloy for pressurized oxygen gas service has positive implications for critical component applications in wet oxidation waste treatment processes and in oxidative ore leach (hydrometallurgical) processes, where oxygen ignition and related safety concerns have thwarted titanium use in the past.

Conclusions

An overview of the comparative resistance corrosion resistance of beta alloys to conventional α and α/β titanium alloys in common industrial and aerospace environments generally reveals attractive behavior by the beta alloys. Optimized use of the beta alloys in each environment depends on alloy composition and, often, metallurgical condition. Expanded performance windows are especially noted for the molybdenum-rich beta alloys, particularly in regard to resisting reducing acids, stress corrosion, and high temperature localized chloride attack, along with hydrogen and oxidation resistance. These beta alloy performance enhancements stem from their very high alloy content and inherent beta crystal (bcc) structure properties, offering a means to expand titanium use into non-traditional service environments.

References

1.   D. R. Klink and R. W. Schutz, "Engineering Incentives for Utilizing Ti-3Al-8V-6Cr-4Zr-4Mo Titanium Tubulars in Highly Aggressive Deep Sour Gas Wells," Paper No. 63, presented at NACE Corrosion '92, Nashville, TN, April 27-May 1, 1992.

2.   D. Dunlap and R. W. Schutz, "Utilization of Beta-C Titanium Components in Downhole Service," Denver '93, 122nd TMS Annual Meeting, Denver, CO, Feb. 21-25, 1993.

3.   W. W. Love, et. al., "The Use of Ti-38644 Titanium for Downhole Production Casing in Geothermal Wells," Sixth World Conference on Titanium Proceedings, Cannes, June 6-9, 1988, Société Française de Métallurgie, Les Ulix Cedex, France, pp. 443-448.

4.   K. Faller and W. Edmonds, "The Incentives for Titanium Alloy Fasteners in the Navy/Marine Environment," presented at the 1991 Ship Maintenance in the 21st Centruy Symposium, Virginia Beach, Virginia, Feb. 21-22, 1991, American Society of Naval Engineers, Alexandria, VA.

5.   Metals Handbook, 9th ed., Vol 13, "Corrosion," ASM International, 1987, pp. 669-706.

6.   R. W. Schutz and J. S. Grauman, "Fundamental Characterization of High-Strength Titanium Alloys," ASTM STP 917, ASTM, Philadelphia, 1986, pp. 130-143.

7.   R. W. Schutz and M. Xiao, "Enhancing Corrosion Resistance of the Ti-3Al-8V-6Cr-4Zr-4Mo Alloy for Industrial Applications," Proceedings of the Seventh World Conference

on Titanium, San Diego, CA, June 1992.

8.   R. W. Wood, Beta Titanium Alloys, Report #MCIC-72-11, Metals and Ceramics Information Center, Battelle-Columbus Laboratories, Sept. 1972.

9.   M. Stern and C. Bishop, "The Corrosion Resistance and Mechanical Properties of Titanium-Molybdenum Alloys Containing Noble Metals," Transactions of the ASM, Vol. 54, 1961, p. 287.

10.   R.W. Schutz and M. Xiao, unpublished lab data, RMI Technical Center, RMI Titanium Company, Niles, OH, Dec. 1992.

11.   R. W. Schutz and J. S. Grauman, "Compositional Effects on Titanium Alloy Repassivation Potential in Chloride Media," Advances in Localized Corrosion, Proceedings of the Second International Conference on Localized Corrosion, June 1-5, 1987, Orlando, FL, National Association of Corrosion Engineers, Houston, TX, 1990, pp. 335-337.

12.   J. S. Grauman, "Beta-21S: A New High Strength, Corrosion-Resistant Titanium Alloy," Proceedings of the Technical Program from the 1990 International Conference, Titanium Development Association, 1990, p. 290.

13.   R. W. Schutz, M. Xiao, and J. W. Skogsberg, "Stress Cracking and Crevice Corrosion Resistance of Pd-Enhanced Ti-38644 Titanium Alloy Products In Deep Sour Gas Well Environments," 12th International Corrosion Congress, Houston, TX, Sept. 1993, NACE, Houston, TX.

14.   M. Ueda, et.al., "Corrosion Behavior of Titanium Alloys in a Sulfur-Containing H_2S-CO_2-CL^- Environment," Paper No. 271, NACE Corrosion '90, April 23-27, 1990, Las Vegas, NV.

15.   Stress-Corrosion Cracking - Materials Peformance and Evaluation, ASM International, Materials Park, OH, July 1992, pp. 265-297.

16.   B. F. Brown, "Stress Corrosion Cracking in High Strength Steels and in Titanium and Aluminum Alloys," Naval Research Laboratory, Washington, D.C., 1972.

17.   R. J. H. Wanhill, "Aqueous Stress Corrosion in Titanium Alloys," Br. Corros, J., Vol. 10 (No. 2), 1975, p. 69-78.

18.   R. P. Gangloff, et.al., "Environmentally Assisted Cracking of High Strength Beta Titanium Alloys," Annual Report - Grant No. N0014-91-J-4164, SEAS Report No. UVA/525464/MSE93/101, Oct. 1992, University of Virginia.

19.   L. M. Young, "Environment Assisted Cracking in β-Titanium Alloys," Denver '93, 122nd TMS Annual Meeting, Denver, CO, Feb. 21-25, 1993.

20.   I. Azkarate and A. Pelayo, "Hydrogen Assisted Stress Cracking of Titanium Alloys in Aqueous Chloride Environments," INASMET, San Sebastian, Spain, 1992.

21.   D. M. Aylor, "A Hydrogen Embrittlement Evaluation of High Strength, Non-ferrous Materials for Fastener Application," Naval Surface Warfare Center, Annapolis Detachment, Carderock Div., Code 2813, Annapolis, MD, 1992.

22.   L. H. Wolfe, et.al., "Hydrogen Embrittlement of Cathodically Protected Subsea

Bolting Alloys," Paper No. 288, NACE Corrosion '93 Annual Conference, NACE, Houston, TX, 1993.

23.   B. Bavarian, "Corrosion Behavior of Beta-21S Ti Alloy In Chloride-Containing Environments," Paper No. 284, NACE Corrosion '93 Annual Conference, NACE, Houston, TX, 1993.

24.   D. E. Thomas and S. R. Seagle, "Stress Corrosion Cracking Behavior of Ti-38-6-44 in Sour Gas Environments, 'Titanium Science and Technology, Proceedings of the 5th International Conference on Titanium, Munich, Germany, Sept. 10-14, 1984, Deutsche Gesellschaft für Metallkunde, E.V., pp. 2533-2540.

25.   R. W. Schutz, M. Xiao and T. A. Bednarowicz, "Stress Corrosion Behavior of the Ti-3Al-8V-6Cr-4Zr-4Mo Titanium Under Deep Sour Gas Well Conditions," Paper No. 51, presented at NACE Corrosion '92, Nashville, TN, April 27-May 1, 1992.

26.   D. E. Thomas, et.al., "Beta-C: An Emerging Titanium Alloy for the Industrial Marketplace," ASTM STP 917, ASTM, Philadelphia, PA, 1986, pp. 144-163.

27.   S. Kitayama, et.al., "Effect of Small Pd Additions on the Corrosion Resistance of Ti and Ti Alloys in Severe Gas and Oil Environment," Paper No. 52, NACE Corrosion '92 Annual Conference, NACE, Houston, TX, 1992.

28.   W. F. Czyrklis and M. Levy, "Stress Corrosion Cracking Susceptibility of β Titanium Alloy 38-6-44," Corrosion, Vol. 32, No. 3, March 1976, p. 99.

29.   R. W. Schutz and M. Xiao, "Stress Corrosion Cracking Behavior of Ti-38644 Titanium Alloy Products in Methanol Solutions," Paper No. 148, NACE Corrosion '93 Annual Conference, NACE, Houston, TX, 1993.

30.   H.-E. Krugmann and J. K. Gregory, "Microstructure/Property Relationships in Titanium Alloys and Titanium Aluminides," eds. R. R. Boyer and Y. Kim, TMS-AIME, Warrendale, PA (1991) in press.

31.   G. R. Yoder, et.al., "Corrosion-Fatigue Resistance of Ti-10V-2Fe-3Al Alloy in Salt Water," Sixth World Conference on Titanium Proceedings, Cannes, June 6-9, 1988, Société Française de Métallurgie, Les Ulix Cedex, France, pp.1741-1746, Les Ulis Cedex, France, June 1988.

32.   N. E. Paton and J. C. Williams, "Effect of Hydrogen on Titanium and Its Alloys," Titanium and Titanium Alloys - Source Book, ASM International, Materials Park, OH, pp. 185-207.

33.   J. E. Costa, et.al., "Hydrogen Effects in β-Titanium Alloys," Beta Titanium Alloys in the '80's, AIME Publication, AIME, Warrendale, PA, 1984, p. 69.

34.   G. A. Young and J. R. Scully, "Effects of Hydrogen on the Mechanical Properties of a Ti-Mo-Nb-Al Alloy," Scripta Metallurgica., Vol. 28, 1993, pp. 507-512.

35.   J. J. DeLuccia, "Electrolytic Hydrogen in Beta Titanium," Report No. NADC-76207-30, Final Report, Naval Air Development Center, Warminster, PA, June 10, 1976.

36.   J. R. Wood and M. L. Bogensperger, "Effects of Hydrogen On the Structure and Properties of Beta-C™ (Ti-38644) Sheet," Proceedings of the Seventh World Conference

on Titanium, San Diego, CA, June 1992.

37.   H. G. Nelson, "Hydrogen-Induced Ductile-to-Brittle Fracture Transition in Beta Titanium Alloys," Proceedings Summary of the 5th Workshop on Hydrogen-Material Interactions, NASP Workshop Publication, April 1993.

38.   D. S. Schwartz, "Effect of Internal Hydrogen on Microstructure and Mechanical Properties of Beta-21S and Ti-15-3," Denver '93, 122nd TMS Annual Meeting, Denver, CO, Feb. 21-25, 1993.

39.   B. Bavarian, et.al., "High Temperature Corrosion Behavior of Beta-21S Titanium Alloy," Paper No. 243, NACE Corrosion '93 Annual Conference, NACE, Houston, TX, 1993.

40.   P. J. Bania and W. M. Parris, "Beta 21-S: A High Temperature, Metastable Beta Titanium Alloy," Presented at the TDA 1990 International Conference on Titanium Products and Applications, Orlando, FL.

41.   Z. Liu and G. Welsch, Metallurgical Transactions A, Vol. 19A, April 1988, p. 1121.

42.   W. M. Parris and P. J. Bania, "Oxygen Effects on the Metallurgical Properties of TIMETAL®21S," Presented at the Seventh World Conference on Titanium, San Diego, 1992.

43.   T. A. Wallace, R. K. Clark, K. E. Wiedemann, "Oxidation of Beta-21S in Air in the Temperature Range 600 to 800°C," Proceedings of the Seventh World Conference on Titanium, San Diego, CA, TMS/Titanium Development Association, 1992.

44.   T. A. Wallace, R. K. Bird, K. E. Wiedemann, "The Effect of Oxidation on the Mechanical Properties of TIMETAL-21S," Presented at the 1993 TMS Annual Meeting: Beta Titanium Alloys, Denver, CO.

45.   K. E. Wiedemann, et.al., "Mechanical Properties of Coated Titanium Beta-21S After Exposure to Air at 700 and 800°C," NASA Technical Memorandum 104220, June 1992.

46.   V. G. Anderson and B. A. Manty, "Titanium Alloy Ignition and Combustion," Report No. NADC 76083-30, United Technologies Corp., Pratt & Whitney Aircraft Group, West Palm Beach, FL, Jan. 15, 1978.

47.   T. R. Strobridge, "Titanium Combustion in Turbine Engines," Report No. FAA-RD-75-51/NBSIR 79-1616, National Engineering Laboratory, National Bureau of Standards, Boulder, CO, July 1979.

48.   D. M. Berczik, U.K. Patent Application No. GB 2238057A, "High Strength Nonburning Beta Titanium Alloys," May 22, 1991.

49.   P. Krag and R. Henson, "An Ignition Resistant Titanium Alloy for Acid Pressure Oxidation Applications," Proceedings of the RANDOL Gold Forum, Vancouver, B.C., 1992, p. 201.

A SOL-GEL-DERIVED COATING FOR PREVENTING ENVIRONMENTAL

DEGRADATION OF TITANIUM BETA-21S AND ITS EFFECT ON PROPERTIES

K. E. Wiedemann
Analytical Services & Materials, Inc.
107 Research Drive, Hampton VA 23666

R. K. Bird, T. A. Wallace, R. K. Clark
NASA Langley Research Center
Hampton VA

ABSTRACT

Mechanical and metallurgical results are presented that show the effectiveness of a thin, lightweight coating when used to protect a titanium alloy from oxidation in high-temperature applications. Interstitial oxygen and, possibly, nitrogen that enter titanium alloys during reactions with air at high temperatures deleteriously affect ductility and fatigue life. A coating that protects titanium alloys by preventing the formation of interstitials was demonstrated using Ti-15Mo-3Al-2.7Nb-0.2Si (Beta-21S) foil, sheet, and metal matrix composite. The demonstration involved studies of ductility, microstruture, fracture surfaces, fatigue behavior, and response to thermal cycling. The coating comprised a 2µm reaction barrier topped by a 1.5µm diffusion barrier. It was found that the coating reduced interstitial levels, extended the life of Beta-21S in high-temperature environments, and effectively prevented the cracking of Beta-21S-matrix composites caused by the combination of matrix embrittlement and thermal stresses.

Beta Titanium Alloys in the 1990's
Edited by D. Eylon, R.R. Boyer and D.A. Koss
The Minerals, Metals & Materials Society, 1993

INTRODUCTION

The strength and fabricability of beta titanium alloys makes them attractive materials for aerospace applications. In particular, Ti-15Mo-3Al-2.7Nb-0.2Si (Beta-21S) can be cold rolled into foil and heat treated to high strengths, making it an attractive material for many uses including the fabrication of titanium-matrix composites.

All titanium alloys undergo life-limiting reactions with air at elevated temperatures. This is caused by interstitial oxygen, and possibly nitrogen, that enter the metal through unprotected surfaces. This process is limited only by diffusion, and consequently, proceeds rapidly at high temperatures For most applications up to 400°C, titanium alloys can be used unprotected. Oxide formation is the dominant process, and the thin oxide that forms is protective because very little diffusion of interstitials occurs (Ref. 1 & 2). Above 400°C, diffusion is much more rapid and the formation of interstitials becomes the dominant process. Useful life times at 600°C may be measured in terms of hours, and at 800°C in terms of minutes.

Conventionally, diffusion coatings 25 to 100-μm thick have been used for protecting titanium alloys exposed to high temperatures. Alternatively, overlay coatings of various types including MCrAlY and glass-ceramics have also been used. Although these coatings are effective in reducing oxidation there is, due to their thicknesses, a significant reduction in fatigue life and a weight penalty that abrogates the advantages of using materials with high specific strengths in thin sections.

This paper presents mechanical and metallurgical results that show the effectiveness of a thin, lightweight coating when used to protect a titanium alloy from oxidation in high temperature applications. Coated and uncoated Beta-21S foil and Beta-21S matrix composite specimens were tension tested after isothermal and thermal-cycling exposures at temperatures to 800°C. The effect of the coating on the fatigue life of sheet material is also presented.

EXPERIMENTAL PROCEDURE

Foil, sheet, and composite material were studied. The foil material was cold-rolled 0.11-mm thick Beta-21S, and the sheet material was mill-annealed 0.635-mm thick Beta-21S, which were cut into 38.1 x 127-mm strips with the longer dimension parallel to the rolling direction. The composite was a Beta-21S matrix [0°/90°]-symmetric 4-ply composite reinforced with 30 vol% SiC fibers.

A coating comprising a 2 μm reaction barrier and a 1.5 μm glass diffusion barrier and sealant (see Figure 1) was applied to foil, sheet, and composite Beta-21S strips. The strips were first detergent cleaned, then rinsed with acetone followed by methanol, and blotted dry. A 1 μm-layer of aluminum was applied to both sides of the strips using electron beam evaporation. The strips were then heat treated for 8 h at 620°C to convert the aluminum layer to $TiAl_3$. The sol-gel-derived glass layers (Ref. 3) were then applied and cured at 650°C for 5 min.

The isothermal exposures were conducted in air at atmospheric pressure by introducing the specimens into a furnace preheated to the test temperature, and by air cooling at the end of the test exposure. The thermal-cycling exposures were also conducted in air at atmospheric pressure. The specimens were loaded into a thermal-cycling apparatus that lifted the specimens into a tube furnace maintained at 800°C. After 5 min. the apparatus lowered the specimens into a dewar of liquid nitrogen (-196°C) for 1 min. The cycle was repeated 144 times to yield a net exposure at 800°C of 12 h.

After exposure, specimens were cut from each strip. The foil tensile specimens were cut three from each strip and were machined to ASTM specification: the width of the reduced

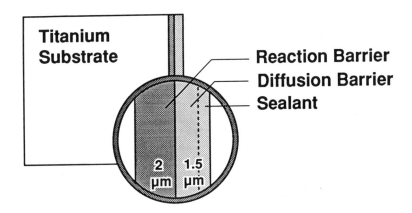

Figure 1. Schematic coating design showing the relationship of the aluminide reaction barrier and the glass diffusion barrier and sealant layer.

section was 6.35 mm and the length was 31.75 mm (Ref. 4). Fatigue specimens were cut two from each strip: the length was 127 mm, the radius was 101.6 mm, and the minimum width was 6.35 mm The composite tensile specimens were machined with straight sides to a width of 6.35 mm.

The tensile specimens were loaded in tension until failure. Back-to-back extensometers with a 25.4-mm gage length were used to measure strain. During each test, a 0.127 mm/min crosshead-deflection rate was used until 2% total strain when the crosshead-deflection rate was increased to 1.27 mm/min. Yield strengths for the un-reinforced material were determined by the 0.2%-offset method and elastic moduli for the un-reinforced material by linear regression over the portion of the stress-strain curve between 0 and 0.5% total strain.

The fatigue specimens were cycled until failure or a reasonable runout condition had been reached. An R value of 0.1 and a frequency of 15 Hertz were used.

Metallurgical specimens were prepared by sectioning and then mounting in a thermosetting plastic. They were then ground using silicon-carbide papers and attack-polished using Kroll's reagent and a colloidal-silica slurry.

RESULTS AND DISCUSSION

Although Beta-21S was developed with oxidation resistance in mind, it is not immune to oxidation effects. This is easily seen from the data given in Figure 2 where the room-temperature properties of uncoated 0.11 mm Beta-21S foil are plotted against exposure time. Useful mechanical properties are lost in less than 30 minutes at 800°C. The most rapidly degraded property was ductility, measured as plastic strain, which was sharply reduced after only 1 minute of exposure.

The thinness of the foil contributed to the speed of degradation. Thicker material would have lasted considerably longer. In this sense, the foil-gauge material provides the most sensitive test of coating effectiveness.

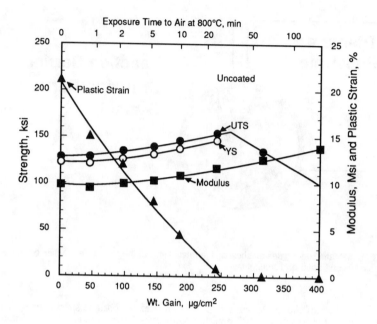

Figure 2. Room-temperature properties of uncoated 0.11 mm Beta-21S foil after exposures to air at 800°C.

Ductility of Coated Foils

The effect of the coating on ductility was shown using 0.11 mm Beta-21 foil that was coated and subjected to exposures in air at 600, 700, and 800°C (see Figure 3). At 800°C, the coated material after 50h exposure had a ductility comparable to the uncoated material after 10 minutes. This represented an increase in useful life of roughly 30,000%.

At each exposure temperature, the ductility of the coated Beta-21S showed an initial rise. This was due to changes in microstructure and was not the result of oxidation or the coating. The as-coated material had a microstructure consisting of very fine accicular alpha in a beta matrix that was a result of the thermal treatments used to cure the coating. The test exposures were, in effect, further thermal processing, and they resulted in a gradual modification of the as-coated microstructure. The nature of the change was a function of temperature.

The microstructural changes occurred most quickly and most completely at 800°C. They were also the most dramatic: ductility rose from the as-coated value of 8% to 16% after 12h. At this temperature, the alloy was very close to its beta transus and the diffusion of alloy constituents was rapid. At the lower temperatures, the changes were slower, and less dramatic.

The ductility data does suggest that the coating is effective in preventing the embrittlement caused by oxidation. Only at 800°C after 48h is there any indication that the properties of the coated foils have been degraded.

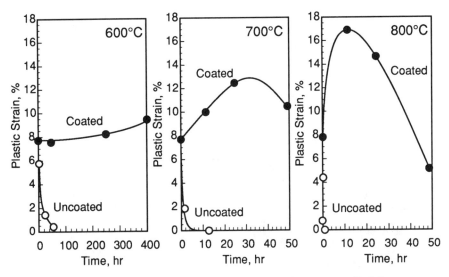

Figure 3. Room-temperature ductilities of 0.11 mm Beta-21S foil after exposure. (R=0.1)

Microstructure and Fracture Behavior

Two distinct microstructures evolved in the coated Beta-21S: one characteristic of exposures at 600 and 700°C, the other characteristic of exposures at 800°C. In the former exposures, the effect of interstitials on the microstructure was difficult to see. In the latter exposures, there was a more pronounced effect and distinct differences between the coated and uncoated material could be seen. At all temperatures, the fracture surfaces showed ductile features for coated Beta-21S that extended to the outer edge of the material next to the coating. Brittle features were seen in the fracture surfaces of uncoated Beta-21S that began near the edges and, often, covered the complete fracture surface.

At 600 and 700°C, the microstructure consisted of transformed beta with interior, accicular alpha and some grain-boundary alpha. The amount of grain-boundary alpha tended to increase with time and temperature. The same microstructure was observed in the coated and uncoated material. The only noticeable difference was that the interior alpha was coarser in the uncoated material (see Figure 4). Nevertheless, the fracture surfaces of the coated and uncoated materials were quite different indicating that the fracture behavior was not being controlled by microstructure, but by some other factor such as embrittlement by interstitials.

At 800°C, uncoated Beta-21S developed heavy, continuous grain-boundary alpha and coarse accicular alpha in many of the beta grains (see Figure 5). The coated Beta-21S had much less alpha both on the grain boundaries and in the interior of the beta grains. The fracture surface of the coated material showed ductile features, and the uncoated, brittle. Distinguishing the microstructure's contribution to the fracture behavior was difficult. It was clear that interstitials were the ultimate cause of the embrittlement, but it was difficult to show whether the mode of embrittlement was reduced toughness of the microstructure or intrinsic embrittlement of the metallic phases.

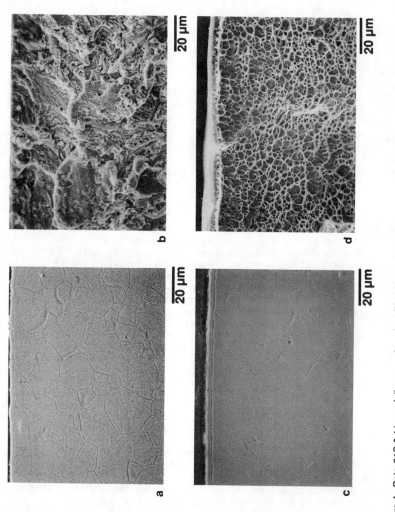

Figure 4. Beta-21S 0.11 mm foil exposed to air at 700°C. (a) Cross section and (b) fracture surface of uncoated material after 16h; (c) cross section and d) fracture surface of coated material after 24h.

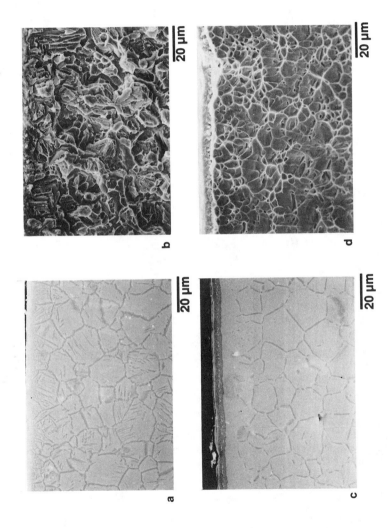

Figure 5. Beta-21S 0.11 mm foil exposed to air at 800°C. (a) Cross section and (b) fracture surface of uncoated material after 2h; (c) cross section and d) fracture surface of coated material after 24h.

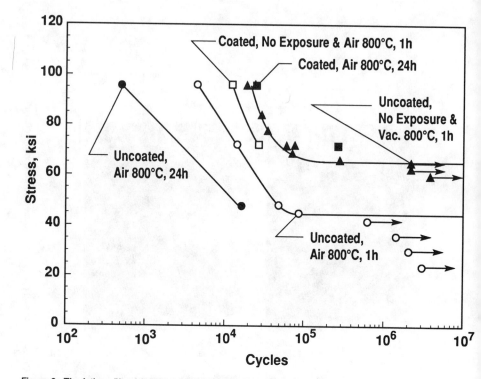

Figure 6. The fatigue life of 0.635-mm thick Beta-21S sheet.

Fatigue Behavior

From Figure 6, it can be seen that exposure to 800°C air significantly reduced the fatigue life of uncoated Beta-21S. After only 1 hour, low cycle fatigue life had decreased by a factor of 5, and the fatigue limit had decreased by 25 ksi. It was observed that the fatigue behavior was not significantly affected by 1h at 800°C in vacuum. The change in fatigue behavior, therefore, was due to the level of oxygen interstitials, and not the thermal treatment.

After coating, the Beta-21S had a shorter fatigue life by about a factor of 2. There was no further decrease, however, during exposure to air for 1h at 800°C. After 24 h, fatigue life increased to the level of the uncoated Beta-21S after 1h at 800°C in vacuum. This showed that the coating was effective in preventing the interstitial contamination that shortened the fatigue life of the uncoated material, and that the coating was beneficial to the fatigue life of Beta-21S exposed to air at high temperatures.

Thermal Cycling of Composites

Thermal cycling imposes on composites the rigors of oxidation damage coupled with the effects of mechanical strains. The difference in thermal expansion between the Beta-21S matrix (CTE = 5.28 x 10-6/°F at 1000°F, Ref. 5) and SiC fibers (CTE = 2.7 x 10-6/°F, Ref. 6) leads to elastic strains of several tenths of a percent along the length of the fibers. In this

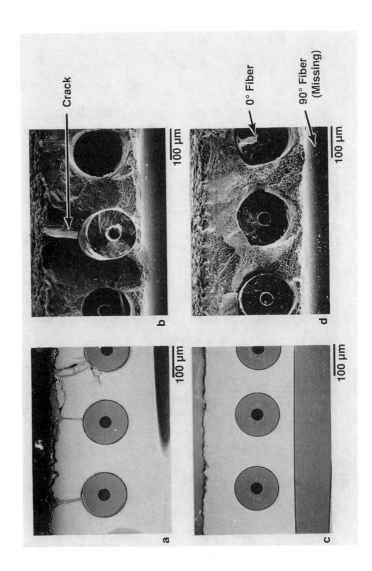

Figure 7. Beta-21S [0/90]$_s$ SiC composite thermally cycled in air between 800 and -196°C for 144 cycles with 5 min. spent at 800°C each cycle. (a) Cross section and (b) fracture surface of uncoated composite; (c) cross section and d) fracture surface of coated composite.

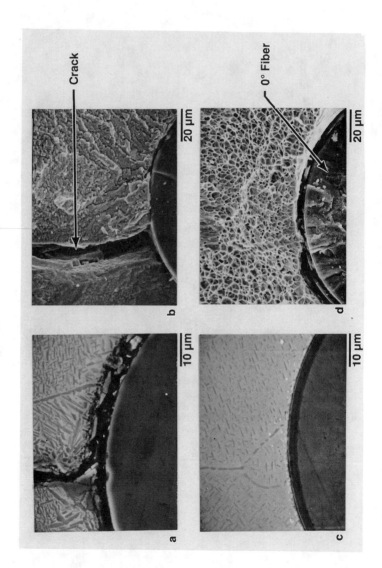

Figure 8. Detail showing the microstructure and fracture behavior near the fiber-matrix interface of a Beta-21S $[0/90]_s$ SiC composite thermally cycled in air between 800 and -196°C for 144 cycles with 5 min. spent at 800°C each cycle. (a) Cross section and (b) fracture surface of uncoated composite; (c) cross section and d) fracture surface of coated composite.

situation, problems such as spalling of the coating, mechanical rupturing, and fatigue may be expected to contribute to the degradation of the coating and the composite.

In the uncoated composite, thermal cycling produced extensive cracking across the entire width and length of the parts with the cracks extending at least to the outer plies and frequently to the center plies (see Figure 7). This was attributed to interstitial embrittlement of the Beta-21S, evidence of which can be seen in both the microstructure and fracture surfaces (see Figure 8). The coated composite did not crack, and it is evident from both the microstructures and fracture surfaces that the level of interstitials was low.

Synergistic effects between oxidation damage and thermal-cycling damage evidently exist. In the uncoated composite, oxidation had reduced the ductility of the matrix, and consequently, cracking occurred during thermal cycling. No cracking occurred in the coated composite, because the ductility of the matrix was maintained through exclusion of interstitials.

CONCLUSIONS

When exposed to air at high temperatures, titanium Beta-21S (Ti-15Mo-3AL-2.7Nb-0.2Si) becomes embrittled and its fatigue life is shortened. As a matrix material for composites, Beta-21S cracks under the combined effects of oxidation and stresses induced by thermal cycling.

A coating comprising a 2 μm reaction barrier and 1.5 μm glass overlayer was shown to be effective in preventing the degradation of Beta-21S when it is exposed to air at high temperatures. This was shown through studies of microstructure, fracture surfaces, ductility, and fatigue behavior.

The coating was shown to prevent the changes in microstructure and fracture behavior that are associated with oxygen interstitials. The coating was effective in preserving the ductility and fatigue life of Beta-21S when exposed to air at high temperatures. The coating also proved effective in preventing the degradation of titanium matrix composites where Beta-21S was the matrix material.

REFERENCES

1.   J. Stringer, "The oxidation of Titanium in Oxygen at High Temperatures," Acta Met., vol 8, pp. 758-766.

2.   P. Kofstad, High Temperature Corrosion, Elsevier Applied Science, London, 1988, pp. 289-299.

3.   K. E. Wiedemann, P. J. Taylor, R. K. Clark, T. A. Wallace, "Thin Coatings for Protecting Titanium Aluminides in High-Temperature Oxidizing Environments," Environmental Effects on Advanced Materials, Eds: R. H. Jones, R. E. Ricker, The Minerals, Metals, and Materials Society, Warrendale, PA, 1991, pp 107-121.

4.   ASTM E8-85b, "Tension Testing of Metallic Materials," 1986 Annual Book of ASTM Standards, vol. 03.01, American Society of Testing and Materials, 1986, pp 124-125.

5.   P. J. Bania, W. M. Parris, "Beta-21S: A High Temperature Metastable Beta Titanium Alloy," Titanium 1990, Products and Applications, Vol. II, Proceedings of the 1990 International Conference, Titanium Development Association, Dayton Ohio.

6.   Mittnick, Melvin A., McElman, John, "Continuous Silicon Carbide Fiber Reinforced Metal Matrix Composites," The 13th Conference on Metal Matrix, Carbon, and Ceramic Matrix Composites, NASA Conference Publication 3054, part 2, 1990.

EVALUATION OF THE ENVIRONMENTALLY ASSISTED CRACKING OF BETA-21S TITANIUM ALLOY

Behzad Bavarian, Vipin Wahi
California State University, Northridge
Northridge, CA 91330

Guido Canzona
Rockwell International
Rocketdyne Division
Canoga Park, CA 91303

Mehrooz Zamanzadeh
PTL, PSI
Pittsburgh, PA 15220

Abstract

Susceptibility of Beta 21S (Timetal 21S) to stress corrosion cracking, corrosion fatigue, and hydrogen embrittlement was investigated in several different corrosive environments. Electrochemical studies of Beta-21S alloy in sodium chloride solution showed it was not susceptible to any localized corrosion, but addition of more than 400 ppm fluoride ions to sodium chloride solution resulted in intergranular attack. Corrosion fatigue tests performed in different fluoride containing solutions exhibited intergranular cracking at initiation stage which changed into transgranular during the crack propagation stage. No susceptibility to SCC was observed in any of the exposed environments. The results showed this alloy to have extreme corrosion resistance and possesses better performance compare to other titanium alloys. While no susceptibility to hydrogen reaction embrittlement was observed due to high hydrogen solubility limit of the alloy, severe hydrogen environment embrittlement(HEE) was observed whenever the alloy was exposed to high pressure high temperature hydrogen gas.

Introduction

Metastable beta titanium alloys are cold rollable with an excellent formability, weldability and age hardening characteristics, rapidly becoming an attractive materials for high strength, corrosive conditions, and light weight applications. However, their limitation is often temperature capability, generally due to poor creep resistance, and oxidation behavior. A new metastable beta alloy, designated Beta-21S (Ti-15%Mo-2.8%Nb-3%Al-0.2%Si), has been developed which circumvents many of these common shortcomings of metastable beta alloys. It also possesses excellent oxidation resistance and elevated temperature mechanical properties for a metastable beta alloy[1-3].

In the aged condition the material shows very good stability, even when highly loaded. The 537°C(1000°F) aged material shows some drop in ductility but minor increases in strength when exposed under high stress at elevated temperature (up to 450°C)[1-3]. Beta-21S has superior corrosion resistance as compared to other metastable beta alloys such as Beta C. Beta-21S provides[8-9] superior corrosion resistance at all concentrations of HCl when compared with grade 2 titanium and beta C. Beta-21S performs equally well in other reducing environments. Tests indicate a chloride crevice corrosion threshold for Beta-21S between pH 0.5-1.0.[8-11].

Stress corrosion cracking resistance of beta titanium alloys is highly dependent on the alloy composition and metallurgical condition. Beneficial alloying elements including Mo, V, Nb, and Pd clearly improve alloy SCC resistance[7]. Mo-rich beta alloys have been demonstrated excellent SCC resistance in seawater, sweet and sour brine mainly due to localized corrosion inhibition of Molybdate formation. Applied anodic or cathodic potential on beta titanium alloys also showed no corrosion attacks in seawater solution[9]. Corrosion fatigue of beta titanium alloys has been studied extensively. However, very limited investigations indicate no significant reduction in alloy endurance limit in seawater.

Titanium alloys have been employed in the aerospace vehicles for weight savings. Unlike the steel and nickel base alloys, titanium is a hydride former and can absorb up to two atoms of hydrogen for every titanium atom to form a very brittle titanium hydride phase[4]. Even without the formation of the hydride phase, titanium alloys are susceptible to hydrogen environment assisted crack growth. With hydrogen present in the metal lattice, failures occur at stresses below that in which step wise crack extension and eventual failure occurs in the absence gaseous hydrogen[5]. The suppression of the omega phase formation by hydrogen tends to support the hypothesis that the decrease in strength and modulus of beta titanium alloys with hydrogen may be due to the decreased volume fraction of omega phase[6]. Nucleation of the omega phase is retarded by the presence of at least 4000 ppm hydrogen, which stabilizes the beta phase[6-7]. However, aging of beta alloys to higher strengths increases volume fraction of alpha phase, mainly along the grain boundaries which reduce the hydrogen tolerance for the alloy(alpha phase becomes brittle at 100 ppm hydrogen)[7-8].

The objective of this investigation was to study the corrosion behavior of Beta 21-S titanium alloy in aqueous chloride/fluoride containing solutions. These investigations consist of electrochemical studies, stress corrosion cracking(using the slow strain rate technique and constant load tests), corrosion fatigue, and hydrogen embrittlement.

Experimental Procedure

The Beta-21S(Mo 15%, Al 3%, Nb 2.8% and Si 0.2%) alloy used in this investigation was supplied by TIMET Corp. The material supplied was 0.172" hot rolled plate and 0.5" rolled bar stock. All samples were machined and heat treated to STA condition (solution treated at 980°C for 10 minutes and aged at 540°C for 8 hours). The mechanical properties of the STA alloy were as followed: tensile strength 202 ksi(1400 MPa), yield strength 180 ksi(1260 MPa), total elongation about 6%, and elastic Modulus was 17×10^6 psi for the aged samples at 540°C.

Electrochemical Studies

The electrochemical studies were conducted per ASTM G5 standard. Hokto model HA-301 potentiostat/galvanostat with a high impedance electrometer system was used. A thick plate of 0.254 cm (0.10 in) was cut into several pieces of 1.6 cm^2 (0.25 in^2) area, to prepare electrochemical samples. The surface of the specimens were finely polished upto 0.05 μm by γ-alumina powder.

A scanning rate of 10 mV/second was used within the potentials range of -1400 mV to 1500 mV and reversed. This range of potentials was predefined in order to get only the areas of interest on the plot. The corrosion potential (E_{corr}), pitting potential (E_p) and the passive potential range was defined for each specimen in variety of solutions. General corrosion tests were performed in several sodium chloride solutions containing different level of fluoride ions. Corrosion rates were generated under controlled potential for Beta-21S alloy, and compared with several other titanium alloys such as Ti 15-3 (15%V, 3%Al, 3%Sn, 3%Cr), Gamma TiAl(Ti-50% Al-2%Cr), and Alpha2 Ti$_3$Al(Ti-25% Al-10%Nb-3%V-1%Mo). Specimens of 2.5 x 2.5cm. were exposed to each solution under an applied potential of + 500 mV$_{SCE}$ for about 72 hours. The corrosion rate was measured in form of average current density when a steady state was achieved for each alloy.

Stress Corrosion Cracking and Corrosion Fatigue

Stress corrosion cracking tests were carried out using two different techniques, namely, slow strain rate technique (SSRT) and a constant load specimen for these materials in alternate immersion. Slow strain rate technique (SSRT) was used to evaluate the stress corrosion susceptibility in the longitudinal direction. Several sodium chloride solutions were used with different level of fluoride ions addition.

The smooth specimens were prepared from a heat treated plate with the thickness of approximately 0.16 inch (0.4 cm), and machined into 0.6 inch (1.5 cm) wide strips with a 6.0 inch (15.24 cm) length. All the specimens were sanded down to a 600 grit finish.

A slow strain rate of 5×10^{-7} (cm/cm)/sec. was used for all of these tests. Some of the samples were tested at open circuit potentials, others were charged anodically or cathodically either by applying potential or current using a potentiostat/galvanostat.

The constant load (U-bend) specimens were prepared from a heat treated plate with the thickness of about 0.16 inch (0.4 cm), and machined into 0.6 inch (1.5 cm) wide strips with an 8.0 inches (20 cm) length. All the specimens were sanded down to a 600 grit finish. A stress of about 100 ksi(700 MPa) was applied on each specimen by a proper bending and then the specimens were subjected to the alternate immersion test (10 minutes wet and 50 minutes dry cycles) in different chloride and fluoride solutions for 1000 cycles.

Corrosion fatigue tests were performed using a rotating beam (R = -1) and a tension-tension (R = 0.1) in 3.5% NaCl solution with different level of fluoride ion. The S-N curve for each solution was established.

Hydrogen Embrittlement

Hydrogen reaction embrittlement(HRE) were studied using the slow strain rate test under the cathodically hydrogen charging condition. Hydrogen charging was performed in a low pH solution (pH 2.0) under the controlled potential of -1500 mV$_{SCE}$ while it was strained until failure occurred. Hydrogen environment embrittlement was performed by exposing tensile specimens to 34.5 Mpa (5000 psi) hydrogen gas at different temperatures (18°C, 120°C, and 260°C)[12]. To accomplish this, the specimens were polished to 600 grit to remove any oxide or scale from their surfaces, followed by degreasing by isopropyl alcohol and acetone. After cleaning process, samples were placed in a vessel (autoclave) which was heated to the charging temperature, while pressurized with 17.2 Mpa (2500 psi) hydrogen gas. After reaching the charging temperature, the hydrogen pressure was adjusted to 34.5 Mpa (5000 psi), and the vessel was held at the charging temperature for various durations. The samples were then removed and stored in liquid nitrogen to prevent hydrogen outgassing until they were tensile tested at 5x10-7 (cm/cm)/sec.

Results and Discussion

Titanium Beta-21S alloy(STA) was tested in chloride-containing solutions to measure the electrochemical properties and to determine its susceptibility to the localized corrosion, stress corrosion cracking, corrosion fatigue and hydrogen embrittlement.

Polarization studies were carried out on the samples of Beta-21S in aerated 3.5% NaCl solutions containing different ppm levels of HF acid. The polarization curves indicated hydrogen evolution (cathodic potential range), active behavior, and passivity at relatively intermediate potentials without any breakdown potentials. Figure 1 demonstrates the summary of the electrochemical behavior of this alloy and its comparison with Ti 6-4 alloy. These results indicated that Beta-21S alloy is more corrosion resistant than the Ti 6-4 alloy. Lack of breakdown potential of Beta-21S alloy also indicates that it is less susceptible to any localized corrosion and possesses a wider range of passivity, while pits were formed on the Ti 6-4 alloy whenever it was subjected to an applied potential above its breakdown potentials.

General corrosion tests that perform in several aggressive corrosive

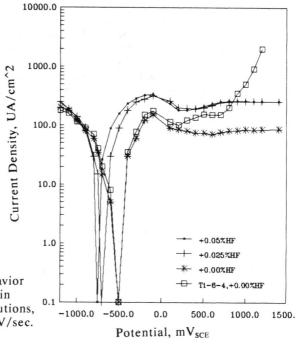

Figure 1: Polarization behavior of Beta-21S titanium alloy in 3.5% NaCl +x.xx% HF solutions, pH3.0. Scan rate was 10 mV/sec.

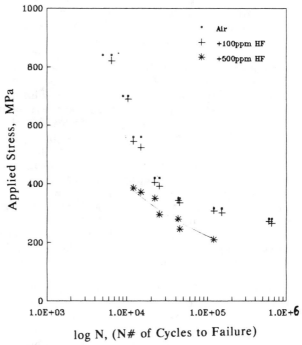

Figure 2: Corrosion fatigue behavior of Beta-21S titanium alloy.

environments using a potentiostatic technique (applied potential of $+500\,mV_{SCE}$) did not show any corrosion attack on the exposed surfaces in any salt solution which contained less than 300 ppm HF acid. An addition of more than 300 ppm HF acid is required to breakdown the alloy passive film. On the other hand, formation of molybdate and its combination with the corrosive species results in the loss of mobility and neutralization of corrosive species, which enhance the stability of the alloy passive film. Therefore, the breakdown of the passive film is extremely difficult in Beta-21S alloy.

Corrosion fatigue test results are shown in Figure 2, which indicates no corrosion fatigue susceptibility for Beta-21S unless the level of HF acid in the solution exceeds 400 ppm. However, above 500 ppm HF in the testing solution, this alloy showed to be extremely susceptible to intergranular attack which acts as the crack initiation site for any fatigue cracking. Figure 3 shows the crack initiation of this alloy in 3.5% NaCl solution which contains 500 ppm HF acid. As indicated, fatigue crack initiation is an intergranular cracking. It appears that above a certain level of aggressive species (F⁻), the protective passive film lose its stability and starts to breakdown. The breakdown of the passive film produces some passages for corrosive species to penetrate the passive film and reach the alloy surface to cause intergranular attacks. However, presence of a stable film on the surface of Beta-21S provides a fatigue life similar to specimen which fatigue tested in air.

Figure 4 summarizes the results of the SCC performed on titanium Beta-21S alloy and its comparison with titanium aluminide(25-10-3-1) using the slow strain rate technique under controlled applied potentials[8]. The plastic strain at failure when tested in the solution as compared to plastic strain at failure when tested in air was primarily measure to estimate the SCC susceptibility. These results show that Beta-21S is not susceptible to SCC. Fractography of the fractured sample under up to $+1000\,mV_{SCE}$ in 3.5% NaCl containing 250 ppm HF, showed completely ductile overload failure. The U-bend tests also showed no SCC susceptibility of this alloy after two months(1000 cycles of alternate immersion) exposure to several aggressive solutions.

The mechanism of passivity and excellent corrosion resistance of the Beta-21S alloy can be explained based on the MacDonald Solute-Vacancy Interaction Model[14]. The presence of molybdenum in this alloy results in formation of highly charged solute in the passive film, such as $Mo^{6+}(MoO_4^=)$. This highly charged solute can form big complexes with negatively charged cation vacancies which decreases concentration of free cation vacancies, neutralize the corrosive species, reduce mobility of corrosive species and improve passivation. Therefore, the breakdown of the passive film is extremely difficult which results in an excellent corrosion resistance for Beta-21S alloy in most corrosive environments.

Cathodically hydrogen charging into Beta-21S did not promote any susceptibility to hydrogen reaction embrittlement. No sign of reduction in its mechanical behavior under the cathodic applied potentials was observed. Figure 4 shows no changes in mechanical properties of Beta-21S alloy tested under -1000 or -1500 mV_{SCE} cathodic applied potentials during the slow strain rate tests. Each test lasted for about 50 hours. Analysis of the level of absorbed hydrogen in this alloy after each test showed to be much lower than alloy hydrogen solubility limit (the hydrogen tolerance level of about 3000-4000 ppm[7,13]). The excellent hydrogen absorption resistance exhibited by Beta-21S can be attributed to extremely protective film improved by presence of molybdenum and niobium

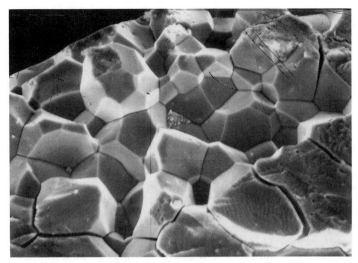

Figure 3: Fractograph of Beta-21S titanium alloy(STA), failed in tension-tension corrosion fatigue test in 3.5% NaCl + 0.05% HF solution (R = 0.1, Max. stress 55 ksi(385 MPa)) after 16300 cycles, showing an intergranular crack initiation which transform into transgranular during crack propagation, x500.

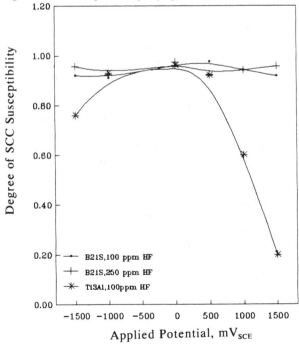

Figure 4: Degree of stress corrosion susceptibility of Beta-21S titanium alloy in 3.5% NaCl+x.xx% HF solutions and its comparison with alpha 2 (25-10-3-1)titanium aluminide using the slow strain rate test. Strain rate was 5×10^{-7} 1/sec. Degree od SCC susceptibility= Total Strain in Solution/Total strain in air.

which acts as a strong barrier for hydrogen penetration[15]. Therefore, hydrogen reaction embrittlement resistance of this alloy is very high.

Hydrogen environment embrittlement tests were carried out on Beta-21S titanium alloy in 34.5 Mpa (5000 psi) hydrogen gas at several different temperatures, and Figure 5 presents the effect of temperature on the strain ratio (strain to failure of exposed alloy/strain to failure of unexposed alloy). Figure 5 illustrates that alloy ductility starts to drop as the exposed temperature and holding time increase. The most severe effects are seen at temperature above 260°C. It should be reported that Beta-21S alloy exhibit a very fine line between high and low susceptibility to hydrogen embrittlement, which it can be seen in figure 5. Figure 6 shows the alloy which was subjected to 34.5 Mpa hydrogen gas for 9 hours at this temperature. As it can be seen the specimen converted to small brittle chips. The formation of titanium hydride is reported to be the primary mechanism of embrittlement of titanium[12-13]. Titanium hydride is larger than the host phase thus can introduce considerable residual stresses as a result of its formation which produces severe internal cracking (chips formation). However, high pressures are generally not required for the formation of titanium hydride, but Beta titanium alloy can hold a large amount of hydrogen in its solid solution (3000-4000 ppm, about 40 a/o hydrogen). Above 4000 ppm hydrogen content, hydrides form causing residual stress high enough to shatter the material.

Conclusions

Corrosion behavior of Beta-21S titanium alloy(STA) was studied to establish its susceptibility to any environmentally assisted cracking. From results of the investigation, the following conclusions can be drawn:

1. Beta-21S is not susceptible to localized corrosion (pitting, crevice corrosion or SCC) in sodium chloride solutions. High Mo content provides a very good corrosion inhibition system for this alloy.
2. Intergranular cracking was observed in corrosion fatigue tests only in the aqueous environment containing more than 400 ppm HF acid. Below this minimum level of HF acid, presence of a very protective film on the alloy surface retards any corrosion attacks.
3. No susceptibility to hydrogen reaction embrittlement was observed for Beta-21S alloy under cathodic applied potential in any short term exposure (duration of a slow strain rate test).
4. Beta-21S alloy is susceptible to hydrogen embrittlement, if it is exposed to a high pressure(34.5 Mpa) hydrogen gas above 250°C. Above 250°C, the formation of titanium hydride is believed to be the primary mechanism of embrittlement of the alloy.

Acknowledgements

The authors wish to acknowledge thanks to Dean of School Engineering and Computer Science of California State University, Northridge for the financial support. Special Thanks to Dr. P. Bania and Dr. T. Wardlaw, TIMET, Henderson Tech. Lab. for providing us with the materials.

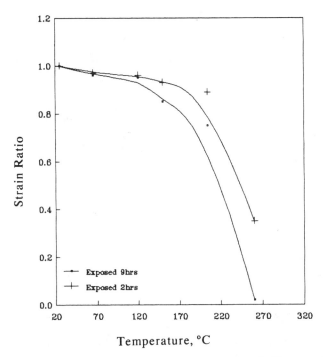

Figure 5: Hydrogen environment embrittlement susceptibility(strain of exposed/strain of unexposed) of Beta-21S titanium alloy, exposed to 34.5 Mpa hydrogen gas at different temperatures.

Figure 6: Scanning electron micrograph of Beta 21S titanium alloy subjected to 34.5 Mpa hydrogen gas for 9 hours at 260 °C. A dogbone tensile sample was converted to numerous small titanium hydride flakes, x100.

References

1. P. J. Bania ,W. M. Parris, "Beta-21S: A High Temperature Metastable Beta Titanium Alloy", Titanium Development Alloys International Conference, 1990, Orlando Florida.

2. P. J. Bania, W. M. Parris, "Beta 21S: A High Strength Metastable Beta alloy'119th TMS annual meeting, February 1990, Warrendale, PA.

3. E. W. Collins, The Physical Metallurgy of Titanium Alloys(Materials Park Ohio, American Society for Metals,International, 1984), pp34-45.

4. W.T. Chandler and R.J. Water, "testing to Determine the Effect of High pressure Hydrogen environments on the mechanical properties of Metals", ASTM STP 543, ASTM, 1974, pp 170-179.

5. R. R. Boyer, Journal of Metals, 5(1992), p.14.

6. I.M. Bernstein, A. W. Thompson, Hydrogen Effects in Metals, the Metallurgical Society, 1980, pp272-277.

7. R. W. Schutz, " An Overview of Beta Titanium Alloy Environmental Behavior" 1993 TMS Annual conference, February 1993.

8. B. Bavarian, S. Dastmalchi, S. Hartouni, M. Zamanzadeh, J. of Materials Science and Engineering, March, 1992, pp34-41.

9. J. S. Grauman, " Beta 21 S : A New High Strength Corrosion Resistant Alloy", Titanium Development Alloys International Conference, 1990, Orlando Florida.

10. R.W Schutz and J. S. Grauman,"Compositional Effects of Repassivation Potential in Chloride Media", International Conference on Localized Corrosion(NACE 1987).

11. S. H. Fores ,H. B. Bomberger, Journal of Metals,7(1985), p22.

12. H. W. Rosenberg, Journal of Metals, 11(1983), p 37.

13. R.J. Walter, R.p. Jewett, "Effects of Hydrogen on Behavior of Materials" I.M. Bernstein and A.W.Thompson (eds), ASM International, Materials Park, 1981, pp 819-827.

14. M.G. Fontana, Corrosion Engineering, (McGraw Hill, 1986), pp265-267.

15. B. Bavarian, V. Wahi, G. Canzona, M. Zamanzadeh, Corrosion/93, paper #93243,(Houston, NACE March 1993).

THE EFFECT OF OXIDATION ON THE MECHANICAL PROPERTIES

OF BETA-21S

T. A. Wallace[*], R. K. Bird[*], and K. E. Wiedemann[**]

[*] NASA Langley Research Center
Hampton, VA 23681-0001

[**] Analytical Services and Materials, Inc.
107 Research Drive
Hampton, VA 23666

ABSTRACT

The effect of oxidation exposure at 600°C, 700°C and 800°C on the room temperature mechanical properties of both 0.0635-cm Beta-21S sheet and 0.0114-cm Beta-21S foil were investigated. The plastic elongation was found to be the most sensitive to small oxidation weight gains, but the tensile strength and elastic modulus were also degraded by oxidation exposure. The amount of decrease in plastic elongation was also found to depend on exposure temperature. Equivalent oxidation weight gains produced a larger decrease in the plastic elongation as the exposure temperature was decreased. Examination of the fractured specimens showed a band of brittle fracture at the metal surface. The depth of this band increased with oxidation exposure. This brittle fracture was due to diffusion of oxygen into the metal, which led to the formation of grain boundary alpha, as well as increased amounts of alpha in the grain interiors.

Beta Titanium Alloys in the 1990's
Edited by D. Eylon, R.R. Boyer and D.A. Koss
The Minerals, Metals & Materials Society, 1993

INTRODUCTION

Current designs for future hypersonic vehicles call for the use of titanium-matrix composites (TMC) for many high temperature components, consituting as much as fifty percent of the structural weight. Utilizing TMC for these applications will require a titanium matrix alloy with good mechanical properties and oxidation resistance from 600°C to 800°C, and that is also easily producible as foil. These requirements have led to the development of new titanium alloys with an good balance of these properties. One such alloy is the metastable β-Ti alloy Beta-21S, Ti-15Mo-2.7Nb-3Al-.2Si (wt. %).[1]

Previous studies showed that although Beta-21S has better oxidation resistance than some other titanium alloys, significant oxidation still occurs in the temperature range 600°C to 800°C.[1-3] The oxidation was found to proceed by two mechanisms; the formation of a compact surface oxide of TiO_2, and diffusion of oxygen into the metal substrate. It is this oxygen diffusion that could be considered the most damaging since increases in oxygen content are known to embrittle titanium alloys.[3-5]

The purpose of this study, therefore, was to examine the effect of oxidation in the temperature range 600°C to 800°C on the room temperature mechanical properties of Beta-21S. Specimens were machined from both 0.0114-cm foil and 0.0635-cm sheet Beta-21S and exposed to laboratory air at 600°C, 700°C, and 800°C for times up to 1000 h. The room temperature mechanical properties, including plastic elongation, modulus, ultimate and yield strengths, of the oxidized specimens were then determined. Metallurgical and fractographic analyses were used to correlate microstructure, fracture morphology, and properties.

EXPERIMENTAL PROCEDURE

Tensile specimens were prepared from both 0.0635-cm thick Beta-21S sheet, and 0.0114-cm Beta-21S foil. The chemical composition, in weight percent, of both the sheet and foil was measured by the manufacturer to be 15.8 Mo, 2.95 Al, 2.88 Nb, 0.23 Si, 0.026 C, 0.114 O, and 0.005 N with the balance being Ti.

Tensile specimens were produced from both the foil and sheet material with dimensions in accordance with ASTM Standard E8[6]. The length and width of the reduced section were 3.175 cm (1.25 in) and 0.635 cm (0.25 in), respectively. Before oxidation exposure, the tensile specimens were chemically cleaned to remove surface contamination.

The specimens were exposed to laboratory air at 600°C, 700°C, and 800°C for times up to 1000 h. Oxidation weight gain was determined by weighing the specimens before and after exposure. Before exposures at both 600° and 700°C, the specimens were annealed in vacuum at the test temperature for 8 h to stabilize the microstructure. This was not done for the 800°C exposures because the as-received material was beta-annealed, and further annealing at 800°C has little effect on this microstructure.[1]

After exposure the tensile specimens were tensile tested at room temperature using a constant cross-head-deflection rate of 0.0042 mm/s (0.010 in/min). Back-to-back extensometers with a 2.54 cm

(1 in) gage length were used to measure strain. Yield strengths were determined by the 0.2% offset method and elastic moduli by linear regression over the portion of the stress-strain curve between 0 and 0.25% total strain.

Selected specimens were chosen for fractography and metallography. Fracture surfaces were cleaned in acetone and examined using scanning electron microscopy. Cross-sections of the tensile specimens were cut remote from the fracture surface, mounted in epoxy, polished, and lightly etched to produce metallographic specimens. These specimens were examined using light microscopy and scanning electron microscopy (SEM).

RESULTS AND DISCUSSION

Oxidation Weight Gain. Figure 1 shows the weight gain data per unit surface area for the sheet material versus the square root of time for exposures at 600°C, 700°C, and 800°C. The data show the typical parabolic weight gain relationship, $w^2 = k_p t$, where w is the increase in weight per surface area, t is the time of exposure, and k_p is the parabolic rate constant. Parabolic weight gain occurred over the entire temperature range investigated, which is consistent with oxidation data obtained from thermogravimetric analysis (TGA) of this alloy.[1,2]

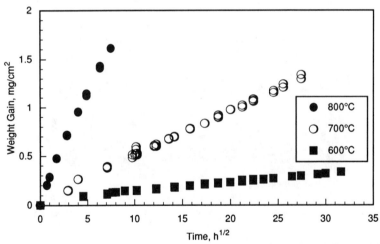

Figure 1 - Weight gain versus exposure time for 0.0635-cm Beta-21S exposed at 800°C, 700°C, and 600°C.

Figure 2 compares the weight gain data for exposure at 800°C for both the sheet and foil tensile specimens. The weight gains measured for the foil samples were greater during the first hour of exposure; but, the weight gains measured for longer exposures were close to those measured for the sheet specimens. The higher oxidation rate for the foil material was related to a thin porous layer, approximately 3 μm thick, at the surface of the as received foil (figure 3). When this layer was machined off, the weight gains measured for the foil specimens were equivalent to those measured for the sheet specimens.

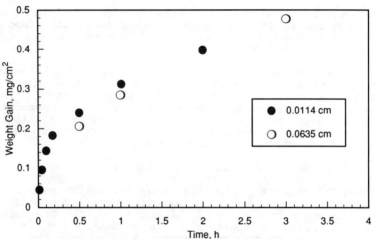

Figure 2 - Weight 2 μm after exposure at 800°C in air for
0.0114 ⊢━┥ d 0.0635-cm Beta-21S.

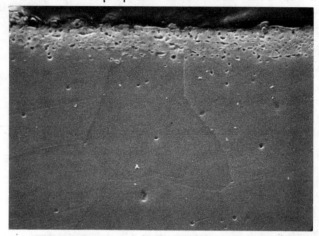

Figure 3 - SEM micrograph showing microstructure of as-
received 0.0114-cm Beta-21S foil.

Mechanical Property Data. Table I lists the resulting mechanical
properties after various exposures at 600°C, 700°C, and 800°C for
both the sheet and foil materials. It should be noted that the
starting properties (0 h exposure) for the three exposure
temperatures are different due to the initial vacuum heat
treatments at 700°C and 600°C. Although all mechanical properties
were affected by oxidation exposure, plastic elongation was the
most sensitive to small oxidation weight gains.

The properties of the foil were degraded after much shorter
exposure times than were those of the sheet. For instance, after
0.5 h at 800°C the foil material had less than 1% plastic
elongation while the sheet material had roughly 17%. This
difference is due to the much larger volume of the sheet specimen
compared to the foil. Oxygen would be expected to diffuse at the

Table I Mechanical properties of Beta-21S sheet and foil after oxidation exposures at 600°C, 700°C, and 800°C.

Temp, °C	Time, h	Weight Gain, %	e_B, %	UTS, MPa	YTS, MPa	E, GPa
		Results for 0.0114-cm Beta-21S foil				
800[1]	0	0	21.1	883.9	848.1	67.6
	50 s	0.171	15.1	887.4	975.0	67.6
	160 s	0.367	12.1	928.8	870.1	67.6
	180 s[2]	0.267	4.1	925.3	864.6	73.1
	350 s	0.552	8.0	963.2	905.3	71.0
	632 s	0.700	4.4	1003.9	946.0	75.2
	0.5	0.921	0.8	1057.7	1009.4	82.1
	1.0	1.198	0.0	926.0	926.0	86.9
	2.0	1.526	0.0	690.9	690.9	95.8
700[3]	0	0	14.9	997.7	932.2	92.4
	1.0	0.344	1.9	1044.6	1003.2	91.7
	3.0	0.537	0.0	895.0	895.0	89.6
	8.0	0.852	0.0	572.3	572.3	97.9
	16.0	1.149	0.0	513.0	513.0	107.6
600[4]	0	0	9.7	1160.4	1122.5	100.0
	12.0	0.480	4.4	1069.4	982.5	107.6
	20.7	0.600	1.5	1201.1	1137.0	106.9
	50.0	0.684	0.4	1083.2	1025.3	104.1
	201.6	0.959	0.2	842.6	842.6	112.4
		Results for 0.0635-cm Beta-21S sheet				
800[1]	0	0	22.4	855.0	849.5	69.0
	0.5	0.147	17.5	892.9	872.2	72.4
	1.0	0.201	9.0	907.4	882.6	75.8
	3.0	0.335	1.7	925.3	906.0	77.2
	8.0	0.507	0.2	885.3	881.9	82.7
	16.0	0.675	0.0	726.7	726.7	86.2
	24.0	0.798	0.0	748.8	748.8	88.3
	40.0	1.010	0.0	618.5	618.5	91.0
700[3]	0	0	20.6	1000.5	935.0	100.0
	8.7	0.106	7.2	970.1	932.2	92.4
	16.0	0.186	2.4	896.4	890.1	91.7
	104.6	0.366	0.6	917.0	908.1	102.7
	500.5	0.760	0.0	425.4	425.4	108.9
	751.7	0.935	0.0	394.4	394.4	108.9
600[4]	0	0	12.0	1299.0	1241.8	108.9
	80.3	0.102	2.6	1168.7	1135.6	108.3
	500.0	0.178	2.3	1066.7	1030.1	108.9
	999.4	0.238	2.0	1029.4	1029.4	110.3

[1] No prior vacuum heat treatment.
[2] Specimens machined prior to exposure to remove porosity layer.
[3] Initial vacuum heat treatment at 700°C for 8 h.
[4] Initial vacuum heat treatment at 600°C for 8 h.

same rate in both the foil and sheet; therefore, the oxygen would diffuse to the same depth in both for an equivalent exposure time. Since the foil is much thinner than the sheet material, a larger percentage of the foil would be damaged by oxygen.

Normalizing the oxidation weight gain by the specimen volume allows direct comparison of the effect of oxidation exposure on the mechanical properties of the sheet and foil. This normalization was accomplished by calculating the percentage increase in specimen weight after oxidation exposure. This measurement can also be used to compare the effect of oxidation exposure at different temperatures for the same thickness material.

Figure 4 shows plastic elongation versus the percentage weight increase due to oxidation exposure at 800°C for both the sheet and foil. While the plastic elongation of both decreased with oxidation exposure, the sheet material became completely embrittled (exhibited no plastic elongation) at a lower percentage weight gain of oxygen than did the foil. The sheet material exhibited no plastic elongation at weight gains of 0.51 percent and greater. Zero plastic elongations were not measured on the foil material, however, until the weight gain was greater than 1 percent. This difference was shown to be due to the surface porosity present in the foil. When this porosity layer was removed from the foil material before oxidation exposure, the resulting average plastic elongation fell more in line with the sheet material than the foil.

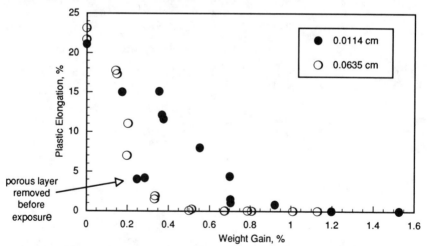

Figure 4 - Effect of oxidation weight gain on the plastic elongation of 0.0114-cm and 0.0635-cm Beta-21S.

Figure 5 shows the room temperature mechanical properties versus the percentage weight increase after exposure at 600°C, 700°C, and 800°C for the sheet material; figure 5(a) shows plastic elongation, figure 5(b) ultimate tensile strength, and figure 5(c) elastic modulus. Although the plastic elongation decreases after exposure at all temperatures, it decreases much quicker with exposure for lower temperature exposures. For instance, after oxidation exposures producing roughly 0.1 percent increase in weight, the plastic elongation of the specimen exposed at 800°C is now 17.5% compared to the starting value of 22.4%, the specimen exposed at 700°C has dropped from 20.6% to 7.2%, and the specimen exposed at 600°C has dropped from 12.0% to 2.6%.

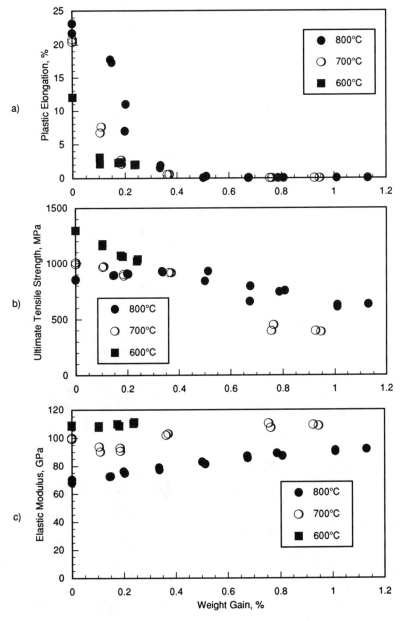

Figure 5 - Effect of oxidation weight gain on the a) plastic elongation, b) ultimate tensile strength, and c) elastic modulus of 0.0635-cm Beta-21S sheet.

The ultimate tensile strength and elastic modulus were also found to be affected by oxidation exposure. For oxidation exposures at 700°C and 800°C, no degradation in the ultimate tensile strength was measured until the weight gain exceeded roughly 0.5 percent. For exposures at 600°C, however, even small oxidation weight gains were shown to degrade the ultimate tensile strength. The modulus was found to increase with increasing exposure at all exposure temperatures.

Metallography and Fractography. Prior to any oxidation exposure, the specimens to be exposed at 700°C and 600°C were vacuum annealed for 8 h at the projected test temperature. This was not done for

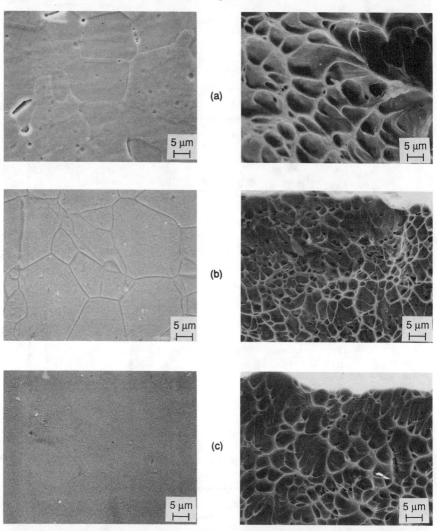

Figure 6 - SEM micrographs of the microstructure and fracture surface for 0.0635-cm Beta-21S a) as-received, and after 8 h vacuum anneal at b) 700°C, and c) 600°C.

specimens to be exposed at 800°C because the as-received material was already beta-annealed, and further exposure at 800°C would not be expected to alter the microstructure.[1] This means that the starting microstructure and mechanical properties of the Beta-21S specimens depended on their projected exposure temperature.

Figure 6 shows the micostructure and corresponding fracture surface of the sheet material in the as-received condition, and after 8 h vacuum exposure at 700°C, and 8 h vacuum exposure and 600°C. The as-received microstructure is almost entirely equiaxed beta. Although the prior beta grain boundaries are still visible after vacuum exposure at 700°C and 600°C, the interior of these grains was converted to an acicular alpha and beta mixture. These alpha and beta particles are smaller at 600°C than at 700°C. The corresponding fracture surfaces show that specimens in all three conditions fail by ductile microvoid coalescence.

Figure 7 shows the effect of oxidation on these three sheet microstructures and the corresponding fracture surfaces. Exposures at 800°C have the most dramatic effect on the microstructure. The increased oxygen content was found to cause the formation of grain boundary alpha as well as the precipitation of alpha in the interior of the beta grains near the surface where the oxygen content is highest. Analysis of the fracture surface shows that the embrittled area consists of two regions. In the surface region, brittle fracture initiated both at the alpha at the grain boundaries as well as the grain interiors. The lower region was a mixture of intergranular fracture, initiated at the grain boundary alpha, and ductile fracture of the beta grains.

Exposures at 700°C and 600°C produce less noticeable results in the microstructure; however, the fracture surfaces are characterized by an obviously oxygen embrittled zone. Examination of the microstructure shows that after exposure at both temperatures, small amounts of alpha have formed along the grain boundaries and the alpha and beta particles have also coarsened. The increase in elastic modulus encountered with oxidation exposure (figure 5) also suggests that there is an increase in the volume percent in alpha since α-Ti has a higher modulus than β-Ti.[4,7] After exposure at 700°C the oxygen affected zone is characterized by multiple initiation sites of brittle fracture corresponding to the alpha both at the grain boundaries and in the grain interiors. After exposure at 600°C the oxygen damaged fracture surface was flat and exhibited river patterns radiating from the surface, indicating brittle cleavage fracture. Due to the similarities in microstructure to the 700°C exposed specimens, fracture would be expected to initiated at the alpha particles. It would appear, however, that once fracture initiates, the finer structure encountered at 600°C is unable to blunt the crack and the entire embrittled region fails by cleavage.

In all but the most oxidized samples, the observed embrittled zone did not extend through the entire specimen thickness. The center of the specimen still exhibited ductile fracture which appeared identical to the fracture surface of the unexposed specimen. The depth of this embrittled zone, however, increased with oxidation exposure until enough of the cross-section was embrittled to cause the measured plastic elongation to be zero. Measurement of the depth of this embrittled zone provides a good measure of the depth of oxygen diffusion into the metal substrate.

123

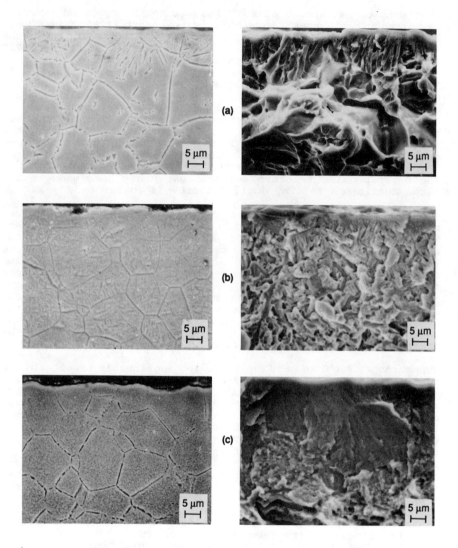

Figure 7 - SEM micrographs of the microstructure and fracture surface for 0.0635-cm Beta-21S exposed for a) 1 h at 800°C, b) 16 h at 700°C, and c) 1000 h at 600°C.

Figure 8 shows the depth of the embrittled zone, measured on selected sheet specimens from all three temperature exposures, against the measured oxygen weight gain. Depths on specimens exposed at 600°C and 700°C were measured from the surface to the beginning of ductile fracture. On the 800°C specimens, both the brittle surface layer and the transition region containing both ductile and intergrandular failure were considered to be included in the embrittled zone.

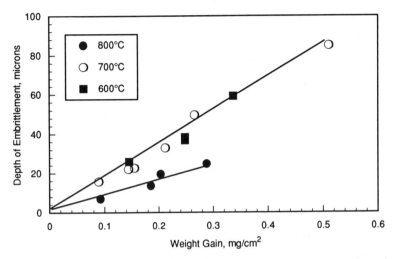

Figure 8 - Depth of oxygen embrittlement versus oxidation
weight gain for 0.0635-cm Beta-21S exposed at
800°C, 700°C, and 600°C.

The results of this analysis can be used to explain why the
decrease in plastic elongation is less for an equivalent percent
increase in weight after exposure at 800°C as compared to
exposures at 600°C and 700°C. The depths measured on specimens
exposed at 600°C and 700°C followed the same straight line
relationship with weight percent. While the depths measured after
exposures at 800°C were also found to follow a straight line
relationship with weight percent, the depths measured at this
temperature were in all cases less than those for equivalent
weight gains at 600°C or 700°C. These results are due to the fact
that exposures at 600°C and 700°C were found to produce similar
mixtures of alpha and beta in the microstructures, while the
microstructures after exposures at 800°C were mostly beta. The
diffusion of oxygen in these two differing microstructures is
different, and this would affect the depth of oxygen penetration
for equivalent weight gains.

The difference in the decrease in plastic elongation after
equivalent weight gain exposure at 600°C and 700°C, however, can
only be explained by the difference in fracture mode produced by
the oxygen embrittlement. After exposure at 600°C brittle
fracture was found to initiate at the surface and proceed by
cleavage fracture across the emrittled layer. The sharp crack
introduced by this process would act as a large stress riser, and
lead to premature failure of the specimen. The multiple initiation
sites of brittle fracture evident after exposure at 700°C,
however, indicate that fracture of the embrittled zone introduces
a group of small cracks. Although these cracks also lead to
premature failure of the specimen, their smaller size introduces
less of an increase in stress than the single crack formed at
600°C. This means that specimens exposed at 700°C would experience
less of a degradation in properties than if exposed for an
equivalent weight percent increase at 600°C

CONCLUSIONS

Examination of the effects of oxidation on the microstructure and mechanical properties of Beta-21S over the temperature range 600°C to 800°C show that the degree of embrittlement produced is very dependent not only on amount of oxidation weight gain, but also on the temperature of exposure. It was shown that equivalent oxidation weight gains (normalized to specimen thickness using weight percent) produces a much greater change in ductility at 600°C than at 800°C (figure 5). These differences can be accounted for by two effects; the lesser depths of embrittlement for oxdiation at 800°C compared to the lower temperatures, and the differing fracture modes produced by oxidation damage at the three temperatures.

REFERENCES

1. P. J. Bania and W. M. Parris, "Beta-21S: A High Temperature Metastable Beta Titanium Alloy," Proceedings of the 1990 International Conference on Titanium Products and Applications (Dayton, OH: Titanium Development Association, 1990).

2. T. A. Wallace, R. K. Clark, and K. E. Wiedemann, "Oxidation Characteristics of Beta-21S in Air in the Temperature Range 600 to 800°C" (NASA TM 104217, 1992).

3. J. Niemann and J. McAfee, "Development of Titanium Matrix Composites" (Paper No. 181, Tenth National Aero-Space Plane Technology Symposium, April 23-26, 1991).

4. G. Welsch and W. Bunk, "Deformation Modes of the α-Phase of Ti-6Al-4V as Function of Oxygen Concentration and Aging Temperature," Metallurgical Transactions A, 13A (1982), 889-899.

5. L. A. Glikman, V. I. Deryabina, N. N. Kilgatin, I. A. Bytenskii, V. P. Teodorovich, and N. S. Teplov, "The Influence of Gas-Saturated Layers on the Strength and Plastic Properties of Titanium Alloys," Titanium and Its Alloys, ed. I. I. Kornilov (Jerusalem: Israel Program for Scientific Translations, 1966), 122-136.

6. E8-85b, "Tension Testing of Metallic Materials," 1986 Annual Book of ASTM Standards, vol. 03.01 (American Society for Testing and Materials, 1986), 124-145.

7. T. W. Duerig and J. C. Williams, "Overview: Microstructure and Properties of Beta Titanium Alloys," Beta Titanium Alloys in the 80's, ed. R. R. Boyer, H. W. Rosenbery (Warrendale, PA: The Metallurgical Society, 1984).

EFFECTS OF AIRCRAFT HYDRAULIC FLUID ON *TIMETAL*®21S

James S. Grauman

TIMET
P.O. Box 2128
Henderson, NV 89009

ABSTRACT

TIMETAL®21S, a recently developed high strength metastable beta titanium alloy, exhibits unsurpassed corrosion resistance for this class of alloy. In particular, the alloy is very resistant to corrosion and hydrogen absorption in hot reducing acids. This corrosion behavior led to testing of the alloy in aircraft hydraulic fluid, which is known to be corrosive to titanium when allowed to evaporate on the metal surface. Initial tests of *TIMETAL*®21S and several other aerospace alloys in hydraulic fluid are reported herein. The resistance of *TIMETAL*®21S to high temperature hydraulic fluid was found to be unsurpassed for a titanium alloy. The new temperature guideline for exposure of *TIMETAL*®21S to hydraulic fluid will allow the use of titanium in areas of aircraft previously restricted to other metals.

Beta Titanium Alloys in the 1990's
Edited by D. Eylon, R.R. Boyer and D.A. Koss
The Minerals, Metals & Materials Society, 1993

INTRODUCTION

The advent of fuel efficient aircraft has led to increased demands for engine and airframe weight reductions. Manufacturers are turning to titanium as one means of achieving these reductions.

However, use of titanium for certain warm (400-700°F) (204-371°C) engine components on commercial aircraft has, in part, been limited because of the risk of exposure to aircraft hydraulic fluid, commonly referred to as Skydrol™. These phosphate-based fluids can decompose above about 350°F (177°C) to form concentrated phosphoric acid solutions.

Hot reducing acids, such as phosphoric acid are very aggressive on most titanium alloys. As such, it is no surprise that titanium alloys have been shown to be susceptible to corrosion and hydrogen embrittlement when contacted by this hot hydraulic fluid.[1] This problem doesn't exist for military aircraft since different types of hydraulic fluids are used that are not subject to acid formation on decomposition.

The fear of catastrophic failure of titanium components due to corrosion and embrittlement by hydraulic fluid has led to most manufacturers stipulating strict temperature guidelines for titanium usage. However, the benefit of weight savings offered through the use of titanium has led to continued Skydrol testing by certain manufacturers.

TIMETAL®21S is a recently developed, high strength beta titanium alloy containing 15 wt. % molybdenum. Although originally developed as an oxidation resistant alloy for the NASP Project, laboratory testing showed it to have excellent corrosion resistance properties. The recognition of *TIMETAL*®21S as a high strength, corrosion resistant alloy led to corrosion screening tests in aircraft hydraulic fluid. In comparison tests with Ti-6-4, *TIMETAL*®21S exhibited very encouraging results (see Table I).

This paper describes the results of comparative testing of one nickel and four titanium alloys in aircraft hydraulic fluid. Differences in corrosion resistance are discussed, as well as new applications for titanium that have occurred as a result of the corrosion behavior of *TIMETAL*®21S.

Table I.   Initial Skydrol Test Results

Test Temp. °F (°C)	Alloy	Corrosion Pit Depth mils (mm)	Hydrogen Pickup (ppm)
350 (177)	TIMETAL®21S	No Attack	12
350 (177)	Ti-6-4	0.4 (0.010)	7
450 (232)	TIMETAL®21S	0.2 (0.005)	97
450 (232)	Ti-6-4	100 (2.54)	>1000
550 (288)	TIMETAL®21S	1.4 (0.036)	138
550 (288)	Ti-6-4	28 (0.711)	>1000
650 (343)	TIMETAL®21S	1.2 (0.031)	83
650 (343)	Ti-6-4	21 (0.533)	>1000

Data courtesy of Boeing Commercial Aircraft
Test Conditions: 48 hr immersion with fluid
replenished every 24 hrs.

EXPERIMENTAL

Table II lists the nominal compositions of the alloys included in this study. The titanium alloys were all tested in the STA condition. Test specimens were cut from mill-produced strip or sheet product into approximately 3"x5" (76mm x 12.7mm) coupons. The titanium test coupons were given a short (5-10 min) immersion in an ambient 35 vol. % HNO_3/5 vol. % HF pickle solution to ensure uniform surface condition. The alloy 625 specimens were lightly abraded with 240 grit silicon carbide paper to prepare the surface for testing. All specimens were given a final rinse in ASTM Type II purified water, and thoroughly dried prior to exposure.

Table II. Nominal Composition of Test Alloys

ALLOY	COMPOSITION (wt%)							
	Al	Cr	Fe	Mo	Ni	V	OTHER	Ti
TIMETAL®21S	3.0		0.15	15.0			2.7Nb,0.2Si	BAL
Ti-6-4	6.0		0.2			4.0		BAL
Ti-6242	6.0		0.2	2.0			2.0Sn,4.0Zr	BAL
Ti-38644	3.0	6.0	0.2	4.0		8.0	4.0Zr	BAL
Alloy 625		22.0	5.0	9.0	BAL		3.5Nb	

The test media was aircraft hydraulic fluid at ambient temperature. Specifically, the hydraulic fluid was Chevron Hyjet IV.

The test apparatus consisted of variable temperature hot plates, situated at a 45 degree angle to allow any excess hydraulic fluid to run off the test specimen. Test specimen temperatures were monitored using thermocouple attachments so as to keep the temperature within about 5°F (2.8°C) of the desired test temperature. A precision metering pump was utilized to deliver the hydraulic fluid onto the specimen. The rate of metering was controlled at one drop every 3 ± 1 minutes. Test temperatures included 400°F (204°C), 450°F (232°C), 500°F (260°C), and 600°F (316°C). Duration of each test was 96 hours.

Post test examination required a light (5 second) sandblasting of the surface to remove the tenacious hydraulic fluid decomposition products. Macro and stereo photography were utilized to illustrate the surface etching/pitting, while transverse section photomicrographs were used to determine pit depths. Hydrogen analysis was performed via the hot vacuum extraction method.

RESULTS

Table III represents a compilation of all test results. Corrosion is reported in terms of depth of attack and visual observation, rather than a corrosion rate since the attack was localized to the area of impingement. Transverse section mounts were prepared for the one specimen from each alloy group exhibiting maximum attack to allow accurate depth of attack measurements. Hydrogen absorption was also measured at the site of hydraulic fluid impingement.

The results revealed that each alloy had a certain temperature zone, within the boundaries of the test, where it exhibited greater susceptibility to corrosion. Generally, temperatures outside of each alloy's specific zone had very little effect. Figures 1-10 illustrate the surface etching observed on the test specimen from each alloy group that exhibited greatest attack. The first figure in each group represents a macrograph of the alloy specimen taken at 1.1X magnification, highlighting the impingement and associated run-off areas with dash marks. The second figure of each grouping is a higher magnification view of just the impingement zone. Figures 11 and 12 illustrate maximum corrosion and hydrogen absorption observed for each alloy over the test temperature range.

Table III. Results of Hydraulic Fluid Dripwise Exposure Testing

ALLOY	TEMPERATURE °F (°C)	NET HYDROGEN ABSORPTION (ppm)	DEPTH OF ATTACK mils (mm)	COMMENTS
TIMETAL®21S	400 (204)	0	-	Very slight etching at 450°F (232°C). No attack at other temperatures.
TIMETAL®21S	450 (232)	0	<1 (<0.03)	
TIMETAL®21S	500 (260)	0	-	
TIMETAL®21S	600 (316)	0	-	
Ti-6-4	400 (204)	16	-	Severe etching at 600°F (316°C). Moderate attack at other temps.
Ti-6-4	450 (232)	0	-	
Ti-6-4	500 (260)	0	-	
Ti-6-4	600 (316)	750	43 (1.09)	
Ti-6242	400 (204)	18	-	Moderate etching at temperatures above 400°F (204°C).
Ti-6242	450 (232)	41	6 (0.15)	
Ti-6242	500 (260)	0	-	
Ti-6242	600 (316)	0	-	
Ti-38644	400 (204)	0	-	Moderate attack at 500°F (260°C). Slight etching at 600°F (316°C).
Ti-38644	450 (232)	0	-	
Ti-38644	500 (260)	15	5 (0.13)	
Ti-38644	600 (316)	13	-	
ALLOY 625	400 (204)	-	-	Slight etching visible at all temperatures.
ALLOY 625	450 (232)	-	1 (0.03)	
ALLOY 625	500 (260)	-	-	
ALLOY 625	600 (316)	-	-	

Note: (-) indicates no measurement.

M4717   1.1X

M4720   16X

Figures 1 and 2 - *TIMETAL®*21S specimen exposed to dripping hydraulic fluid at 450°F (232°C) for 96 hours. Stamp marks on specimen were used to outline the area of hydraulic fluid drip exposure prior to sandblasting.

M4691 1.1X

M4694 8X

Figures 3 and 4 - Ti6Al-4V specimen exposed to dripping hydraulic fluid at 600°F (316°C) for 96 hours.

M4710 1.1X

M4713 8X

Figures 5 and 6 - Ti-6242 specimen exposed to dripping hydraulic fluid at 450°F (232°C) for 96 hours.

4697   1.1X   M4699   8X

Figures 7 and 8 - Ti-38644 specimen exposed to dripping hydraulic fluid at
               500°F (260°C) for 96 hours.

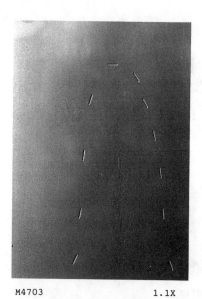

M4703   1.1X   M4707   8X

Figures 9 and 10 - Alloy 625 specimen exposed to dripping hydraulic fluid at
                450°F (232°C) for 96 hours.

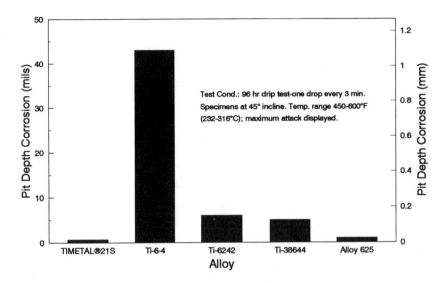

Figure 11 - TIMET Skydrol corrosion test results.

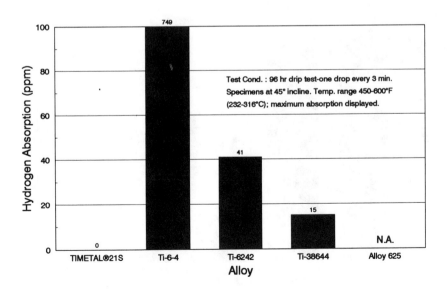

Figure 12 - TIMET Skydrol hydrogen absorption test results.

DISCUSSION

Molybdenum, as an alloying addition to titanium, has been shown to be very effective in enhancing corrosion resistance.[2,3,4] In particular, additions above about 3-4 wt. % dramatically improve general and crevice corrosion resistance in hot reducing acids. Increased molybdenum content above this minimum level directly correlates with improved resistance. Various studies[5,6] have shown that *TIMETAL®21S* exhibits greater than expected resistance (based on molybdenum content) to reducing acid corrosion. This phenomenon has been attributed to an additive effect of molybdenum with niobium in *TIMETAL®21S*, as shown in Figure 13.[6] [Note that neither the Ti-Nb or Ti-Mo binary alloys alone approach the resistance of *TIMETAL®21S*.] This additive effect also explains the excellent hydrogen absorption resistance exhibited by *TIMETAL®21S*.[5,6,7]

The excellent reducing acid corrosion resistance of *TIMETAL®21S* appears to translate to equally good behavior when the alloy is exposed to high temperature hydraulic fluid. This is not particularly surprising since the corrosive medium formed from high temperature chemical breakdown of aircraft hydraulic fluid, phosphoric acid is among the group of reducing acids. Among the group of four titanium alloys tested, *TIMETAL®21S* was clearly superior under conditions of intermittent exposure. Even the one specimen of *TIMETAL®21S* that exhibited slight etching (see Figure 2) required twice the magnification used for the other alloys just to obtain a clear photograph of the corrosion. Overall, it appears the degree of susceptibility for the alloys studied correlates quite well with the molybdenum content of each alloy.

Compared with alloy 625, *TIMETAL®21S* displayed equivalent resistant over the range of temperatures studied. In fact, alloy 625 exhibited slight etching at each test temperature above 400°F (204°C). The corrosion on alloy 625 appeared more uniform in nature, and was observed along the fluid run-off area of the specimen, rather than at the drop impingement site (see Figure 9).

Based on this intermittent hydraulic fluid impingement study, the following ranking can be compiled for corrosion and hydrogen absorption resistance:

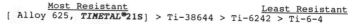

   Most Resistant   Least Resistant
[Alloy 625, *TIMETAL®21S*] > Ti-38644 > Ti-6242 > Ti-6-4

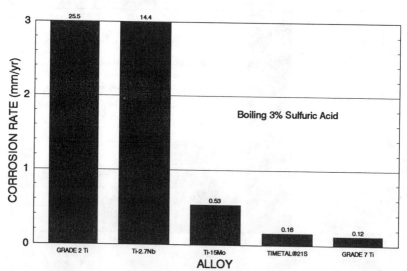

Figure 13 - Effect of molybdenum and niobium additions on the corrosion rate of titanium in sulfuric acid.

134

However, certain types of immersion (as compared to dropwise) exposures appear to be more aggressive towards the titanium alloys, making alloy 625 stand out as the most resistant. Thus, the type of exposure the component could see might well influence the material consideration.

New applications arising from the excellent Skydrol resistance of *TIMETAL*®21S involve warm engine components. Included on this list would be the aft cowling, aft fairing heat shield, and the exhaust assembly (plug and nozzle). The use of *TIMETAL*®21S for these components will result in an estimated 500 pounds (227 kg) weight savings for the new Boeing 777 aircraft.

CONCLUSIONS

1.   In the temperature range of 400°F-600°F (204°C-316°C), *TIMETAL*®21S exhibits only very slight surface etching when exposed to intermittent (dropwise) contact with aircraft hydraulic fluid.

2.   *TIMETAL*®21S exhibits no susceptibility towards hydrogen absorption under the above stated conditions.

3.   When compared with other titanium alloys, *TIMETAL*®21S appears best suited to handle intermittent high temperature hydraulic fluid exposure without risk of component failure.

4.   For resistance to intermittent exposure of high temperature hydraulic fluid, *TIMETAL*®21S appears to be equivalent to alloy 625. However, alloy 625 exhibits greater tolerance under certain conditions involving immersion.

ACKNOWLEDGEMENT

The author wishes to acknowledge the assistance and cooperation of Mr. Rod Boyer, Boeing Commercial Airplane Div.

REFERENCES

1.   Unpublished TIMET data.

2.   R. W. Schutz and J. S. Grauman, "Fundamental Corrosion Characterization of High Strength Titanium Alloys", Industrial Applications of Titanium and Zirconium, 4th Volume, STP 917, American Society for Testing and Materials, (1986) 130-143.

3.   M. Stern and C. Bishop, Transaction ASM, 54, (1961) 286-298.

4.   N. D. Tomashov, et.al., "Studies of the Alloys of the Titanium-Molybdenum-Niobium-Zirconium System With Higher Corrosion Resistance in Acid Solutions", Proceedings of the 3rd International Conference on Titanium, Moscow, (1976) 927-940.

5.   J. S. Grauman, "Beta-21S: A New High Strength, Corrosion Resistant Titanium Alloy", Proceedings of the 1990 International Conference, Titanium Development Association, Dayton, OH, Vol. 1 (1990) 290-299.

6.   J. S. Grauman, "Corrosion Behavior of *TIMETAL*®21S For Non-Aerospace Applications", Presented at the 7th World Conference on Titanium, San Diego, CA, June 1992.

7.   B. Bavarian, V. Wahi, and M. Zamanzadek, "Corrosion Behavior of Beta-21S Titanium Alloy in Chloride - Containing Environments", Paper No. 284, **CORROSION '93**, New Orleans, March 1993, National Association of Corrosion Engineers, Houston, TX.

KINETICS OF HYDROGEN TRANSPORT THROUGH

Ti-15Mo-2.7Nb-3Al-0.2Si (BETA-21S) ALLOY*

Sankara N. Sankaran and Rebecca K. Herrmann
Analytical Services and Materials, Inc.
107 Research Drive, Hampton, VA 23666

Ronald K. Clark and R. A. Outlaw
NASA Langley Research Center
Hampton, VA 23681-0001

ABSTRACT

The kinetics of hydrogen absorption by Ti-15Mo-2.7Nb-3Al-0.2Si (Beta-21S) alloy were studied using an ultrahigh vacuum thermogravimetric analysis apparatus. Hydrogen uptake kinetics were measured over a temperature range of 400°C to 800°C and pressures up to 50 torr. The alloy has a substantial solubility for hydrogen, and the kinetics are significantly influenced by the microstructure of the alloy. The differences in the uptake kinetics of the individual phases (alpha and beta phases) influence the transport characteristics. The beta-stabilizing effect of hydrogen influences the microstructure of the alloy after hydrogen exposure.

* This work was supported by the National Aero-Space Plane Joint Program Office (ASC/NA), Wright-Patterson AFB, Ohio 45433-6503 through Contract No: F33657-91-C-2183 with technical guidance from Dr. H.G.Nelson of NASA Ames Research Center. Partial support was also provided by the NASA Langley Research Center.

Beta Titanium Alloys in the 1990's
Edited by D. Eylon, R.R. Boyer and D.A. Koss
The Minerals, Metals & Materials Society, 1993

137

INTRODUCTION

Ti-15Mo-2.7Nb-3Al-0.2Si (Beta-21S), a metastable β-titanium alloy, is a candidate material for titanium matrix composite structures in hydrogen-fueled hypersonic planes because of its excellent formability and adequate mechanical properties in the 500-800°C temperature range [1]. The alloy is strengthened through the precipitation of fine α particles in the β matrix. The mechanical properties and microstructure are controlled by a solutionizing/ageing heat treatment.

The low solubility of hydrogen in α-titanium results in the precipitation of hydrides that embrittle the metal [2-4]. In Beta titanium alloys, it has been suggested that the large solubility of hydrogen in the beta phase of titanium would preclude the precipitation of hydrides, especially at low hydrogen pressures [5]. However, hydrogen raises the ductile/brittle transition temperatures of Beta-21S to levels above room temperature depending on the hydrogen content [6].

The objective of the present investigation was to determine the hydrogen transport characteristics of the Beta-21S alloy at moderate pressures (0.5 to 50 torr) and at temperatures ranging from 400°C to 800°C. Sorption and desorption measurements using a microbalance technique, also referred to as thermogravimetric analysis (TGA) under ultrahigh vacuum (UHV) conditions was used as the primary technique in this work.

THEORY

The diffusion process in coupon specimens can be approximated to that of a plane sheet of material in which effectively all the diffusing substance enters through the plane faces and a negligible amount through the edges. Diffusion in a planar specimen has been described in Crank [7]. For a sheet of thickness, $2l$ ($-l < x < l$), initially at a uniform concentration, C_0, and with the surfaces maintained at a constant concentration, C_s, solution to the Fick's laws yield:

$$\frac{C - C_0}{C_s - C_0} = 1 - \frac{4}{\pi} \sum_{n=0}^{\infty} \frac{(-1)^n}{2n + 1} e^{-D(2n+1)^2 \pi^2 t/4l^2} \cos \frac{(2n + 1)\pi x}{2l} \tag{1}$$

where, D is the diffusivity of the solute atoms and is assumed to be independent of the concentration of the species, and n is an integer to track the summation. The total amount of the diffusing species, M_t, that has entered the sheet at time t, can be expressed as a fraction of the saturation value, M_∞, as:

$$\frac{M_t}{M_\infty} = 1 - \sum_{n=0}^{\infty} \frac{8}{(2n+1)^2\pi^2} e^{-D(2n+1)^2\pi^2 t/4l^2} \tag{2}$$

The hydrogen uptake by the sheet is depicted in Figure 1. Two regimes can be identified as a function of time. At long times, the exponential terms vanish, and the entire sheet reaches the surface concentration C_s. At this stage, no more species can diffuse into the sheet, and it reaches saturation corresponding to:

$$M_\infty = 2A(C_s - C_0)l \tag{3}$$

where A, is the area of the sheet exposed to the diffusing species. At short times [7], the corresponding solution can be written:

$$\frac{M_t}{M_\infty} = 2\left[\frac{Dt}{l^2}\right]^{1/2} \left\{ \pi^{-1/2} + 2 \sum_{n=0}^{\infty} (-1)^n \text{ierfc} \frac{nl}{\sqrt{Dt}} \right\} \tag{4}$$

It is interesting to note that the slope of the curve in Figure 1 can be further approximated at very short times, $\left[\frac{Dt}{l^2}\right]^{1/2} < 0.5$, to:

$$\frac{M_t}{M_\infty} = \frac{2}{\sqrt{\pi}} \left[\frac{Dt}{l^2} \right]^{1/2} \tag{5}$$

Thus, the uptake kinetics of the diffusant display a square root time dependence during the early stages of the exposure, with a slope that reflects the diffusivity of the solute in the sheet material. On the other end of the spectrum, very long exposures lead to saturation of the species in the material that enables the determination of the maximum solubility, C_s using equation (3).

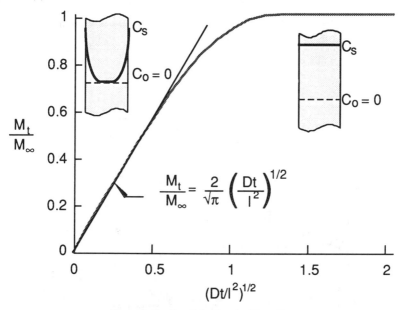

Figure 1. Sorption Behavior of a Plane Sheet.

Monitoring the uptake of the diffusing species from the beginning of the exposure up to saturation provides a means for characterizing the solubility and diffusivity of the material through a single experiment. The saturation values can be used to determine the pressure dependence of the concentration of the species in the material being studied assuming that the experiments are performed with a gaseous species, and diffusion of the species is the rate-controlling step. A corresponding experiment is the desorption of species from a saturated sample. The calculations are identical but the boundary conditions are changed.

EXPERIMENTAL PROCEDURE

Specimen Characteristics

The specimens were in the form of coupons, 2.0 cm × 0.8 cm × 0.1 cm in dimension with a surface finish of 0.25 μm. They were ultrasonically cleaned with soap solution, rinsed in deionized water, and finally acid-cleaned in a mixture containing 68% deionized H_2O, 30% HNO_3, and 2% HF maintained at 54°C. Following acid-cleaning, the specimens were again rinsed in deionized water and hot air dried. In order to ensure the same starting microstructure prior to hydrogen exposure, the specimens were given a controlled heat-treatment consisting of heating to 900°C and holding for 30 minutes to expel any initial hydrogen and solutionize the sample in the β-phase region, rapid cooling to room temperature, and ageing at 640°C. Some of the earlier samples were aged at this temperature for 4 hours, but the later samples were aged for 12 hours to insure homogeneity of the microstructure. As will be described later, the entire heat-treatment was carried out inside the TGA system under UHV conditions ($< 10^{-9}$ torr) and the sample was never re-exposed to atmospheric air prior to hydrogen exposure. This heat-treatment procedure was also followed between runs in samples that were subject to multiple exposures: this procedure helped to remove the previously absorbed hydrogen and to restore the starting microstructure.

Apparatus

A schematic of the TGA apparatus is shown in Figure 2. The heart of the system is the microbalance which is located inside a large vacuum chamber evacuated by an ion-pump, and isolated from the pump through a gate valve. The sample is suspended from the microbalance into the reaction chamber through a long inter-connecting tube. The sample temperature is maintained by a tubular furnace, and the hydrogen exposure pressure is regulated through a pressure controller, and monitored by capacitance gages. The reaction zone is evacuated by a separate ion pump.

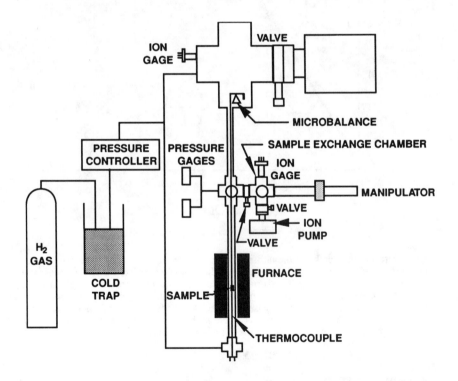

Figure 2. Schematic of the Thermogravimetric Apparatus

Additional features include a sample exchange chamber to enable remote transfer of sample into the reaction zone without breaking the UHV conditions, and a magnetically coupled sample lifting facility to enable rapid cooling of the sample. Diagnostic equipment include ion-gages to monitor the pressure, and a quadrupole mass-spectrometer to monitor the composition of the residual gases in the system. All the equipment in the system are provided with interfaces to enable communication and data acquisition through a computer.

Testing Procedure

Prior to sample introduction, the system was baked and brought to UHV conditions. The specimens for testing were introduced into the system only through the sample exchange chamber, and the main system was always maintained under UHV conditions. Between sample exchanges, cold trapped argon or helium gas was used to backfill the sample exchange chamber. These precautions were taken to minimize contaminants into the system that have been shown to undermine the characterization of hydrogen absorption in titanium alloys [8]. Prior to hydrogen exposure, the samples were subjected to the control heat-treatment which also helped in dissolving any native surface films that may interfere with the hydrogen uptake. Data were then taken using three different methods.

In the first method, weight gain measurements were made using a flowing system with the hydrogen gas maintained at the test pressure using a pressure controller and capacitance pressure gages. Flow was established by exhausting a portion of the gas from the chamber into an auxiliary pump at a flow rate of 30 sccm. This method was chosen to help alleviate any barrier layer effects from contaminants in the gas [8]. However, even the small buoyancy changes associated with the flow control produced considerable noise in the weight readings, especially at the early stages of exposure. Therefore, this technique was not adequately sensitive for monitoring weight gain rates. Nevertheless, this technique was found to be the best one for unambiguously characterizing the equilibrium concentrations of hydrogen in the alloy at conditions requiring long-term exposures where barrier layer effects pose a problem.

In the second method, the sample was withdrawn from the hot zone of the reaction chamber after the conclusion of the heat-treatment cycle, and the system was backfilled with hydrogen gas to the test pressure while maintaining the reaction chamber at the test temperature. The gas source was then valved off. After ensuring the stabilization of the pressure in the system, the sample was returned to the reaction chamber and the pressure decay and weight gain were monitored to estimate the hydrogen uptake. Following saturation, the specimen was quickly withdrawn from the hot zone of the furnace after hydrogen charging while simultaneously pumping out the gas. The quick withdrawal of the specimen from the hot zone after hydrogen charging was designed to quench in the absorbed hydrogen in the sample so that it can be used for microstructural characterization.

In the third method, the specimens were subjected to a multiple exposure cycle. In this method, the specimen was heated to 800°C after the heat treatment cycle, and was then allowed to equilibrate with hydrogen gas at the test pressure. The temperature was then progressively decreased to 400°C in 50°C intervals ensuring equilibrium at each of the intervening temperatures. The tests were conducted at pressures ranging from 1 torr to 50 torr.

RESULTS AND DISCUSSION

Equilibrium Hydrogen Concentration

The concentration of hydrogen in equilibrium with the Beta-21S alloy as a function of temperature at various pressures of hydrogen gas is summarized in Figure 3 in the form of an Arrhenius plot. The data covers a temperature range of 400-800°C, and hydrogen pressures ranging from 1 to 50 torr. The open symbols represent data taken during single exposures to the indicated test temperature and pressure. The closed symbols represent data from multiple exposures. At the outset, it can be noted that the alloy has a substantial solubility for hydrogen; concentrations as high as 2 weight percent were observed. Also, the hydrogen concentration displays a positive slope, indicative of a negative heat of solution for hydrogen in the alloy consistent with other titanium alloys [9].

In an ideal material that obeys Sievert's law, the equilibrium concentration of hydrogen in the alloy for a given exposure condition can be expressed as:

$$C_s = SP_{H_2}^{1/2} \qquad (6)$$

where, S is the solubility of hydrogen in the alloy, and P_{H_2} is the hydrogen pressure. For a dilute solution (Henry's law regime), this equation predicts that S is only a function of temperature, and the concentration curves at different pressures must be straight lines parallel to each other with a slope that is representative of the heat of solution of hydrogen in the alloy. This ideal

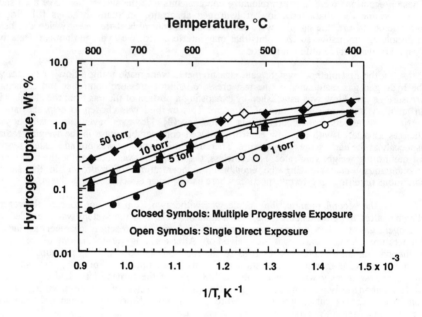

Figure 3. Temperature Dependance of Hydrogen Concentration in Beta-21S

behavior is followed by the Beta-21S alloy only at the higher temperatures, but at lower temperatures, the hydrogen concentration departs from a log-linear relationship. Further, the linear behavior extends over a larger temperature range at lower exposure pressures (e.g., 1 torr), and the deviation from linearity occurs at higher temperatures for higher hydrogen pressures. These deviations may be due to deviation from Sievert's law, i.e., the solubility is not a constant at higher hydrogen pressures. A change in solubility due to phase transformations can also produce a similar effect.

Effect of Hydrogen Pressure

Since the deviation of solubility from log-linear relationship with respect to hydrogen pressure in the Beta-21S alloy was most prominent at the lower temperatures (Figure 3), a more extensive investigation of the pressure dependence was carried out at 500°C. This temperature was selected for this investigation because the data at this temperature seemed to follow linear behavior at lower pressures, and display deviations from linearity at higher pressures, suggesting the existence of a transition region. The data covering a pressure range of ≈ 0.5 torr to 50 torr are presented in Figure 4. These measurements were made on multiple samples. The excellent repeatability in the measurements is evident from the data. The exposure conditions for each specimen was selected in such a manner that samples representing specific pressure regimes were available for microstructural characterization studies. The data shows that the hydrogen uptake follows Sievert's law up to a pressure of about 1 torr, but deviates from the one-half power dependence at higher pressures. However, a sharp discontinuity in the pressure dependence, which is typical of phase transitions like the formation of a hydride phase, is not present. The behavior noted in Figure 4 is indicative of a change in the dominant rate controlling step, probably caused by microstructural changes in the alloy.

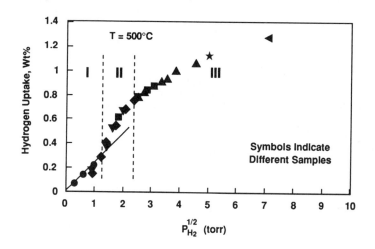

Figure 4. Pressure Dependence of Hydrogen Uptake in Beta-21S

Photo-micrographs of samples representing the three distinct regions in the pressure dependence data are shown in Figure 5. The microstructural characterization was done on the cross-section of the hydrogen exposed samples using light microscopy. At low hydrogen pressures (P < 1 torr), the base microstructure of the alloy consisting of a fine distribution of α precipitates in a β matrix seems to remain undisturbed by the hydrogen as indicated in Figure 5a.

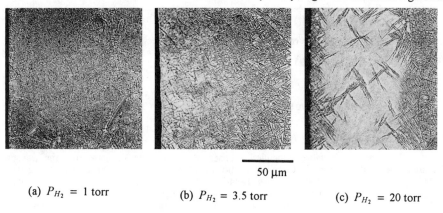

50 µm

(a) $P_{H_2} = 1$ torr   (b) $P_{H_2} = 3.5$ torr   (c) $P_{H_2} = 20$ torr

Figure 5. Microstructural Effects of the Hydrogen Exposure

At higher pressures (P > 10 torr), substantial enrichment of the β phase can be observed, particularly at regions close to the surface of the specimen (see Figure 5c). In this pressure region, though the kinetics indicate equilibrium uptake, there is a distinct β-phase case adjacent to the sample surface. Even very long exposures at these conditions produced the same effect suggesting that complete stabilization of the β-phase does not extend to the entire sample cross-section. It is speculated that the kinetics of movement and redistribution of the other alloying elements is inhibiting the full scale conversion of the β phase. At intermediate pressure ranges (1 torr < P < 10 torr), more of the β-phase is present as compared to lower pressures, but large islands of β-phase are not observed.

Kinetics of Hydrogen Uptake

The TGA data depicting the kinetics of the hydrogen uptake in Beta-21S at 500°C over the pressure range of 1 to 50 torr are shown in Figure 6. The rate of uptake, slope of the curves in Figure 6(a), is influenced by the hydrogen pressure: equilibrium seems to be attained earlier at higher hydrogen pressures. This feature is very conspicuous in Figure 6(b) where the weight gain has been normalized to the final concentration at each pressure. For the experimental conditions typical of the TGA, Figure 1 and equation (5) predict that all of the above curves should overlap since the tests were conducted at the same temperature. Also, the initial slopes, indicative of the diffusivity should be a constant.

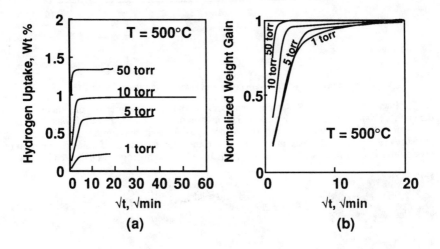

Figure 6. Kinetics of Hydrogen Uptake in Beta-21S

The type of pressure dependence observed in the data suggests that the transport characteristics of hydrogen in the alloy are not indicative of diffusion through a homogeneous material. In the heat treated condition, Beta-21S alloy has a two phase structure consisting of precipitates of α phase in a metastable β matrix. Hydrogen is a β-phase stabilizer in titanium alloys and it also has a higher diffusivity in the β-phase compared to the α-phase. Therefore, it is to be expected that the relative proportion of the α and β phases in the Beta-21S alloy will be altered by the presence of hydrogen, i.e., conversion of the starting α+β microstructure to all β microstructure. However, the kinetics of this process will be regulated by the transport behavior of each of these phases, not only with respect to hydrogen, but the redistribution of the other elements in the alloy.

The overall kinetics can be explained through the following model. At low hydrogen pressures, i.e., at hydrogen activities lower than that required for the transformation of the α-phase to β-phase, the starting microstructure will be retained. The kinetics will reflect multiple mechanisms: at short times, the phase with higher permeability (presumably the β-phase in this alloy) will dominate the uptake behavior, while at longer times, the phase with the lower permeability will control the kinetics until saturation is achieved. At higher hydrogen pressures, i.e., at hydrogen activities that are sufficient to stabilize the β-phase, the kinetics will be dominated by the conversion process. The rate of this process is limited by the flux balance at the α/β interface. The flux of hydrogen at the interface depends on the activity gradient of hydrogen in the β-phase in the vicinity of the interface. At a constant temperature, the higher the hydrogen activity in the beta phase, the faster the conversion process. In other words, at higher hydrogen pressures, this effect is manifest in the form of fast conversion of the α phase because the hydrogen flux into the interface from the gas phase through the β phase is significantly enhanced.

CONCLUDING REMARKS

Uptake measurements in Beta-21S alloy show that the hydrogen transport is very rapid over the temperature range of 400-800°C, and the solubility for hydrogen is high even at low to moderate pressures (up to 50 torr) of hydrogen exposure. Like the other titanium alloys, the hydrogen solubility increases with decreases in the temperature of exposure. The equilibrium hydrogen concentration in the alloy, expressed as a function of the hydrogen pressure, shows deviations from Sievert's law at higher pressures. These deviations are probably associated with microstructural changes in the alloy caused by hydrogen: the proportion of beta phase is increased by partially converting the alpha precipitates that were produced during the heat-treatment cycle. The microstructural changes, which occur during hydrogen exposure, are reflected in the kinetics of the hydrogen uptake.

REFERENCES

1. P.J. Bania and W.M. Parris, "Beta21S: A High Temperature Metastable Beta Titanium Alloy", Proceedings of the 1990 International Conference on Titanium Products and Applications, (Dayton, OH: Titanium Development Association, 1990).

2. A.D. McQuillan, "An Experimental and Thermodynamic Investigation of the Hydrogen-Titanium System", Proceedings of the Royal Society, 204 (1951), 309-323.

3. A.D. McQuillan, "The Titanium-Hydrogen System for Magnesium-Reduced Titanium", The Journal of the Institute of Metals, 79 (1951), 371-379.

4. G.A. Lenning, C.M. Craighead and R.I. Jaffee, "Constitution and Mechanical Properties of Titanium-Hydrogen Alloys", Transactions of the AIME, 200 (1954), 367-376.

5. J.E. Costa, D. Banerjee and J.C. Williams, "Hydrogen Effects in β-Titanium Alloys", Beta Titanium Alloys in the 80's, (Warrendale, PA: TMS, 1984), 69-84.

6. H.G. Nelson, "Hydrogen Induced Ductile-to-Brittle Fracture Transition in Beta Titanium Alloys", (Paper presented at the 5th Workshop on Hydrogen-Materials Interactions, Scottsdale, AZ, September 23-25, 1992).

7. J. Crank, The Mathematics of Diffusion, (London, United Kingdom: Oxford University Press, 1970), 42-45.

8. S.N. Sankaran, R.A. Outlaw and R.K. Clark, "Surface Effects on Hydrogen Permeation Through Ti-14Al-21Nb Alloy", (NASA TP-3109, 1991).

9. W.M. Mueller, "Titanium Hydrides", Metal Hydrides, ed. W.M. Mueller, J.P. Blackledge and G.G. Libowitz, (New York, NY: Academic Press, 1968), 337.

THE EFFECTS OF HYDROGEN ON THE ROOM TEMPERATURE MECHANICAL

PROPERTIES OF Ti-15V-3Cr-3Al-3Sn AND Ti-15Mo-3Nb-3Al

George A. Young Jr. and John R. Scully

Materials Science & Engineering
The University of Virginia
Charlottesville, VA 22903

Abstract

The effects of electrochemically introduced hydrogen on the room temperature mechanical properties of two β titanium alloys, TIMETAL 15-3 (Ti-15V-3Cr-3Al-3Sn, wt%) and TIMETAL 21S (Ti-15Mo-3Nb-3Al, wt%) are compared. Solution annealed, peak aged (538°C, 8h), and duplex aged (440°C, 20h, 538°C, ½h) conditions are investigated. Bridgman notched tensile bars are employed to quantify the degree of embrittlement both by reduction in the maximum longitudinal stress developed at the centerline of the notch and the average effective plastic strain across the notch diameter at maximum load. Fracture paths are correlated with the slip behavior observed in solution annealed material. Possible hydriding of the α and β phases is investigated through x-ray diffraction. Results show that TIMETAL 21S is more susceptible to hydrogen embrittlement than TIMETAL 15-3 as evidenced by reductions in the longitudinal stress, plastic strain, and changes in fracture mode at hydrogen concentrations above 1000 wt. ppm. Possible hydriding of a large volume fraction of the α or β phases was not observed over the range of hydrogen concentrations investigated. The increased susceptibility of TIMETAL 21S to hydrogen embrittlement is attributed to a high temperature, long time, solution treatment which removed heterogeneous nucleation sites from the grain interiors. Subsequent aging occurs preferentially on β grain boundaries and lastly in the grain interiors, resulting in fine intragranular precipitates. These fine α plates are readily sheared and promote planar slip. In contrast, a lower temperature, shorter duration solution treatment for TIMETAL 15-3, results in a material with more homogeneous, larger α precipitates, which, in turn, promote wavy slip. Results indicate that persistent planar slip exacerbates both hydrogen embrittlement and aqueous environmentally assisted cracking in metastable β titanium alloys.

Beta Titanium Alloys in the 1990's
Edited by D. Eylon, R.R. Boyer and D.A. Koss
The Minerals, Metals & Materials Society, 1993

Introduction

Modern ß titanium alloys are candidates for many hydrogen environment applications due to their desirable mechanical properties, protective oxide film, and high hydrogen solubility [1-4]. However, ß titanium alloys are susceptible to hydrogen embrittlement. Both ß and α titanium form a brittle hydride phase and recent reports suggest that the bcc ß phase is intrinsically embrittled by hydrogen [5-9]. The present study seeks to compare and contrast the effects of a range of electrochemically charged hydrogen concentrations on the room temperature mechanical properties of the metastable ß titanium alloys TIMETAL 15-3 and TIMETAL 21S in both the solution annealed and ß+α aged conditions. Duplex aging was performed in an attempt to promote a finer, more homogenous α distribution and avoid preferential precipitation on ß grain boundaries. Preferential α precipitation has previously been associated with increased susceptibility to stress corrosion cracking, hydrogen embrittlement, and intergranular cracking [9-12].

Experimental

Cross rolled plate, nominally 10 mm thick, of TIMETAL 15-3 (Ti-15V-3Cr-3Sn-3Al) and TIMETAL 21S (Ti-15Mo-3Nb-3Al) was received in the solution annealed condition. Solution annealed (SA), single step aged (referred to hereafter as peak aged (PA)), and duplex aged (DA) heat treatments were investigated. Table I details the heat treatments and corresponding hardness of each condition. Both PA and DA 21S exhibited preferential nucleation of α along ß grain boundaries while α nucleated intragranularly in 15-3 after one hour at 538°C as shown in Figure 1.

Table I. Heat treatments and corresponding hardnesses of the conditions investigated.

Condition: Material	Heat Treatment	HRC
Solution Annealed: TIMETAL 15-3	816°C, 0.5 h → Air Cool	25.2 ± 0.9
TIMETAL 21S	871°C, 8 h → Air Cool	28.7 ± 1.2
Peak Aged: TIMETAL 15-3	538°C, 8 h → Air Cool	38.9 ± 0.6
TIMETAL 21S	538°C, 8 h → Air Cool	42.1 ± 1.2
Duplex Aged: TIMETAL 15-3	440°C, 20 h → Air Cool → 538°C, 0.5 h → Air Cool	41.4 ± 0.7
TIMETAL 21S	440°C, 20 h → Air Cool → 538°C, 0.5 h → Air Cool	48.8 ± 0.8

Figure 1. Optical micrographs showing the aging response of (a) TIMETAL 15-3 and (b) TIMETAL 21S after aging for 1 hour at 538°C.

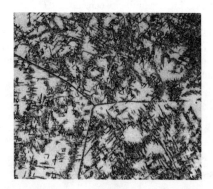 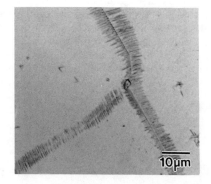

Electrochemical pre-charging of hydrogen was conducted in a solution of 10cc H_2SO_4, 1000cc H_2O and 0.8 g $Na_4P_2O_7$ at 90°C [13]. Previously machined and heat treated tensile specimens were cathodically polarized to 100 A/m^2 for various times to promote hydrogen uptake and were tested in air, as described elsewhere [9]. Upon removal from the charging bath, the oxide which forms in air is an effective barrier to hydrogen egress. Hydrogen concentrations reported for each tensile specimen were obtained from a section of the tensile bar adjacent to the notch and represent an average total concentration for the volume of metal tested. Note that the hydrogen uptake rate of 15-3 was approximately 5-6 times greater than that of 21S for all the heat treatments investigated. This increase in hydrogen uptake is attributed, in part, to a higher hydrogen fugacity on the surface of 15-3. At 25°C, galvanostatic measurements in the charging solution indicate that 15-3 develops a potential approximately 200 mV cathodic to 21S at an equivalent current density of 100 A/m^2.

Circumferentially notched "Bridgman" tensile bars were employed to quantify the effects of hydrogen on the mechanical properties of the alloys investigated [14]. Degree of embrittlement was quantified by determining the maximum longitudinal stress developed at the centerline of the notched region and the effective plastic strain across the notch diameter at maximum load following the procedures of Hancock et al. [15-16]. All tensile tests were conducted at a crosshead displacement rate of 1.5 x 10^{-2} mm/min. The effect of constraint on the failure stress and strain in PA material was determined as a function of hydrogen concentration at four different initial constraint levels (0.52, 0.62, 1.03, 1.43) where the triaxial constraint is defined as the ratio of mean to effective stress, $(\sigma_m/\bar{\sigma})$ [14-16]. These constraint levels correspond to notch radii of 7.9mm, 4.8mm, 1.6mm, and 7.9mm respectively at a constant initial diameter across the notch of 6.4mm. Additional tensile tests, conducted at the constraint level of 1.43 (0.33 = uniaxial tension, 2.50 = sharp notch), compared the effects of hydrogen on SA, PA, and DA heat treatments.

The slip behavior of each alloy was investigated by deforming electropolished cubes of SA material (approximately 1 cm^3) in compression and observing the surface slip lines via optical microscopy. X-ray diffraction experiments were performed with a Scintag automated diffractometer utilizing copper K-α radiation, which was continuously scanned over 30-80° 2θ at a rate of 1° per minute. Both the heat treated and heat treated + hydrogen charged conditions were investigated. Diffraction spectra of electrochemically charged plate were taken both at the charged surface and well into the specimen interior by sectioning and grinding.

Results

Mechanical Testing

Constraint and Hydrogen   The longitudinal stress increased and the plastic strain decreased as level of constraint increased for both uncharged, PA 15-3 and 21S as shown in Figure 2. The longitudinal stress and plastic strain developed in 15-3 decreased linearly with increasing hydrogen concentration. TIMETAL 21S, however, exhibited a sharp decrease in longitudinal stress and plastic strain at hydrogen concentrations greater than 1000 wt. ppm. Note that the largest reductions in plastic strain for 21S occur at the highest constraint.

Figure 2. The effects of constraint and hydrogen on the stress and strain developed in peak aged TIMETAL 15-3 and TIMETAL 21S. σ_{YIELD} is the uniaxial yield strength.

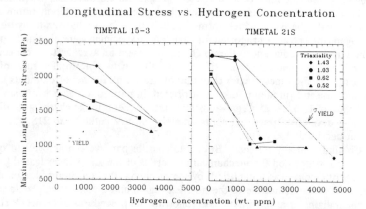

Longitudinal Stress vs. Hydrogen Concentration

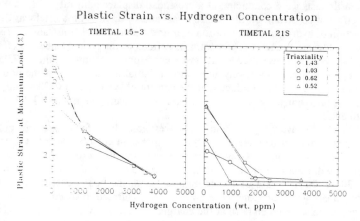

Plastic Strain vs. Hydrogen Concentration

Hydrogen Concentration (wt. ppm)

Heat Treatment   Concerning SA material at the highest constraint level investigated (1.43), the failure stress and strain of 15-3 are unaffected by hydrogen concentrations up to 3000 wt. ppm. In contrast, 21S is embrittled at less than 3000 wt. ppm as shown in Figure 3a. Although the failure stresses for each material in the SA condition are nearly equal, 15-3 displays approximately twice the plastic strain of 21S and nearly three times the plastic strain at high hydrogen concentrations. In the PA condition (Fig. 3b), both 15-3 and 21S exhibit a decrease in strength and ductility at approximately 1000 ppm H. This decrease is more pronounced for 21S. Duplex Aged 15-3 also exhibits greater resistance to hydrogen embrittlement than DA 21S (Fig. 3c). Comparison of material at equivalent hardness levels (as in the case of DA 15-3 versus PA 21S) shows that 15-3 still exhibits superior resistance to hydrogen embrittlement.

Deformation Mode   Compression tests on SA material revealed that both alloys are prone to planar slip at low plastic strains (3%) while at strains on the order of 8%, extensive cross slip occurred in 15-3 but not in 21S. Figure 4 compares the surface slip observed in SA 15-3 and 21S at approximately 8% plastic strain and illustrates the difference.

150

Figure 3. The effects of hydrogen on the stress and strain developed in TIMETAL 15-3 and TIMETAL 21S in the (a) solution annealed, (b) peak aged, and (c) duplex aged conditions.

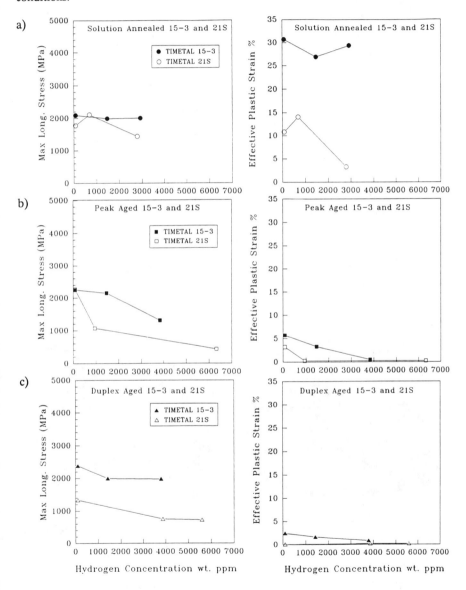

Figure 4. Surface slip lines observed in solution annealed (a) TIMETAL 15-3 and (b) TIMETAL 21S deformed to approximately 8% plastic strain in compression.

a) 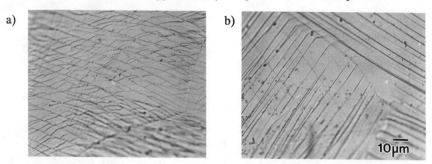 b)

Fractography

The fracture mode of SA 15-3 was relatively insensitive to hydrogen concentration. At all hydrogen levels investigated SA 15-3 failed by microvoid coalescence. In PA material, the fracture mode of 15-3 was also microvoid coalescence at all constraint and hydrogen levels investigated. The only noticeable effect of increasing hydrogen and constraint was to decrease the depth and width of the microvoids. The fracture mode of 15-3 consists of fine equiaxed microvoids up to its highest hardness and hydrogen concentration (DA to R_C 41, 3793 ppm H) where the fracture mode changed to transgranular ductile tearing as shown in Figure 5.

Figure 5. Fracture surfaces from TIMETAL 15-3 (a) uncharged, solution annealed and (b) duplex aged, 3700 wt. ppm hydrogen.

a) 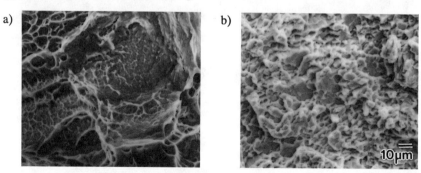 b)

In contrast to 15-3, the fracture mode of SA, PA, and DA 21S changed as the hydrogen concentration was increased (Fig. 6). The fracture mode of SA 21S progressed from large microvoids at approximately 100 ppm hydrogen (Fig. 6a), to small microvoids and ductile tearing features at 680 ppm H, and finally at 3000 ppm to flat fracture characterized by three distinct fracture modes; 1) fine voids and tearing features similar to the 680 ppm hydrogen level, 2) flat transgranular fracture, and 3) transgranular fracture displaying parallel markings (Fig. 6b). These parallel markings typically extend across an entire grain diameter and are consistent with hydrogen induced slip band fracture which has been observed in hydrogen charged bcc steels, pure nickel, and nickel & iron base alloys, [17-20]. The difference in fracture mode between individual grains is believed to be caused by differences in slip system orientation relative to the tensile axis.

The fracture mode in PA 21S is strongly influenced by both level of constraint and hydrogen concentration as detailed elsewhere [9]. At the highest constraint levels, peak aged 21S exhibits a duplex microvoid structure in the uncharged condition, which progresses to a mixture of intergranular and transgranular fracture as H concentration is increased (Fig. 6c). The flat featureless areas of 4664 ppm PA 21S (Fig. 6d) indicate that some slip plane decohesion has occurred. Duplex aged 21S fails by ductile tearing at 75 ppm H progressing to flat fracture characterized by elongated, but microscopically ductile voids at 3800 ppm (Fig. 6e), and finally by cleavage (Fig. 6f) at 5600 ppm.

Figure 6. Fracture modes produced in TIMETAL 21S at a constraint level of 1.43 and different hydrogen levels and heat treatments (a) SA, 100 ppm, (b) SA, 3000 ppm, (c) PA, 1000 ppm, (d) PA, 3500 ppm, (e) DA, 3800 ppm, and (f) DA, 5600 ppm.

a)   b)

c)   d)

e)   f)

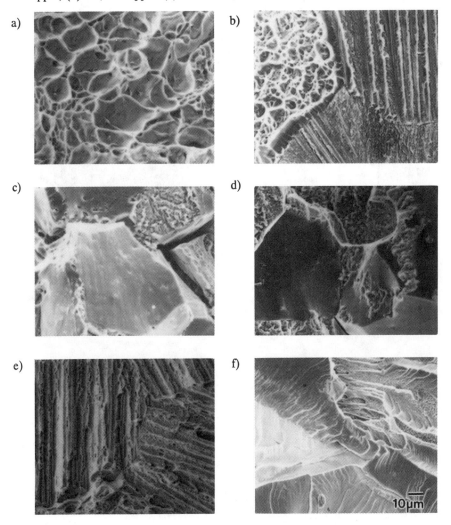

X-ray Diffraction

Diffraction spectra taken from the surface of as charged PA plate indicated that surface exposed α of 21S spalled off while that of 15-3 did not (Fig. 7). Spalling of the α phase has previously been reported by Nakasa in Ti-6Al-4V during electrochemical hydrogen charging [21]. TIMETAL 15-3, however, exhibited some evidence of hydriding as shown by diffraction spectra of Figure 8. Hydriding appears to be a phenomena associated with surface exposed α for both alloys since no hydride peaks were detected in the interior of either material when diffraction spectra were taken after serial grinding of the charged surface. Instead, partitioning of hydrogen to the β phase was suspected as indicated by large changes in the β lattice parameter. However, this does not preclude the possibility of deformation assisted hydriding or localized hydriding of the α/β interfaces as discussed by Boyd [22].

Figure 7. (a) Diffraction spectra from the surface of peak aged TIMETAL 21S charged 64 hours in H_2SO_4 solution and (b) SEM micrograph of the charged surface illustrating the spalled α phase.

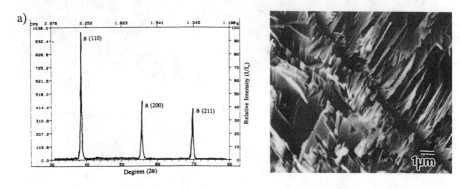

Figure 8. Diffraction spectra taken from the surface of PA TIMETAL 15-3, charged 24 hours showing the α, β, and δ phase peaks.

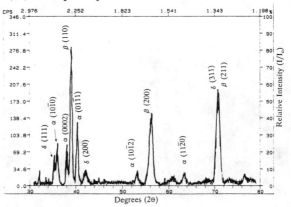

Embrittlement appears to be exacerbated by persistent planar slip in 21S. Slip behavior is dependent on both alloy composition and precipitate morphology. The fact that alloying additions affect stacking fault energy (SFE) and subsequent slip behavior is well known but the authors know of no reports of the effects of alloying additions on SFE in ß-Ti alloys. Furthermore, hydrogen affected fracture modes such as intergranular cracking appear to be primarily affected by heat treatment. Alloying elements may play a secondary role. Both L.M. Young and Meyn & coworkers have produced intergranular cracking in 15-3 which was heat treated above the ß transus for longer times than investigated here (e.g. 2 hours) [10,23].

Heat treatment, specifically solution treatment and its effects on the resulting microstructure and slip behavior may dominate hydrogen embrittlement resistance in metastable ß titanium alloys. Solution treatment has been identified as the controlling factor in hot salt stress corrosion cracking susceptibility of the metastable ß alloy Beta-III but the mechanism of embrittlement was not discussed [12]. Okada, Banerjee, and Williams studied the transfer of slip from ß to α as a function of precipitate morphology in Ti-15V-3Cr-3Al-3Sn. They demonstrated that slip initiates in the ß phase and that parallel α plates of the same variant (colony type structure) allow dislocations to shear the α phase while α plates of differing orientations promote homogeneous slip [24]. While TEM/SAD experiments have shown that both 15-3 and 21S exhibit Burger's α (i.e. $(110)_B$ ∥ $(0001)_\alpha$, $[111]_B$ ∥ $[11\bar{2}0]_\alpha$) [25], high magnification SEM of PA metallographic specimens (Figure 9) show a difference in the size and orientation of the α, especially at the grain boundaries [10].

Figure 9. High magnification SEM micrographs of α in peak aged (a) TIMETAL 15-3 and (b) TIMETAL 21S (from ref 10).

a) b)

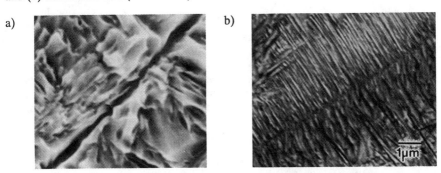

The parallel, linear features shown on the fracture surfaces of hydrogen charged 21S in the SA and DA conditions (Figures 6b and 6e) and previously observed in hydrogen charged PA material [9] suggest an interaction between hydrogen, slip, and fracture mode. Hydrogen segregation to dislocations and transport along slip lines is well documented in bcc metals [26-27]. Recall that hydrogen has been shown to partition to, and is relatively mobile in, the bcc lattice which provides a readily available source for hydrogen pickup by dislocations and possible deposition of this hydrogen at grain boundaries [9,28].

It is plausible that rather than crystallographic orientation, physical alignment at grain boundaries and coarseness of α in the 15-3 (which has nucleated first and grown for a longer time) is a more effective slip barrier than the fine α of the 21S. A greater impedance to planar slip promotes more homogeneous deformation (i.e. cross slip). In turn, less hydrogen is transported to ß grain boundaries and the locally high hydrogen concentrations along persistent planar slip bands is reduced. This explanation is well supported by the fracture modes observed in this study and is consistent with the work of Albrecht, Thompson, and Bernstein in aluminum alloys [29-32].

TIMETAL 21S which is prone to planar slip (at the investigated heat treatment) undergoes slip line, slip plane, and intergranular cracking depending on microstructure, degree of constraint and hydrogen concentration. In contrast, 15-3, in which wavy slip is easier to induce, exhibits a microvoid rupture fracture mode until the highest strength and hydrogen level investigated where the fractographic features are on the size and order of the deformation structure seen in SA material as shown by a comparison of Figures 4a and 5b. Correlation of the different fracture modes with the mechanical properties exhibited by each alloy indicate that deformation mode strongly influences the hydrogen effected fracture paths and subsequent mechanical properties.

Hydrogen has previously been suggested as the embrittling species in aqueous chloride testing of PA ß titanium alloys [10-11]. Hydrogen-slip interactions may explain the superior EAC resistance of PA 15-3 vs. PA 21S in aqueous sodium chloride solution [10-11]. L.M. Young and R.P. Gangloff have suggested that dislocation motion is a requisite for EAC in 21S based on J-integral resistance curve testing conducted at varying load-line displacement rates and "ripple" loaded tests of PA 15-3 and 21S in aqueous saltwater [10]. This statement fits well with the observations made in this paper. Consider the production of hydrogen at an acidified crack tip and absorption of this hydrogen into the fracture process zone. Under a "ripple" load designed to rupture the surface oxide film (but well below the fatigue ΔK threshold for crack growth in moist air) hydrogen enters the metal and diffuses through the lattice, partitioning primarily to the ß phase where it is highly soluble. Persistent planar slip band formation does not occur, however, below the moist air fatigue ΔK threshold.

Under a rising load test, however, where dislocation motion is occurring, hydrogen is transported by dislocations and deposited at the dislocation sinks (i.e. grain boundaries) where it is trapped and promotes intergranular separation. The difference in the EAC resistance between 15-3 and 21S is attributed to the tendency of 21S toward planar slip (which promotes hydrogen transport over long distances to grain boundaries). Cross slip, which has been shown to occur more readily in SA 15-3 than 21S, retards the transport and deposition of hydrogen to grain boundaries and concurrently lowers local hydrogen concentrations as dislocations transport hydrogen to newly activated slip systems.

Conclusions

1. Hydriding of a large volume fraction of the α and ß phases is not required for embrittlement to occur in ß titanium. Reduction in the longitudinal stress and plastic strain developed in solution annealed TIMETAL 21S at hydrogen concentrations < 3000 wt. ppm suggest that the ß phase is intrinsically embrittled by hydrogen especially when deformation occurs by persistent planar slip.

2. Embrittlement is a function of hydrogen concentration, constraint, and yield strength for material (21S) which has a susceptible microstructure. Susceptible microstructures are caused by excessive solution treatment temperatures which remove heterogeneous nucleation sites from grain interiors, delaying the onset of intragranular α precipitation. The resulting fine α precipitates are readily sheared by dislocations, promoting planar slip, concentration of hydrogen along persistent planar slip bands, and transport of hydrogen to grain boundaries.

3. The relationships between hydrogen transport, microstructure, and deformation behavior described in this paper accurately account for the observed effects on specimens which were precharged with hydrogen and tested in air as well as specimens which were simultaneously polarized and tested in aqueous chloride environments. This correlation supports a hydrogen embrittlement/dislocation transport mechanism for aqueous saltwater stress corrosion cracking of metastable β titanium alloys.

Acknowledgements

This research was supported by the Office of Naval Research (Grant N00014-91-J-4164) with Dr. A. John Sedriks as Scientific Monitor and by the Virginia Center for Electrochemical Science and Engineering at the University of Virginia. The material used in this study was graciously donated by TIMET. The authors wish to acknowledge the invaluable assistance of R.P. Gangloff, R.J. Kilmer, S.S. Kim, D.G. Kolman, B.P. Somerday, and L.M. Young.

References

1. H.G. Nelson, First Thermal Structures Conference, E. Thornton, ed., Nov 13-15, The University of Virginia, Charlottesville, VA, (1990) 301-311.
2. P.J. Bania, G.A. Lenning and J.A. Hall, Beta Titanium Alloys in the 80's, R.R. Boyer and H.W. Rosenberg eds., TMS-AIME, Warrendale, PA, (1983) 209-237.
3. P.J. Bania and W.M. Parris, "Beta-21S: A High Temperature Metastable Beta Titanium Alloy", paper presented at the Titanium Development Association Conference, Orlando, FL (1990).
4. R.W. Schutz and D.E. Thomas, "Corrosion of Titanium and Titanium Alloys," Metals Handbook, 9th ed. 13, ASM, Metals Park, Ohio, (1987) 669-706.
5. J.E. Costa, J.C. Williams, and A.W. Thompson, "The Effects of Hydrogen in Mechanical Properties in Ti-10V-2Fe-3Al" Met. Trans. A, (18A) (1987), 1421-1430.
6. J.J. DeLuccia, Report No. NADC-76297-30, Naval Air Development Center, Warminster, PA (1976).
7. K. Nakasa and J. Liu, "Bending Strength of Hydrogen Charged Ti-13V-11Cr-3Al alloy", J. Japan Inst. Metals, 55 (9) (1991) 922-927.
8. D.S. Shih and H.K. Birnbaum, "Evidence of FCC Titanium hydride Formation in β Titanium Alloy: An X-ray Diffraction Study," Scripta Met., (20) (1986), 1261-1264.
9. G.A. Young Jr. and J.R. Scully, "Effects of Hydrogen on the Mechanical Properties of a Ti-Mo-Nb-Al Alloy", Scripta Met., (28) (1993) 507-512.
10. L.M. Young, "Hydrogen Environment Embrittlement of Beta Titanium Alloys," to be published in Proceedings of the Seventh World Conference on Titanium, F.H. Froes and I.L. Caplan, eds., TMS-AIME, Warrendale, PA (1993).
11. L.M. Young, "Environment Assisted Cracking in β-Titanium Alloys," Master's Thesis, The University of Virginia, Charlottesville, VA (1993).
12. J.B. Guernsey, V.C. Petersen, and F.H. Froes, "Discussion of Effect of Microstructure on the Strength, Toughness, and Stress-Corrosion Cracking Susceptibility of a Metastable β Titanium Alloy" Met. Trans. A, (3) (1972) 339-340.
13. Z.A. Foroulis, "Factors Affecting Absorption of Hydrogen in Titanium from Aqueous Electrolytic Solutions," Ti'80, Science and Technology, H. Kimura and O. Izumi eds., TMS-AIME, Warrendale, PA, (1980) 2705-2711.
14. P.W. Bridgman, Studies in Large Plastic Flow, McGraw-Hill Inc., New York, NY (1952) 9-37.

15. J.W. Hancock and A.C. Mackenzie, "On the Mechanisms of Ductile Failure in High-Strength Steels Subjected to Multi-Axial Stress-States," J. Mech. Phys. Solids, (24) (1976) 147-169.
16. A.C. Mackenzie, J.W. Hancock and D.K. Brown, "On the Influence of State of Stress on Ductile Failure Initiation in High Strength Steels," Engineering Fracture Mechanics, (9) (1977) 167-188.
17. J. Eastman et al., "Hydrogen Effects in Nickel--Embrittlement or Enhanced Ductility," Hydrogen Effects in Metals, I.M. Bernstein and A.W. Thompson eds., The Metallurgical Society of AIME (1980) 397-409.
18. R.E. Stoltz and A.J. West, "Hydrogen Assisted Fracture in FCC Metals and Alloy," Hydrogen Effects in Metals, I.M. Bernstein and A.W. Thompson eds., The Metallurgical Society of AIME (1980) 541-553.
19. N.R. Moody and F.A. Greulich, "Hydrogen Induced Slip Band Fracture in an Fe-Ni-Co Superalloy," Scripta Met., (19) (1985) 1107-1111.
20. N.R. Moody, R.E. Stolts, and M.W. Perra, "The Effect of Hydrogen on Fracture Toughness of the Fe-Ni-Co Superalloy IN903," Met Trans A, (18A) (1987) 1469-1482.
21. K. Nakasa and J. Liu, "Surface Peeling of Ti-6Al-4V Alloy Specimens during Hydrogen Charging" J. Japan Inst. Metals, 54 (11) (1990) 1261-1269.
22. J.D. Boyd, "Precipitation of Hydrides in Titanium Alloys" Trans ASM, (62) (1969) 977-988.
23. D.A. Meyn and P.S. Pao, "Slow Strain Rate Testing of Precracked Titanium Alloys in Salt Water and Inert Environments", to be published in Slow Strain Rate Testing: Research and Engineering Applications.
24. M. Okada, D. Banerjee, and J.C. Williams, "Tensile Properties of Ti-15-V-3Al-3Cr-3Sn Alloy," Titanium Science and Technology, G. Lütjering, U. Zwicker, and W. Buck, eds., (3) (1980) 1835-1842.
25. K.R. Lawless, Unpublished Research. University of Virginia, Charlottesville, VA (1993).
26. J.P. Hirth, "Effects of Hydrogen on the Properties of Iron and Steel", Met Trans A, (11A) (1980) 861-890.
27. C.J. MacMahon Jr., "Effects of Hydrogen on Plastic Flow and Fracture in Iron and Steel," Hydrogen Effects in Metals, I.M. Bernstein and A.W. Thompson, eds., The Metallurgical Society of AIME (1980) 219-233.
28. R.L. Shulte and P.N. Adler, "Stress-Induced Hydrogen Redistribution in High Purity Ti-31V Alloy," Hydrogen Effects in Metals, I.M. Bernstein and A.W. Thompson eds., The Metallurgical Society of AIME (1980) 177-185.
29. J.K. Tien et al., "Hydrogen Transport by Dislocations," Met. Trans. A, (7A) (1976) 821-829.
30. J. Albrecht, I.M. Bernstein, and A.W. Thompson, "Evidence for Dislocation Transport of Hydrogen in Aluminum," Met. Trans. A, (13A) (1982) 811-820.
31. D.A. Hardwick, A.W. Thompson, and I.M. Bernstein, "The Effect of Copper Content and Microstructure on the Hydrogen Embrittlement of Al-6Zn-2Mg Alloys," Met. Trans. A, (14A) (1983) 2517-2526.
32. D. Nguyen, A.W. Thompson, and I.M. Bernstein, "Microstructural Effects on Hydrogen Embrittlement in a High Purity 7075 Aluminum Alloy," Acta Met., (35) 10 (1987) 2417-2425.

EFFECTS OF INTERNAL HYDROGEN ON MICROSTRUCTURES

AND MECHANICAL PROPERTIES OF ß21S and Ti-15-3

R. J. Lederich,[†] D. S. Schwartz,[†] and S. M. L. Sastry[††]

[†]McDonnell Douglas Corporation
P.O. Box 516
St. Louis, MO 63166

[††]Washington University
Box 1185, One Brookings Drive
St. Louis, MO 63130

ABSTRACT

The effects of elevated temperature exposures of Ti-15Mo-2.7Nb-3Al-0.2Si (ß21S) and Ti-15V-3Al-3Sn-3Cr (Ti-15-3) to gaseous hydrogen on microstructures and mechanical properties were studied. In both of the alloys hydrogen is absorbed into the ß phase and the amount of ß phase present is greatly increased. With increasing hydrogen concentration, the ß phase becomes more highly strained, and the ductile-to-brittle transition temperatures of both alloys increase. Prior aging has little effect on ductility although the microstructures are affected. Increased oxygen levels further increase the ductile-to-brittle transition temperature.

Beta Titanium Alloys in the 1990's
Edited by D. Eylon, R.R. Boyer and D.A. Koss
The Minerals, Metals & Materials Society, 1993

INTRODUCTION

The elevated temperature properties of titanium alloys render them attractive candidates for monolithic and composite matrix applications in advanced hypersonic aircraft.[1] Since some of these high temperature applications involve exposure to gaseous hydrogen of moderate concentrations of ~ 130 Pa (1 Torr), a knowledge of hydrogen effects in these alloys is essential for usage in a variety of high temperature applications.[2]

This paper reports the effects of elevated temperature hydrogen exposures on the microstructure and mechanical properties of Ti-15Mo-2.7Nb-3Al-0.2Si (ß21S) and Ti-15V-3Al-3Sn-3Cr (Ti-15-3).

EXPERIMENTAL

The ß21S and Ti-15-3 alloys were produced by TIMET Corp. in the form of production lot sheets and were ß annealed after finish rolling and surface grinding. Actual compositions were very close to nominal compositions and the oxygen concentration was 0.13 wppm in both alloys. These alloys were hydrogen charged by heating at 525°C and 600°C for 16 h at atmospheric pressure in flowing helium + 500 Pa (3.8 Torr) hydrogen and helium + 3 kPa (23 Torr) hydrogen. After charging, the samples were removed from the furnace and cooled to 25°C in the hydrogen environment. Charging for times greater than 16 h did not increase the hydrogen concentration, indicating that equilibrium was achieved. To determine non-hydrogen related thermal exposure effects on microstructure and mechanical properties, samples were sealed in diffusion-pump-evacuated quartz capsules and annealed for the same times and same temperatures.

Testing at temperatures other than room temperature was performed using a small insulated chamber which could be either heated by hot air or cooled by vapor from liquid nitrogen. Stress rupture testing was conducted by mounting the specimens in a creep fixture and initially subjecting them to loads corresponding to 80% of their yield stresses. If no failure occurred within 50 h, the stress level was increased by 12% and the sequence continued until failure did occur. The stress rupture specimens were notched to produce a stress concentration factor of 3.3.

RESULTS AND DISCUSSION

Effect on Microstructures

Hydrogen is a potent ß stabilizer and thus strongly modifies the microstructures of these ß titanium alloys as shown in Figure 1 for ß21S. The as-received ß annealed microstructure shown in Figure 1a consists of ß grains approximately 25 μm in diameter. This is the starting microstructure for hydrogen charging. The microstructure in Figure 1b was obtained by annealing in vacuum at 600°C for 16 h and air cooling. It consists of relatively coarse α platelets near the prior ß grain boundaries and finer platelets in the interiors of the grains. Charging this sample in 500 Pa (3.8 Torr) hydrogen at 600°C results in 3600 wppm absorbed hydrogen and the microstructure shown in Figure 1c. The coarse α platelets are present along the prior ß grain boundaries, but the grain interiors are largely devoid of the fine α platelets. Figure 1d shows the microstructure after charging to 7200 wppm hydrogen at 600°C in 3 kPa (23 Torr) hydrogen. Essentially no α phase is present, indicating that the absorbed hydrogen has lowered the ß transus below 600°C.

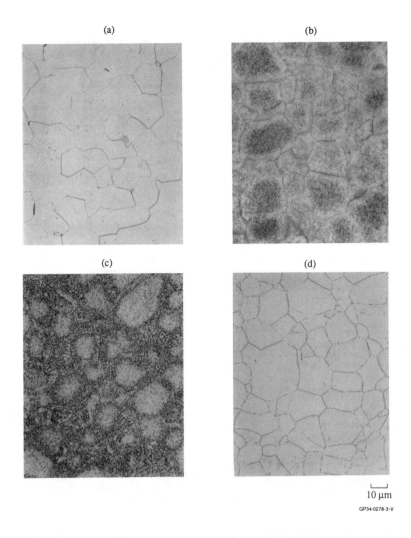

(a)   (b)

(c)   (d)

10 μm

GP34-0278-3-V

Figure 1 - Microstructures of ß21S (a) in as-received (ß annealed) condition (25 wppm H),
(b) after annealing in an inert environment at 600°C/16 h/AC, (c) after heating in
helium + 500 Pa (3.8 Torr) hydrogen at 600°C/16 h/AC (3600 wppm H), and
(d) after heating in helium + 3 kPa (23 Torr) hydrogen at 600°C/16 h/AC
(7200 wppm H).

About 1 vol% of the heavily charged ß21S sample (7200 wppm) has the unique
microstructures shown in Figures 2a and 2b. The irregular shapes of these regions suggest
that extensive partitioning of solute elements occurred as hydrogen was absorbed. However
energy dispersive spectroscopy (EDS) indicated that the only evidence of solute migration is a
slightly higher level of aluminum (3.5% vs. 3.0%) in these regions. The small volume
fraction of these irregularly shaped regions suggests that their effect on mechanical properties
is minimal.

161

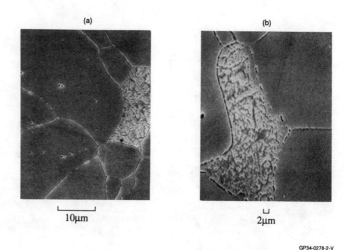

(a)

(b)

10µm

2µm

GP34-0278-2-V

Figure 2 - Scanning electron micrographs of ß21S charged with 7200 wppm hydrogen.

Figure 3 shows the effect of internal hydrogen on the microstructures of Ti-15-3. The ß annealed microstructure (Figure 3a) contains large (25 µm diam) ß grains, and annealing at 600°C produces a somewhat coarse acicular α structure (Figure 3b). Charging ß annealed specimens at 600°C in 500 Pa (3.8 Torr) hydrogen results in a hydrogen concentration of 2500 wppm and a reduced amount of α with the platelets mostly located near the grain boundaries (Figure 3c). Charging in 3 kPa (23 Torr) hydrogen increases the hydrogen concentration to 6800 wppm and results in an all ß phase microstructure (Figure 3d) similar to that of ß21S (Figure 1d). Charging at 525°C in 3 kPa (23 Torr) hydrogen increases the hydrogen level to 14,700 wppm and produces precipitates at prior ß grain boundaries (Figure 3e). EDS shows these precipitates to be a tin-rich compound.

X-ray diffractometry shows that nearly all of the hydrogen is absorbed into the ß phase. Table I indicates that while the α cell size remains constant, the ß phase cell size increases by 6.7% when charged in 0.5 kPa hydrogen, and 8.7% when charged in 3 kPa hydrogen. No evidence of hydride precipitation was seen using transmission electron microscopy. However the ß phase showed strong strain contrast, indicating local straining.

Effect on Mechanical Properties

Figure 4 shows the effects of composition of charging gas and charging temperature on room temperature ductility. Both alloys maintain greater than 10% tensile elongation when exposed to 0.5 kPa (3.8 Torr) hydrogen at 600°C. Whereas Ti-15-3 has 12% tensile elongation at 6800 wppm hydrogen, ß21S becomes brittle at 7200 wppm hydrogen (whether charged at 500 Pa (3.8 Torr) hydrogen at 525°C or 3 kPa (23 Torr) at 600°C). Although both highly hydrogen charged alloys have 100% ß phase (Figures 1d and 3d) ß21S is brittle, whereas Ti-15-3 retains full ductility. However Ti-15-3 is brittle when hydrogen charged to 14,700 wppm at 525°C in 3 kPa (23 Torr) to produce the microstructure shown in Figure 3e.

The effects of internal hydrogen on stress rupture properties shown in Table II indicates that these alloys are not particularly sensitive to notched stress rupture testing when charged in 500 Pa (3.8 Torr) hydrogen at 600°C. Internal hydrogen, which increases the amount of ß phase, reduces the yield strengths by ~ 5%. The notched rupture strengths are lower in the charged specimens than in the uncharged specimens. However the notched rupture strengths of the hydrogen charged alloys are comparable to their yield strengths.

162

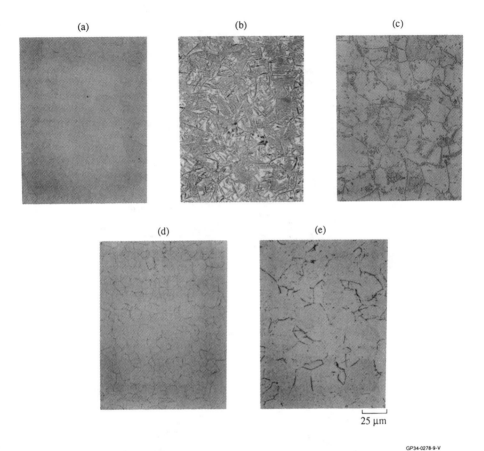

(a)　　　　　(b)　　　　　(c)

(d)　　　　　(e)

25 μm

GP34-0278-9-V

Figure 3 - Microstructures of Ti-15-3 (a) in as-received (ß annealed) condition (25 wppm H), (b) after annealing in an inert environment at 600°C/16 h/AC, (c) after heating in helium + 500 Pa (3.8 Torr) hydrogen at 600°C/16 h/AC (2500 wppm H), (d) after heating in helium + 3 kPa (23 Torr) hydrogen at 600°C/16 h/AC (6800 wppm H), and (e) after heating in helium + 3 kPa (23 Torr) hydrogen at 525°C/16 h/AC (14700 wppm H).

TABLE I. - EFFECT OF HYDROGEN CHARGING ON LATTICE PARAMETERS OF
ß21S AS DETERMINED BY X-RAY DIFFRACTION

Condition	Phase	a(Å)	c(Å)	Cell vol. (Å³)	% increase in cell vol.
Uncharged (25 wppm H)	ß	3.254		34.46	
	α	2.924	4.674	34.61	
Charged with He + 500 Pa H₂	ß	3.325		36.76	6.7
(3600 wppm H)	α	2.927	4.676	34.68	0.2
Charged with He + 3 kPa H₂	ß	3.346		37.46	8.7
(7200 wppm H)	α	2.920	4.674	34.51	0.0

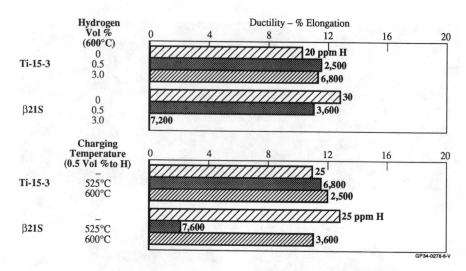

Figure 4 - Effect of charging gas hydrogen content and charging temperature on room temperature ductility.

TABLE II. - EFFECTS OF INTERNAL HYDROGEN ON STRESS RUPTURE

Samples Charged in He + 0.5 vol% H_2 at 600°C/16 h/AC
Stress Rupture Samples Were Notched

Alloy	Yield Stress, MPa (ksi)		Stress Rupture Strength, MPa (ksi)	
	Uncharged	Charged	Uncharged	Charged
ß21S	989 (143.5)	957 (138.8)	1094 (158.6)	946 (137.2)
Ti-15-3	887 (128.6)	841 (121.9)	968 (140.4)	890 (129.1)

To further elucidate the effects of internal hydrogen on mechanical properties, tensile testing was conducted over a range of temperatures to characterize the ductile-to-brittle transition temperature (DBTT) and the results are shown in Figure 5. Both alloys exhibit very sharp transitions, typical for body-centered cubic materials. The transition for uncharged Ti-15-3 occurs at approximately 50°C below that for uncharged ß21S. Increasing amounts of internal hydrogen raise the DBTT to higher temperatures. In general, a charged Ti-15-3 specimen remains ductile at 50°C below a similarly charged ß21S sample.

The fracture morphology of both alloys changes abruptly across the DBTT, as shown in Figure 6 for ß21S samples containing 3600 wppm hydrogen. The brittle sample (Figure 6a) shows a transgranular cleavage fracture at -50°C. However, testing the sample at -25°C results in a ductile failure mode (Figure 6b). Ti-15-3 undergoes similar changes in fracture morphology as the DBTT is crossed (Figures 7a and 7b). Figure 7c shows the fracture morphology of highly charged Ti-15-3 tested at 25°C. The transgranular fracture mode indicates that the Sn rich precipitates at grain boundaries (Figure 3e) do not induce grain boundary fracture at this temperature.

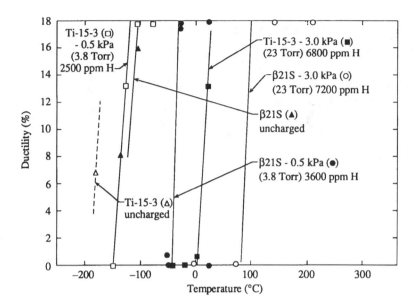

Figure 5 - Ductile-to-brittle transitions in Ti-15-3 and ß21S following hydrogen exposures.

(a)   (b)

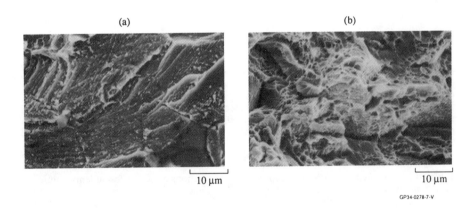

GP34-0278-7-V

Figure 6 - Fracture surfaces of ß21S charged at 600°C/16 h/AC in helium + 500 Pa (3.8 Torr) hydrogen (3600 wppm H) and tested at (a) -50°C and (b) -25°C.

(a) (b) (c)

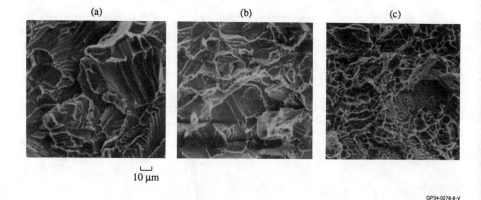

L___J
10 µm

GP34-0278-8-V

Figure 7 - Fracture surfaces of Ti-15-3 (a) uncharged and tested at -190°C, (b) containing 6800 wppm hydrogen and tested at -25°C, and (c) containing 14,700 wppm hydrogen and tested at 25°C.

Effect of Aging Treatment

To determine the effect of aging treatment on hydrogen uptake and mechanical properties, ß21S specimens were given the following aging treatments:

a. 620°C/8 h/AC - This aging treatment results in a good combination of strength and ductility.

b. 760°C/8 h/AC - This treatment results in lower strength and greater ductility than the 620°C age, and produces ß phase with large grain boundary α precipitates.

Figure 8 shows the microstructures after charging in 500 Pa (3.8 Torr) hydrogen at 600°C. Hydrogen charging unaged samples (Figure 8a) and stabilized samples (Figure 8b) results in similar microstructures. However, samples aged at higher temperatures (Figure 8c) retain their grain boundary α after charging. Figure 9 shows the hydrogen induced microstructural changes in ß21S which was initially aged at 760°C prior to charging. The grain boundary α precipitates are largely unaffected by charging in 500 Pa (3.8 Torr) hydrogen (Figures 9a and 9b). However when charged to greater amounts of hydrogen, the grain boundary α apparently undergoes refinement and begins transforming to another phase (Figure 9c).

Table III lists the hydrogen concentrations and % elongations at two different temperatures of ß21S samples which were unaged, and aged at 620°C and 760°C. Aging slightly reduced the amount of hydrogen absorbed, possibly because of the greater amount of α phase initially present. However, its effect on mechanical properties appears to be minor.

Effect of Oxygen Concentration

Both alloys are processed to nominally contain 0.13 wt% oxygen. Oxygen is a very strong α stabilizer with each 0.1 wt% increase in oxygen concentration raising the ß transus temperature by 22°C.[3] Hence, increasing amounts of oxygen could significantly affect hydrogen uptake and interactions. M. Parris has demonstrated that oxygen levels in ß21S can rise to at least 0.23 wt% without adversely affecting room temperature strengths and ductilities.[4] With the objective of determining synergistic effects of oxygen and hydrogen, ß21S sheets containing 0.13, 0.23, and 0.33 wt% oxygen provided by TIMET were evaluated.

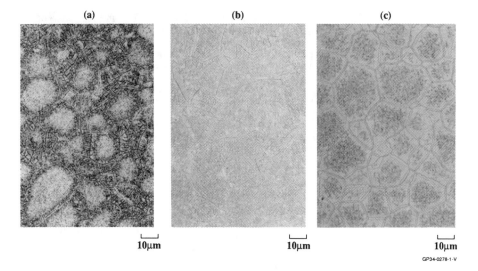

Figure 8 - Microstructures of ß21S after charging in 500 Pa (3.8 Torr) hydrogen after (a) beta annealing (2355 wppm H), (b) beta annealing plus aging at 620°C/8 h/AC (2000 wppm H), and (c) beta annealing plus aging at 760°C/8 h/AC (1975 wppm H).

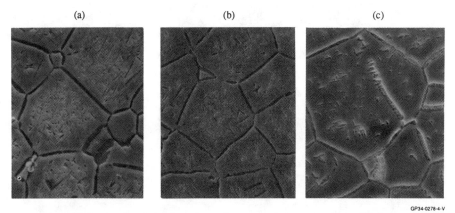

Figure 9 - Scanning electron micrographs of ß21S which was aged at 760°C/8 h/AC and then heated for 16 h at 600°C in (a) helium (25 wppm H), (b) helium + 500 Pa (3.8 Torr) hydrogen (1975 wppm H), and (c) helium + 3 kPa (23 Torr) hydrogen (6500 wppm H).

TABLE III. - EFFECT OF AGING TREATMENT ON
TENSILE DUCTILITIES OF HYDROGEN CHARGED ß21S

Aging Temperature	% Elongation		H Concentration (ppm)
	-50°C	-25°C	
None	5.5	10.6	2355
621°C	10.8	13.8	2000
760°C	9.5	9.7	1975

These samples were aged to provide similar strength levels (Table IV) and then charged to two different levels of hydrogen. Figure 10 shows that the microstructures of the three alloys after aging and hydrogen charging to the higher hydrogen concentration are similar, although the density of α precipitates near the prior ß grain boundaries increases with increasing oxygen concentration. The effects of two different levels of hydrogen concentration on yield strengths and tensile elongations for these three alloys are shown in Table V. At the lower hydrogen level (~ 300 wppm), the effects of hydrogen are believed to be negligible, and nearly all of the ductility losses are attributed to increased oxygen content. The properties of the samples at the higher hydrogen level (~ 2300 wppm) reflect additive effects of both internal hydrogen and oxygen. An increase in the oxygen level from 0.13 to 0.23 results in an increased susceptibility to hydrogen, while an increase to 0.33 results in a nearly complete loss in ductility for both conditions tested.

TABLE IV. - AGING TREATMENT AND RESULTING MECHANICAL PROPERTIES
OF ß21S ALLOYS CONTAINING DIFFERENT CONCENTRATIONS OF OXYGEN

Oxygen concentration (wppm)	ß Transus °C	Aging treatment	Yield stress MPa (ksi)	% Elongation
.13	810	620°C/8 h/AC	912 (131.3)	15.6
.23	834	665°C/8 h/AC	1012 (146.8)	12.4
.33	854	682°C/8 h/AC	1022 (148.2)	12.7

(a)   (b)   (c)

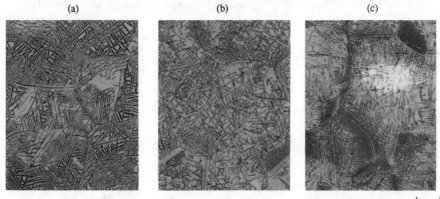

10 µm

Figure 10 - Microstructures of ß21S samples having oxygen contents of (a) 0.13 wt%,
(b) 0.23 wt%, and (c) 0.33 wt%, after aging and hydrogen charging at 600°C/16
h/AC in helium + 3 kPa (23 Torr) hydrogen (2000 wppm H).

TABLE V. - EFFECT OF OXYGEN CONCENTRATIONS ON
ROOM TEMPERATURE TENSILE DUCTILITY OF ß21S
CHARGED TO DIFFERENT CONCENTRATIONS OF HYDROGEN

Oxygen concentration (wt%)	Hydrogen concentration (wppm)	% Elongation at 25°C	Hydrogen concentration (wppm)	% Elongation at -75°C
.13	2665	17.9	355	15.6
.23	2020	10.8	280	5.3
.33	2240	3.9	310	1.5

CONCLUSIONS

Both ß21S and Ti-15-3 retain significant ductility at room temperature at a hydrogen level of 3600 wppm. Ti-15-3 retains room temperature ductility at a hydrogen level of 6800 wppm. In this work hydrides have not been observed; however, as hydrogen is absorbed into the ß phase it becomes locally strained. Both alloys possess sharp ductile-to-brittle transitions, which shift towards higher temperatures as more hydrogen is absorbed and as the ß phases become more highly strained. The DBTT of uncharged Ti-15-3 is 50°C lower than that of ß21S, and this difference is maintained as the hydrogen concentration is increased. Prior aging, while affecting the microstructure of charged samples, has little effect on ductility. Increased oxygen levels increase the ductile-to-brittle transition temperature which is further increased by absorbed hydrogen.

ACKNOWLEDGEMENTS

The authors would like to thank H. G. Nelson for technical discussions. Appreciation is also extended to D. J. Deuser, J. J. Evans, II, and J. D. Keyes for technical assistance, and to M. Parris of TIMET Corp. for providing ß21S material with different oxygen concentrations. This research results from NASA Contract NAS2-13182 as well as the McDonnell Douglas Aerospace Independent Research and Development program.

REFERENCES

1. M. A. Steinberg, "Material for Aerospace," Scientific American, 10 (1986), 66.

2. R. W. Schutz, "An Overview of Beta Titanium Alloy Environmental Behavior," this conference.

3. M. Hansen, Constitution of Binary Alloys (New York, NY: McGraw Hill, 1958), 1069.

4. M. Parris and P. Bania, "Oxygen Effects on the Mechanical Properties of TIMETALß21S," (Paper presented at the Seventh World Conference on Titanium, San Diego, CA, 28 June-2 July, 1992).

MICROSTRUCTURE-PROPERTY RELATIONSHIPS IN BETA TITANIUM ALLOYS

MICROSTRUCTURAL INSTABILITIES IN BETA TITANIUM ALLOYS

--Keynote Lecture--

Alain Vassel

Direction des Matériaux, ONERA, BP 72, 92322 Châtillon Cedex, France.

Abstract

Complex phase transformations occur in metastable beta alloys owing to the large amount of different alloying elements that they contain. Apart from the equilibrium alpha and beta phases, transient phases and brittle compounds may precipitate depending upon alloy composition, processing history and service conditions. This paper describes recent work on microstructural instabilities in several industrial alloys (Ti-10V-2Fe-3Al, Beta-C, Beta-21S, BETA-CEZ,...) and their effect on mechanical properties.

Beta Titanium Alloys in the 1990's
Edited by D. Eylon, R.R. Boyer and D.A. Koss
The Minerals, Metals & Materials Society, 1993

Introduction

Beta titanium alloys offer a unique combination of high strength, low density, deep hardenability, hot and cold processability and corrosion resistance. They contain large additions of beta-isomorphous and beta-eutectoid stabilizers in order to achieve the desired properties. These additions may give rise, depending upon alloy chemistry, to the precipitation of transient phases and intermetallic compounds during processing, heat treatments or service conditions. As the presence of these phases generally have an adverse effect on mechanical properties (ductility, fatigue resistance, fracture toughness), it is important to investigate their domain of stability.

This paper deals with recent work carried out on microstructural instabilities in different metastable beta alloys. The conditions of precipitation of isothermal omega phase, intermetallic compounds, stress induced martensite and non-Burgers alpha phase are successively reviewed and their influence on mechanical properties is discussed.

Isothermal omega phase

Transient precipitation of omega phase

The omega phase forms in lean metastable beta alloys at low ageing temperatures, below 400°C. It has been known for a long time that this phase gives very high strength but with low ductility [1]. In many metallurgical systems, the precipitation of a metastable phase can control the subsequent development of a more stable phase. Thus, in beta titanium alloys, omega particles act as alpha phase nucleation sites in two step ageing treaments [2-4]. The alpha phase grown from omega particles has a strong strengthening effect owing to its extremely fine size and high volume fraction. This precipitation mode generally promotes low ductility values [5].

Most beta titanium alloys are designed to be used in a solution treated plus aged condition. In thick products, the cooling rate after solution treatment and the heating rate to the ageing temperature are very different at the core and near the surface. These parameters will influence the transformation mechanism of the high temperature beta phase and a transient precipitation of isothermal omega phase may appear on slow heating to the ageing temperature.

The influence of the aforementioned parameters on microstructure and tensile properties has been studied on the BETA-CEZ alloy (Ti-5Al-2Sn-4Zr-4Mo-2Cr-1Fe) [6], recently developed by CEZUS [7]. Microstructural investigations were performed on metallographic samples 10mm in diameter. They were either water quenched or air cooled after a solution treatment of 4 hours at 830°C in the alpha+beta field, 60°C below the beta transus temperarure. The corresponding cooling rates are $50°C.s^{-1}$ and $5°C.s^{-1}$ approximately. It is to be noted that the second condition is representative of a water quenching on a 70mm thick titanium alloy pancake. The ageing treatments were performed in a salt bath or in a progammed air furnace. The heating rates that have been selected ($\approx 200°C.min^{-1}$ and $1°C.min^{-1}$) correspond to extreme values encountered in industrial conditions. In both cases, an ageing treatment of 8 hours at 580°C is applied to the samples once the temperature has been reached.

A cooling rate of $5°C.s^{-1}$ after solution treatment induces a decomposition of the high temperature beta phase ; dark field transmission electron micrographs reveal the presence of needle like secondary alpha phase in the beta matrix. On the other hand, the beta phase is retained upon rapid quenching ($50°C.s^{-1}$) to room temperature. Selected area diffraction patterns on $<110>_\beta$ zone exhibit streakings and diffuse athermal omega reflections. These microstructures

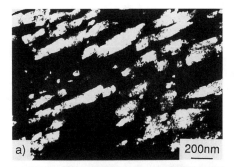

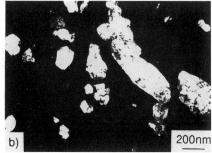

Figure 1 TEM micrographs of transformed beta phase in solution treated (830°C 4h, cooling rate 5°C.s⁻¹) plus aged material. Dark field images showing secondary alpha phase :
a) 1°C.min⁻¹ up to 580°C,
b) 1°C.min⁻¹ up to 580°C + 8h 580°C,
c) 580°C 8h (salt bath).

will behave differently during ageing.

In the case of the lower cooling rate, the secondary alpha phase slightly grows when it is slowly heated to the ageing temperature (Fig. 1a), then it coarsens after 8 hours at 580°C (Fig. 1b). A similar microstructure as the one shown in Fig. 1b is observed when the ageing treatment is carried out in salt bath (Fig. 1c). No other phase apart from alpha and beta has been detected.

As far as the rapid quenching is concerned, weak reflections that may be attributed to iso-thermal omega phase appear on $<110>_\beta$ zone patterns in a sample slowly heated at 1°C.min⁻¹ up to 300°C. The precipitation of that phase is confirmed at 350°C and 375°C (Fig. 2). The size and the volume fraction of omega particles seem to be maximum at 375°C. The omega phase morphology is controlled by the misfit between the precipitate and the beta matrix : it is ellipti-cal in Ti-Nb and Ti-Mo alloys whereas it is cubic in Ti-Cr, Ti-Fe and Ti-V systems [8]. Here, it is difficult to determine the exact morphology due to the small size of the particles. The secon-dary alpha phase begins to precipitate at 400°C ; at that temperature, isothermal omega and se-condary alpha phases coexist and dark field images confirm that omega particles act as nuclea-tion sites for the alpha phase. When a sample is slowly heated up to 450°C, the omega phase disappears and fine needles of secondary alpha phase only are observed.

It is worth comparing the morphology and the distribution of secondary alpha phase after a slow heating up to 580°C when the alloy has been subjected to different cooling rates after solution treatment (Fig. 1a and 3a). It appears that the alpha phase is much finer and more uni-formly distributed when it precipitates from isothermal omega particles instead of the beta ma-trix. The same tendency, although less pronounced, can be observed after 8 hours at 580°C (Fig. 1b and 3b). In the case of ageing in salt bath, i.e. a fast heating rate, the alpha phase di-rectly precipitates from beta and its morphology does not depend on the cooling rate after solu-tion treatment (Fig. 1c and 3c).

Figure 2 TEM micrograph of transformed beta phase in solution treated plus aged material (830°C 4h, cooling rate 50°C.s^{-1}+ 1°C.min^{-1} up to 375°C). Dark field image showing isothermal omega phase.

a)

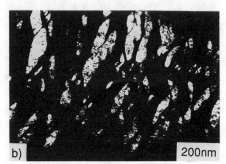

b)

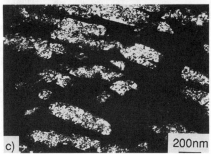

c)

Figure 3 TEM micrographs of transformed beta phase in solution treated (830°C 4h, cooling rate 50°C.s^{-1}) plus aged material. Dark field images showing secondary alpha phase :
a) 1°C.min^{-1} up to 580°C,
b) 1°C.min^{-1} up to 580°C + 8h 580°C,
c) 580°C 8h (salt bath).

Table I Influence of heat treating parameters on tensile behaviour of BETA-CEZ alloy at room temperature [6].

Solution treatment	Ageing treatment	Y.S. (MPa)	U.T.S. (MPa)	El. (%)	R.A. (%)
830°C 4h ↘ 50°C.s^{-1}	1°C.min^{-1} ↗ 580°C	1407	1545	0.7	3
	1°C.min^{-1} ↗ 580°C + 8h 580°C	1302	1389	4.3	6
	580°C 8h (salt bath)	1302	1363	4.8	9
830°C 4h ↘ 5°C.s^{-1}	1°C.min^{-1} ↗ 580°C	1230	1368	4	9
	1°C.min^{-1} ↗ 580°C + 8h 580°C	1185	1261	6	16
	580°C 8h (salt bath)	1253	1316	6.3	14

Following microstructural investigations, tensile tests at room temperature were carried out in some selected conditions (Table I). A high strength and a very low ductility have been measured when the alloy is rapidly quenched after solution treatment and slowly heated up to 580°C. This result is attributed to the extremely fine size and strong strengthening effect of the secondary alpha phase grown from isothermal omega particles (Fig. 3a). The subsequent ageing treatment induces a coarsening of alpha platelets and leads to a decrease in strength and an improvement in ductility. The optimum properties are obtained with a cooling rate of 5°C per second after solution treatment. In that condition, the heating rate to the ageing temperature has hardly no influence on tensile properties and this result is consistent with microstructural observations.

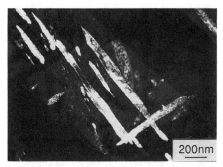

Another interesting feature is the influence of the solution treatment temperature on the transient precipitation of the isothermal omega phase. When decreasing the solution treatment temperature, the beta phase becomes more stable owing to an increased amount of beta stabilizers and there is no formation of isothermal omega on slow heating to the ageing temperature. Accordingly, the equilibrium alpha phase directly precipitates from the beta matrix and its morphology is much coarser (Fig.4) than when it forms from omega particles (Fig. 3a).

Figure 4 TEM micrograph of transformed beta phase in solution treated plus aged material (770°C 4h, cooling rate 50°C.s^{-1}+ 1°C.min^{-1} up to 580°C). Dark field image showing secondary alpha phase.

In conclusion, the transient precipitation of isothermal omega phase in the BETA-CEZ alloy only occurs after a rapid quenching followed by a very low heating rate to the ageing temperature, conditions not encountered in industrial heat treatments.

From a more general point of view, the domain of stability of isothermal omega phase is now well established and heat treating parameters of metastable beta alloys can be adjusted to avoid its precipitation.

Composition of omega phase

The composition of the isothermal omega phase is difficult to determine with accuracy because of its very small size, generally less than 30nm. A quantitative analysis was recently obtained using a field emission microscope coupled with an atom probe [9,10]. Schematically, the atom probe investigation of a material can be described as the analysis of a specimen along a cylinder. The lateral resolution is chosen between 0.5nm and 5nm whereas the depth resolution is

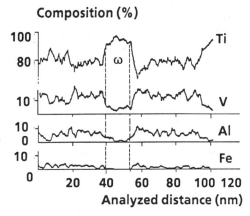

Figure 5 Atom probe composition profile of a Ti-10V-2Fe-3Al specimen solution treated at 780°C and aged for one hour at 300°C [9].

strictly equal to one atomic plane.

The study was carried out on Ti-10V-2Fe-3Al alloy solution treated for 2 hours at 780°C (20°C below the beta transus temperature), water quenched, and aged at 300°C to form isothermal omega phase. Figure 5 shows a composition profile after one hour at 300°C ; the crossing of an omega particle by the analysis cylinder is clearly visible. It can be observed that the omega phase corresponds to titanium enriched zones depleted in all the alloying elements, alpha (Al) and beta (V, Fe) stabilizers. Also, it could be mentioned that there is no chemical segregation at the beta/omega interface. The atom probe data for parent beta phase and omega phase for two ageing times (1 and 8 hours) are shown in Figure 6. This graph illustrates the rejection of each of the alloying elements from the omega phase. After 8 hours at 300°C, the omega phase contains 93.1±0.5% Ti, 4.6±0.4% V, 0.7±0.2% Fe and 1.6±0.2% Al (wt.%). It should be noted that the depletion in the beta stabilizing elements is consistent with the nucleation of the equilibrium alpha phase from omega precipitates. In contrast, a decrease in the aluminium concentration is unexpected.

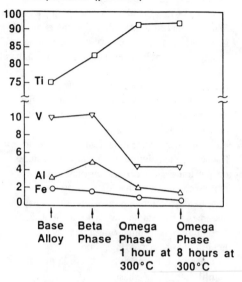

Composition (percent)

Figure 6 Composition (wt.%) of beta and omega phases as deduced from atom probe data [9].

Intermetallic compounds

An important property for all metallic materials is the stability over time of their microstructure and mechanical characteristics. Most metastable beta alloys contain beta-eutectoid stabilizers which may induce the precipitation of intermetallic compounds during processing or service conditions, with a consequent decrease in ductility, fatigue resistance and fracture toughness. Main beta-eutectoid reactions in order of increasing eutectoid temperature are listed in Table II. The eutectoid reaction is either sluggish (Mn, Fe, Cr) or fast (Cu, Si), its kinetics increasing with the eutectoid temperature.

The alloy Ti-13V-11Cr-3Al is one of the first beta titanium alloys that was developed. It exhibits very high strength levels but its applications are limited because of the embrittlement encountered above 400°C. Owing to the high chromium content, The $TiCr_2$ phase forms above that temperature after prolonged exposures [11]. This compound precipitates at beta grain boundaries and thus leads to an intergranular failure mode and a severe drop in ductility. The chromium content in the Ti-3Al-8V-6Cr-4Zr-4Mo (Beta-C) alloy is lower than in the Ti-13V-11Cr-3Al alloy. However, the grain boundary precipitation of $TiCr_2$ has been also observed in that alloy in the temperature range 500-550°C only after 500h ageing. The absence of $TiCr_2$ in samples aged above 550°C indicates a very narrow range of stability for this phase [12].

The BETA-CEZ alloy contains two beta-eutectoid stabilizers, chromium and iron, which

Table II Characteristics of beta-eutectoid reactions.

Element	Alpha composition (wt.%)	Compound	Eutectoid composition (wt.%)	Eutectoid temperature (°C)
Mn	0.5	TiMn	20	550
Fe	0.06	TiFe	15	590
Cr	0.5	$TiCr_2$	14	667
Co	1.0	Ti_2Co	9.9	680
Ni	0.1	Ti_2Ni	5	770
Cu	1.5	Ti_2Cu	7	799
Si	0.45	Ti_5Si_3	0.65	861

Table III Effect of prolonged exposures at operating temperatures on room temperature tensile properties of BETA-CEZ alloy [6].

			Y.S. (MPA)	U.T.S. (MPa)	El. (%)	R.A. (%)
Initial condition			1195	1270	15.0	23
σ=0 (1)	425°C	5000h	1233	1277	13.5	25
	475°C	500h	1226	1274	13.5	25
σ=0 (2)	425°C	5000h	1241	1294	10.2	14
	475°C	500h	1241	1275	10.3	13
σ=250MPa (2)	425°C	1000h	1185	1264	12.3	19

(1) surface removed   (2) surface retained

Table IV Crystal structure of silicides in metastable beta and alpha titanium alloys.

Alloy	Composition	Lattice parameter a (nm)	c (nm)	Space group	Ref.
β-C	$(Ti,Zr)_xSi_y$	0.696	0.365	P6/mmm	[14]
β-CEZ+0.2Si	$(Ti,Zr)_xSi_y$	0.700	0.363	P6/mmm	[15]
IMI 685	$(Ti,Zr)_xSi_y$	0.705	0.372	-	[16]
IMI 685	$Ti_{3.4}Zr_{2.6}Si_3$	0.698	0.365	P$\bar{6}$2m	[17]
IMI 829	$(Ti,Zr)_xSi_y$	0.696	0.365	-	[18]
IMI 829	$(Ti,Zr)_2Si$	0.714	0.374	P$\bar{6}$2m	[19]

may induce the precipitation of TiCr$_2$ and TiFe intermetallic compounds with a consequent decrease in ductility. Also, a possible ordering reaction of the alpha phase has been looked for, bearing in mind that the percentage of alpha-stabilizing elements in BETA-CEZ is low enough so that an extensive precipitation of Ti$_3$Al is unlikely to occur.

Microstructural studies and tensile tests were performed on a cylindrical bar supplied in a fully heat treated condition (830°C 1h, cooling rate 2°C.s^{-1} + 580°C 8h) [6]. The following exposures which correspond to service conditions were considered :

- 425°C up to 5000 hours,
- 475°C up to 500 hours.

Tensile tests at room temperature were carried out after these exposures and compared to the initial condition. In order to separate the influence of structural instability and surface oxidation, the specimens were tested with the surface removed or retained, respectively. Also, some creep tests (250MPa) were performed under similar conditions of temperature and time to detect a possible stress assisted precipitation.

All tensile tests results are listed in Table III. It appears that with no applied stress there is no change in properties when specimens are machined after exposure. Examination on thin foils was unsuccessful to show the existence of heterogeneous precipitation of a titanium compound like TiCr$_2$ or TiFe. Taking into account the very low solubility of chromium and iron in the alpha phase, 0.3 and 0.05 wt% at 500°C respectively [13], and the concentration of these elements in the alloy, the preceding observation suggests that the stabilization of the beta phase by molybdenum, a beta-isomorphous element, allows chromium and iron to remain in solid solution in beta. The only minor microstructural change occurs after 5000 hours at 425°C ; diffuse superlattice spots can be seen on diffraction patterns of the alpha phase indicating the very beginning of Ti$_3$Al formation.

When specimens are tensile tested with the surface retained, a slight decrease in ductility is observed after prolonged exposures and this is attributed to a superficial oxygen contamination. Finally, the application of a stress does not lead to a measurable damage. No stress assisted precipitation was detected by transmission electron microscopy.

The preceding results show that a limited amount of beta-eutectoid stabilizers (Fe, Cr) can be used in beta alloys without inducing any structural instability.

Metastable beta titanium alloys may also contain a small quantity of silicon. For instance, the silicon content is ≤0.1wt.% in the Ti-3Al-8V-6Cr-4Zr-4Mo (Beta-C) alloy and 0.2wt.% in the Ti-15Mo-3Nb-3Al-0.2Si (Beta-21S) alloy. The solubility limit of silicon in beta titanium alloys is very low and an heterogeneous silicide precipitation occurs either on slow cooling from the beta phase field or after long ageing times. It was shown that a silicide phase appears in a solution treated and aged Beta-C alloy at alpha/beta interfaces in the temperature range 550-650°C, and at beta grain boundaries in the range 650-700°C [12].

The crystal structure of silicides has been studied in detail in the Beta-C alloy [14] and in a modified composition of the BETA-CEZ alloy with 0.2wt.% Si [15]. It is worth noting that the silicide phase was found to be the same in both alloys. It possesses an hexagonal structure with a space group that was determined as P6/mmm by convergent beam electron diffraction patterns. The lattice parameters are identical within the experimental scatter (Table IV). Energy dispersive X-ray spectra revealed that the composition is (Ti,Zr)$_x$Si$_y$.

A comparison of the crystal structure of silicides in metastable beta and alpha alloys is of particular interest. Silicon is added to alpha alloys in order to improve the creep resistance and this element may precipitate in the form of a (Ti,Zr)$_x$Si$_y$ phase after a prolonged annealing treatment. Table IV gives a comparison of the crystal structure of silicides in two metastable beta alloys (Beta-C, BETA-CEZ+0.2Si) and two alpha alloys (IMI 685, IMI 829). Although metastable beta and alpha alloys have very dissimilar chemical compositions, Table IV reveals that the silicide phase is nearly the same in all alloys. Only three elements are present in this phase : titanium, zirconium and silicon. The space group was identified as P6/mmm in metastable beta alloys and P6̄2m in alpha alloys, but it should be said that P6/mmm and P6̄2m are very similar.

Schematic equilibrium phase fields of an alpha alloy (IMI 834) and three metastable beta alloys (BETA-CEZ+0.2Si, Beta-C, Beta-21S) were drawn from literature data (Fig. 7). This comparison shows that the temperature of silicide solvus lies below the beta transus in the alpha alloy whereas the silicide solvus is well above the beta transus in metastable beta alloys. The beta+silicide field is wide : 210°C in BETA-CEZ+0.2Si, 260°C in Beta-C and 140°C in Beta-21S. These equilibrium phase fields clearly illustrate the very limited solubility of silicon in metastable beta alloys. It was suggested that the limited solubility of silicon was due to the presence of zirconium, the silicide being strongly enriched in this element [18]. The preceding explanation cannot be put forward in the case of the Beta-21S alloy since it is zirconium free. An investigation of the crystal structure of silicide in Beta-21S would be of interest.

The presence of silicides may affect mechanical properties and the magnitude of the effect will depend on their morphology and distribution. Various thermal and thermomechanical processing procedures were used to change the silicide particles distribution in the Beta-C alloy [14]. A beta matrix with a uniform distribution of silicides exhibits a ductile transgranular failure and a very good tensile ductility. On the opposite, a continuous grain boundary silicide precipitation results in intergranular fracture and a low elongation to rupture.

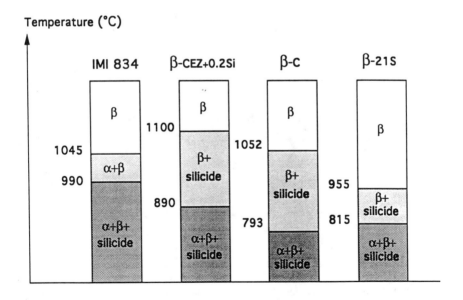

Figure 7 Comparison of equilibrium phase fields of an alpha alloy (IMI 834) and three metastable beta alloys (BETA-CEZ+0.2Si, Beta-C, Beta-21S).

Stress-induced martensite

In a number of lean beta alloys (Transage alloys, Ti-10V-2Fe-3Al,...), the Ms temperature is depressed below room temperature whereas the Md temperature lies above room temperature. In such alloys, a stress-induced martensite appears when deforming a metastable beta phase at room temperature [20-26]. This stress-induced martensite has an orthorhombic structure and is called α''. Orthorhombic representations of the beta and alpha phases can be given and α'' can be easily visualized as an intermediate structure with lattice parameters lying between those of alpha and beta [21,27].

The interesting feature in alloys which exhibit the stress-induced martensitic transformation is that the original shape can be restored in some of these alloys by heating the material rapidly to its As temperature. In almost all the alloys exhibiting the phenomenon, both the parent and martensitic phases are ordered. However, a significant shape memory effect has been observed in lean beta alloys although they do not have an ordered structure. An example of shape memory behaviour in a Ti-12Mo-3Al alloy is given in Figure 8. Rectangular specimens quenched from the beta phase field were bent at room temperature and then heated at 100, 150 and 200°C. A full shape recovery occurs at 200°C for the top specimens whereas the shape recovery is only partial for the bottom specimens which were subjected to a greater deformation. The magnitude of the shape recovery was found to depend on the heating rate and the amount of the initial strain : a perfect recovery is observed up to about 3% strain [21,24].

It is to be noted that the shape memory effect in lean beta alloys exhibits two limitations in view of commercial applications. Firstly, the As temperature was reported to be approximatrely 200°C [21,24]. Unfortunately, a quenched beta phase is unstable above that temperature and tend to decompose to isothermal omega quickly. Secondly, the stress generated on the shape recovery was estimated to be small [24]. So, there will be restrictions to applications which require large forces accompanying the shape memory.

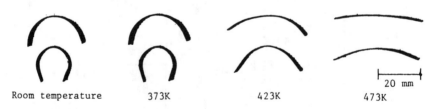

Room temperature   373K   423K   473K

20 mm

Figure 8 Shape memory effect in a Ti-12Mo-3Al alloy [24].

Non-Burgers alpha phase

The existence of two types of alpha phases has been reported in the literature on ageing beta titanium alloys : the well-known Type 1α which is Burgers related to the parent beta and Type 2α which does not obey the Burgers orientation relation [28-30]. Type 2α precipitates from Type 1α at long ageing times and is twin related to Type 1α. The presence of Type 2α is revealed by arced reflections on electron diffraction patterns. Non-Burgers alpha phase has been observed in Ti-10V-2Fe-3Al and Ti-3Al-8V-6Cr-4Mo-4Zr (Beta-C) alloys. Although a direct comparison of Types 1 and 2α could not be made, there does not seem to be any correlation between precipitate type and mechanical properties [29].

Similar arced reflections were observed on $<111>_\beta$ zone electron diffraction patterns of Ti-10V-2Fe-3Al alloy solution treated and aged at temperatures above 375°C [31,32]. Before concluding to the existence of Type 2α phase, all the prismatic and first-order pyramidal poles of the twelve variants of Type 1α which precipitate from a single beta grain were plotted on a (110) stereographic projection of the beta phase (Fig. 9). Then, theoretical $<111>_\beta$ zone diffraction patterns were drawn from this stereographic projection including all possible Burgers

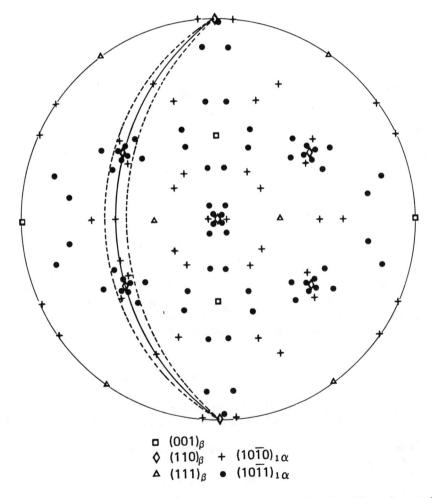

□	$(001)_\beta$	
◊	$(110)_\beta$	+ $(10\overline{1}0)_{1\alpha}$
△	$(111)_\beta$	• $(10\overline{1}1)_{1\alpha}$

<u>Figure 9</u> (110) stereographic projection of the beta phase with all the prismatic and first-order pyramidal poles of the twelve variants of Type 1α phase. The full line depicts <111> beta zone and dotted lines a tilt of ± 5 degrees from this zone.

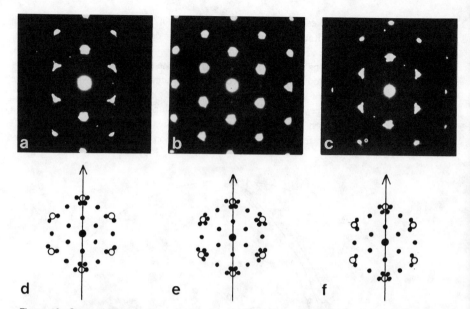

Figure 10 Comparison between experimental and theoretical <111> beta zone electron diffraction patterns for small angular variations. All possible Type 1α reflections are plotted on theoretical patterns [31].

(a) and (d) : +5 degrees from <111> beta zone,
(b) and (e) : <111> beta zone,
(c) and (f) : -5 degrees from <111> beta zone.

oriented alpha reflections. The stereographic projection predicts some particular strong alpha reflections when a ± 5 degrees tilt from the <111>$_\beta$ zone is considered (Fig. 9). Theoretical and corresponding experimental <111>$_\beta$ zone diffraction patterns are shown in Figure 10. The very good correlation between theoretical and experimental diffraction patterns demonstrates that additional reflections around {110}$_\beta$ spots correspond to the Burgers alpha phase [31]. As arced reflections can be perfectly interpreted in terms of the contribution of the twelve variants of Type 1α, we should be very careful before concluding to the existence of a non-Burgers alpha phase.

References

[1] J. C. Williams, B.S. Hickman, H.L. Marcus, Met. Trans., 2 (1971), 1913-1919.

[2] T.W. Duerig, J.C. Williams, Beta Titanium Alloys in the 1980's (AIME, Warrendale, PA, 1984), 19-67.

[3] G.M. Pennock, H. M. Flower, D.R.F. West, Titanium'80 Science and Technology, vol.2 (AIME, Warrendale, PA, 1980), 1343-1351.

[4] M.H. Campagnac, A. Vassel, Designing with Titanium (The Institute of Metals, London, 1986), 261-266.

[5] T.W. Duerig, G.T. Terlinde, J.C. Williams, Met. Trans., 11A (1980), 1987-1998.

[6] A. Henri, A. Vassel, paper presented at the Seventh World Conference on Titanium, San Diego, CA, 28 June - 2 July, 1992.

[7] B. Prandi et al., Proceedings of the 1990 International Conference on Titanium Products and Applications (TDA, Dayton, OH, 1990), 150-159.

[8] J.C. Williams, M.J. Blackburn, Trans. AIME, 245 (1969), 2352.

[9] A. Menand, L. Hadjadj, D. Blavette, Sixth World Conference on Titanium, vol. 3 (Les Editions de Physique, Paris, France, 1989), 1577-1582.

[10] L. Hadjadj, M.H. Campagnac, A. Vassel, A. Menand, Microscopy, Microanalysis, Microstructures, (Les Editions de Physique, Paris, France), to be published.

[11] A. DeLeon, D. Northwood, Microstructural Science, vol.11 (Elsevier Science Pub., 1983), 29-38.

[12] T.J. Headley, H.J. Rack, Met. Trans., 10A (1979), 909-920.

[13] J.L. Murray, Phase Diagrams of Binary Titanium Alloys (ASM, Metals Park, OH., 1987).

[14] S. Ankem et al., Met. Trans., 18A (1987), 2015-2025.

[15] A. Henri, Doctorat Thesis, University of Paris VII, February 1993.

[16] A. Vassel, La Recherche Aérospatiale, 6 (1982), 61-69.

[17] F. Barbier, C. Servant, C. Quesne, P. Lacombe, J. Microsc. Spectrosc. Electron., 6 (1981), 299-310.

[18] D. Banerjee, J.E. Allison, F.H. Froes, J.C. Williams, Titanium Science and Technology, vol.3 (DGM, Oberursel, Germany, 1985), 1519-1526.

[19] A.P. Woodfield, M.H. Loretto, Scripta Metallurgica, 21 (1987), 229-232.

[20] T.W. Duerig, R.M. Middleton, G.T. Terlinde, J.C. Williams, Titanium'80 Science and Technology, vol.2 (AIME, Warrendale, PA, 1980), 1503-1512.

[21] T.W. Duerig, J. Albrecht, D. Richter, P. Fischer, Acta Metallurgica, 30 (1982), 2161-2172.

[22] A.I.P. Nwobu, H.M. Flower, D.R.F. West, Journal de Physique, Colloque C4, Tome 43 (1982), 315-320.

[23] E.S.K. Menon, J.K. Chakravartty, S.L. Wadekar, S. Banerjee, ibid, 321-326.

[24] H. Sasano, T. Suzuki, Titanium Science and Technology, vol.3 (DGM, Oberursel, Germany, 1985), 1667-1674.

[25] A.I.P. Nwobu, H.M. Flower, D.R.F. West, Sixth World Conference on Titanium, vol. 3 (Les Editions de Physique, Paris, France, 1989), 1583-1588.

[26] S. Ishiyama, S. Hanada, O. Izumi, ISIJ International, 31 (1991), 807-813.

[27] K. Mukherjee, M. Kato, Journal de Physique, Colloque C4, Tome 43 (1982), 297-302.

[28] C.G. Rhodes, J.C. Williams, Met. Trans., 6A (1975), 2103-2114.

[29] C.G. Rhodes, N.E. Paton, Met. Trans., 8A (1977), 1749-1761.

[30] J.R. Toran, R.R. Biederman, Titanium'80 Science and Technology, vol.2 (AIME, Warrendale, PA, 1980), 1491-1501.

[31] M.H. Campagnac, A. Vassel, Designing with Titanium (The Institute of Metals, London, 1986), 261-266.

[32] M.H. Campagnac, Doctorat Thesis, University of Paris VI, January 1988.

STRENGTHENING CAPABILITY OF

BETA TITANIUM ALLOYS

Y. Kawabe and S. Muneki

National Research Institute for Metals,
Tsukuba Laboratories
1-2-1, Sengen, Tsukuba-shi,Ibaraki
305,JAPAN

ABSTRACT

The strengthening behavior of beta titanium alloys at room temperature has been investigated under a wide variety of solution treated and aged conditions, and relationships between microstructures and mechanical properties will be discussed.
Beta alloys can be strengthened mainly by the precipitation hardening mechanism, but increases in the strength of smooth, notched and precracked specimens are limited by ductility, notched tensile strength and fracture toughness, respectively. As a result, there is a critical value for strengthening for each specimen.
Tensile ductility is strongly dependent on the beta grain size while fracture toughness is not.
Therefore, the strength-ductility balance can be improved by grain refinement through applying the thermomechanical treatment while the strength-toughness balance cannot.
Accordingly, there is a possibility of raising the critical strength when the application is limited by ductility, but it is very difficult to raise it when it is limited by toughness. The critical strength when limited by notched tensile strength is, at present, about 1700MPa in tensile strength, which is considered to be the maximum strength available for general applications.

Beta Titanium Alloys in the 1990's
Edited by D. Eylon, R.R. Boyer and D.A. Koss
The Minerals, Metals & Materials Society, 1993

1. Introduction

The strengthening and toughening of materials are the most important research subjects for structural materials. However, the strength and toughness of high strength metallic materials are characteristics which are opposed to each other and toughness generally reduces with an increase in strength level. Therefore, strengthening of alloys has to be limited by ductility and toughness at some strength levels.

Beta titanium alloys[1] can be hardened over a considerable range by the precipitation of the α phase in the beta matrix[2], but increases in the strength, as mentioned above, may be limited by ductility and toughness and there is, hence, a critical value for strengthening.

The objectives of this study are to make clear the critical strength at room temperature when limited by ductility and toughness respectively and a possibility for its further strengthening will be discussed. In addition, some examples of the relationship between microstructures and mechanical properties and the comparison of the critical strength with other alloys such as ultrahigh strength steels will be presented.

2. Experimental procedure

Chemical composition of alloys used is listed in Table 1, in which four commercial alloys are included. 5kg ingots were melted by a plasma electron beam furnace and were hot worked into 11mm square bars.

Solution treatment was performed at every 50K interval temperature from 973 to 1173K for 3.6ks to vary the beta grain size. Aging treatment was performed at temperatures between 573 and 923K for 28.8ks to control the aged structure. Mechanical tests were carried out in air for the specimen in the solution treated and aged condition. Tensile test specimen of 3.5mm diameter and 14mm gauge length, notched tensile test specimen with circumferential notch of Kt=3.5, and fracture toughness (K_{Ic}) specimen with 10mm thickness, 10mm width and 55mm length for three points bending were used.

Table 1 – Chemical composition of alloys used (mass%)

Ti–10V–2Fe–3Al *
Ti–10V–2Fe–5Zr–3Al
Ti–11.5Mo–6Zr–4.5Sn *
Ti–8Mo–2Fe–3Al
Ti–15V–3Cr–3Sn–3Al *
Ti–10V–5Mo–2Fe–3Al
Ti–3Al–8V–6Cr–4Mo–4Zr *
Ti–15Mo–2Fe–3Al
Ti–8V–8Mo–2Fe–3Al

* commercial alloy

3. Results and discussion

3.1 Effect of the microstructure on mechanical properties

Fig.1 shows the effect of solution treatment temperature on mechanical properties of a Ti–3Al–8V–6Cr–4Mo–4Zr alloy aged at 773K for 28.8ks. This is an example showing the effect of solution treatment temperature on mechanical properties of various alloys. As clearly shown in this figure, tensile strength is almost independent of the solution treatment temperature but ductility values such as elongation and reduction in area decrease greatly with increasing the solution treatment temperature. Notched tensile strength and fracture toughness are slightly varied with the solution treatment temperature but those variations are not very much compared with the ductility values.

Fig.2 shows the effect of aging temperature on mechanical properties of a Ti–15V–3Cr–3Sn–3Al alloy solution treated at 1073K for 3.6ks. In this alloy, there are two peaks at 623

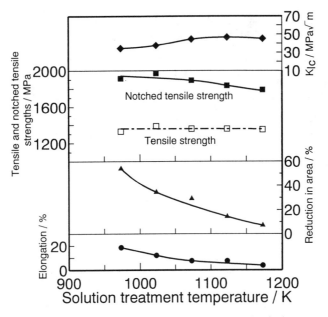

Figure 1 – Effect of solution treatment temperature on mechanical
properties of a Ti–3Al–8V–6Cr–4Mo–4Zr alloy aged at 773K
for 28.8ks.

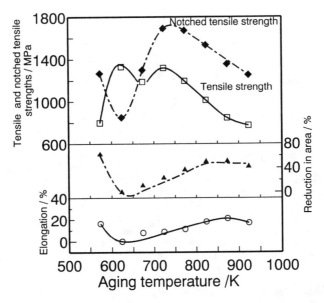

Figure 2 – Effect of aging temperature on mechanical properties of a Ti–
15V–3Cr–3Sn–3Al alloy solution treated at 1073K for 3.6ks.

and 723K. The two peaks indicate the same value of hardness and tensile strength, but the specimen aged at 623K show poor ductility and notched tensile strength compared with the specimen aged at 723K. As clearly shown in Fig.2, the ductility and toughness values are strongly dependent upon the aged structure.

Figs.3 and 4 summarized the effect of beta grain size on tensile properties, notched tensile strength and fracture toughness of three commercial beta alloys aged at 773K. For three alloys, tensile strength is independent of the beta grain size. Both elongation and reduction in area are strongly dependent upon the grain size and reduce with increasing the beta grain size. Fracture toughness is also independent of the beta grain size but notched tensile strength shows the slight decrease with increasing the grain size. The grain size dependence of notched tensile strength show the intermediate behavior between ductility and fracture toughness.

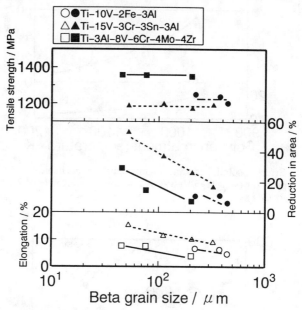

Figure 3 – Effect of beta grain size on tensile properties of three commercial beta alloys aged at 773K for 28.8ks.

3.2 Strength–ductility balance of beta alloys[3])

Fig.5 shows a collection of ductility (reduction in area) values as a function of tensile strength for various beta alloys. Although there is a tendency for the ductility to be reduced with an increase in the strength level, the ductility values can be divided into two groups. The group with higher ductility consists of the results for samples strengthened by aging above 723K together with the fine grained structure. On the other hand, the group with lower ductility consists of the results for samples which have coarse beta grained structure and which are strengthened by aging below 723K. The results clearly show that the strength–ductility balance is strongly dependent on the beta grain size and the aged structure.

It is also worth noting that the ductility values show a remarkably steep decrease above 1500MPa in tensile strength as indicated by the broken line in Fig.5. It has generally been considered that the relationship between strength and ductility shows a tendency to decrease gradually with increasing strength, as indicated by that of an upper limit of Fig.5. Accordingly, the fact that ductility shows a drastic decrease above a certain strength level, is connected to the ocurrence of the grain boundary failure due to the insufficient grain refinement as mentioned below.

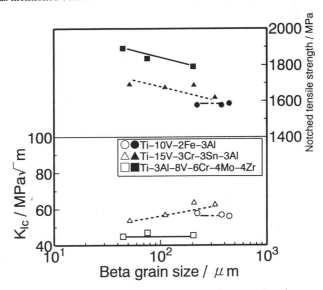

Figure 4 – Effect of beta grain size on notched tensile strength and fracture toughness of three commercial beta alloys aged at 773K for 28.8ks.

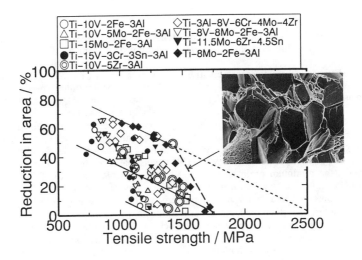

Figure 5 – Relationship between ductility (reduction in area) and tensile strength for various beta alloys, which are subjected to the heat treatment alone.

Fig.6 shows a schematic variation of the strength and ductility as a function of the grain size under constant aging conditions. When the grain size becomes larger than a critical size A, the ductility is reduced due to the occurrence of grain boundary failure. Above a critical size B, the ductility becomes zero, which leads to unstable low stress failure.

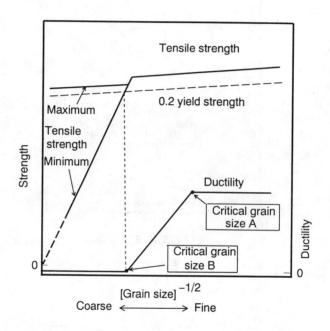

Figure 6 – Schematic variation of strength and ductility as a function of the grain size under constant aging conditions.

Fig.7 shows a schematic representation of the relationship between tensile strength and critical grain size. Those critical grain size A and B for any alloys become smaller with increases in the strength level. Hence, the grain size should be kept smaller, with higher strength levels, in order to avoid drastic decreases in ductility. On the other hand, there is a technological limit for the grain refinement. Applying the thermomechanical treatment can produce smaller grain sizes than applying the heat treatment as the sole mean of microstructure modification. The broken line in Fig.5 shows that the ductility becomes zero at a tensile strength of about 1700MPa. This means that when heat treatment alone is applied to make the minimum grain size about 50 μm, unstable low stress failures occur above 1700MPa and the amount of further strengthening may be limited.

Fig.8 shows schematically the improvement of the strength–ductility balance by microstructure modification[4]. An upper limit of the balance after the heat treatment and the effects of microstructures on strength and ductility are indicated in this figure. Ductility can increase by the grain refinement and homogeneous precipitation of the α phase, whereas strength can increase by the homogeneous and fine precipitation of the α phase and by an increase in dislocation density. The strength–ductility balance can be improved   by combining both effects. The extrapolated line shown by the dotted line in Fig.8 may be an attainable upper limit of the strength–ductility balance. This is a guideline for strengthening of beta titanium alloys.

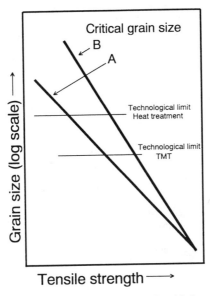

Figure 7 – Schematic representation of the relationship between tensile strength and critical grain size.

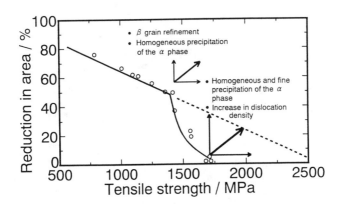

Figure 8 – Schematic representation showing the improvement of the strength–ductility balance by microstructure modifications.

Fig.9 shows the recent improvement of the strength–ductility balance through applying the thermomechanical treatment[5]. Applying the thermomechanical treatment can produce the microstructure with refined beta grain and evenly spaced and small precipitates of the α phase.

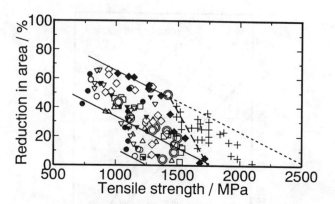

Figure 9 – The improvement of the strength–ductility balance through
applying the thermomechanical treatment (+ mark).

3.3 Strength–toughness balance of beta alloys

Fig.10 shows a collection of notched tensile strength as a function of tensile strength for
various alloys. As mentioned in 3.1 section, notched tensile strength is almost independent
of the grain size but strongly dependent on the aged structure. Therefore, when samples are
aged above 723K, notched tensile strength changes from notch strengthening to notch
embrittlement conditions with increasing strength levels, as indicated by the hatched region

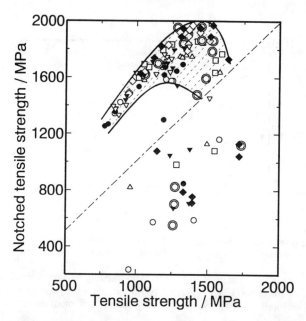

Figure 10 – Relationship between notched tensile strength and tensile
strength for various beta alloys, which are subjected to the
heat treatment alone.
Stress concentration factor Kt is 3.5.

194

in the figure. The notched tensile strength is reduced to the embrittlement condition with increasing strength levels, even if the microstructures are properly controlled from the point of view of improving the strength–ductility balance. This is an example of strengthening being limited by notch toughness. The tensile strength at the point where the ratio of notched tensile strength to tensile strength reaches one, is a critical value for strengthening when it is limited by the notched tensile strength. This critical value is, at present, about 1700MPa.

Fig.11 shows a collection of fracture toughness (K_{Ic}) values as a function of tensile strength for various beta alloys. Fracture toughness decreases monotonously with increasing the strength level. When samples are aged at 773K, fracture toughness is not dependent on the composition nor microstructure and is located within the narrow band shown in the figure. In samples which are strengthened by aging below 723K, fracture toughness reduces slightly to a level below this band. Fracture toughness of alloys decreases with increasing the strength level, but the required toughness for assuring the reliability of alloys should be increased with an increase in the strength level. The tensile strength to intersect both toughness curves mentioned above, is a critical value for strengthening when the application is limited by fracture toughness. This critical strength is, of course, affected by the size of the precrack. 1300MPa in tensile strength shown in the figure, is when the diameter of the precrack is assumed to be about 0.3mm. There is, at present, no guideline for microstructure modification for improving the strength–toughness balance.

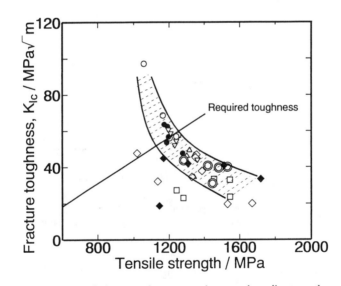

Figure 11 – Relationship between fracture toughness and tensile strength for various beta alloys, which are subjected to the heat treatment alone.
Required toughness is estimated when the diameter of the precrack is assumed to be about 0.3mm.

3.4 Possibility for further strengthening

As mentioned in the previous sections, increases in the strength of smooth, notched and precracked specimens are limited by ductility, notched tensile strength and fracture toughness, respectively. As a result, there is a critical value for strengthening for each specimen. Three critical strengths of beta titanium alloys, as measured by tensile strength, are 2000MPa for smooth, 1700MPa for notched and 1300MPa for precracked specimens, respectively.

These critical strengths are compared with those of ultrahigh strength steels, in order to discuss a possibility for further strengthening. Fig.12 shows the comparison of three critical strengths, as measured by specific strength, between titanium alloys and ultrahigh strength steels. For notched and precracked specimens, critical specific strengths are almost the same values in both alloys. However, for smooth specimen, the critical specific strength of beta titanium alloys is considerably lower than that of ultrahigh strength steels. This is a reason that there is a possibility for further strengthening for the smooth specimen of beta titanium alloys. In addition, ductility values can be improved by grain refinement but toughness values cannot, as mentioned repeatedly. Accordingly, there is a possibility of raising the critical strength when limited by ductility, but it is very difficult to raise it when limited by toughness. Hence, the critical strength when limited by notched tensile strength, about 1700MPa, may be the maximum strength available for general applications.

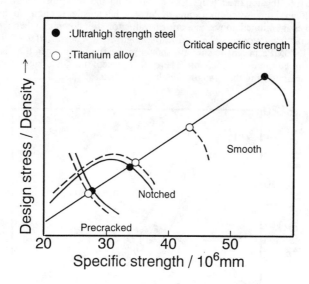

Figure 12 – Comparison of three critical strengths, as measured by specific strength, between titanium alloys and ultrahigh strength steels.

4.Conclusion

The strengthening behavior of beta titanium alloys at room temperature has been investigated under a wide variety of solution treated and aged conditions. As a result, relationships between microstructures and mechanical properties have been made clear and strengthening capability of beta alloys has been discussed. The main results are summarized as follows:

(1) Beta alloys can be hardened by the age hardening behavior, but increases in the strength are limited by ductility and toughness. Hence, there is a critical strength for strengthening.
(2) Tensile ductility is strongly dependent upon the grain size and the aged structure. On the other hand, toughness values are almost independent of the grain size and dependent on the aged structure alone.
(3) Therefore, the strength–ductility balance can be improved by grain refinement through applying the thermomechanical treatment while the strength–toughness balance cannot.
(4) Accordingly, there is a possibility of raising the critical strength when limited by

ductility, but it is very difficult to raise it when limited by toughness.

(5) The critical strength when limited by notched tensile strength, about 1700MPa in tensile strength, may be the maximum strength available for general applications.

References

1) R.R.Boyer and H.W.Rosenberg,ed., Beta Titanium Alloys in the 1980's (New York,NY:TMS–AIME,1984)

2) F.H.Froes,J.C.Chesnutt,C.G.Rhodes, and J.C.Williams, Toughness and Fracture Behavior of Titanium (Philadelphia,PA:ASTM STP651 ASTM,1978),1154

3) Y.Kawabe and S.Muneki, Present Aspects of Titanium Materials Research in Japan, (Tokyo,Japan,ISIJ,1989),147

4) Y.Kawabe and S.Muneki, "Strengthening and Toughening of Titanium Alloys" ,ISIJ International,31(1991),785～791

5) C.Ouchi,H.Suenaga and Y.Kohsaka, 6th World Conf. on Titanium,ed.by P.Lacombe, R.Tricot,G.Béranger (Paris,Franch,Les Editions de Physique,1989),819

IMPROVEMENT OF MECHANICAL BEHAVIOR IN

Ti-3Al-8V-6Cr-4Mo-4Zr BY DUPLEX AGING

L. Wagner[1] and J.K. Gregory[2]

[1] Technical University Hamburg-Harburg, 21073 Hamburg, Germany
[2] GKSS Research Center, 21502 Geesthacht, Germany

Abstract

Solute-rich β-titanium alloys such as Ti-3Al-8V-6Cr-4Mo-4Zr can be heat treated to strength levels ranging from roughly 800 to 1700 MPa by the precipitation of the α-phase in the β-matrix, which occurs at 400-600 °C. Above roughly 500 °C, the α distribution tends to be inhomogeneous, leaving locally weak regions. "Duplex" aging treatments are helpful in producing a more uniform distribution of α-precipitates at strength levels in the range of 1000-1300 MPa in fully recrystallized and partially recrystallized microstructures. The advantages of duplex over simplex aging are that at a given final aging temperature, higher strengths are achieved, and at a given yield stress, higher fatigue limits are obtained (with other mechanical properties essentially unchanged). The choice of the pre-aging temperature is critical for the success of the duplex treatment.

Beta Titanium Alloys in the 1990's
Edited by D. Eylon, R.R. Boyer and D.A. Koss
The Minerals, Metals & Materials Society, 1993

Introduction

Age-hardenable high-strength β-alloys such as Ti-3Al-8V-6Cr-4Mo-4Zr (Beta C) tend to age inhomogeneously at temperatures greater than about 500 °C, leaving regions which are free of α-precipitates (1,2). These aging temperatures normally result in yield strengths of 1000-1300 MPa. Although cold-working prior to aging has been shown to be particularly useful in this alloy, not only in eliminating the precipitate-free regions, but also in obtaining the very high strengths which only this class of titanium alloys offers in reasonable times (3), it is only practical for small cross-sections. Duplex aging, i.e., pre-aging at 425-480 °C followed by a final age at the desired temperature, is also effective in eliminating the precipitate-free regions (2,,4-6) and can be carried out on thick sections. This paper compares the mechanical behavior of duplex-aged with simplex-aged material for various starting conditions by examining the influence of these precipitate-free regions on both primary and secondary mechanical properties, and attempts to provide guidelines for heat treating Ti-3Al-8V-6Cr-4Mo-4Zr to obtain optimum combinations of properties.

Influence of Aging Parameters on α Precipitate Distribution

The distribution of α precipitates in the β matrix depends in general on both aging temperature and prior working history. At low aging temperatures, the transformation presumably occurs via β →β+ β'→β+ α (7), and homogeneous distribution is favored. The precursor to the α phase, β', nucleates homogeneously, and heterogeneous nucleation of the a phase can then occur on this β' precursor. At higher aging temperatures, the direct heterogeneous nucleation (i.e., at grain boundaries and dislocations) of the α phase is favored and is expected to be sensitive to the degree of residual deformation from cold and/or warm working (7,8). For example, aging 8 h at 540 °C (which is underaged) results in the strongly non-uniform distribution seen in Fig. 1a, which consists primarily of grain boundary precipitates. At a lower aging temperature of 440 °C, the distribution of the α precipitates is homogeneous (Fig. 1b). However, not only existing grain boundaries, but also *prior* grain boundaries which existed before the recrystallization treatment can act as preferential nucleation sites. This is illustrated by the sequence of micrographs in Fig. 2. Elongated β grains from hot rolling are shown in Fig. 2a. Recrystallizing at 927 °C resulted in the equiaxed grains (diameter roughly 160 μm) in Fig. 2b which served as the starting microstructure for the aging treatments in Figs. 1a and 1b. Aging 4 h at 455 °C (as opposed to 440 °C) caused precipitation at the prior β grain boundaries (Fig. 2c). Thus, while it is possible in principle to obtain a homogeneous α precipitate distribution for the desired strength level by judicious choice of a low temperature (pre-age) followed by a high temperature cycle (final age), in practice, extreme care must be taken with the temperature for the pre-age. The schematic TTT diagram showing the kinetics of the β →α transformation in Beta-C in Fig. 3 is intended as a guide in determining aging cycles.

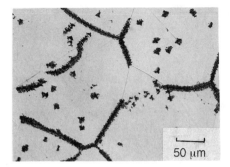

Fig. 1a - Inhomogeneous distribution of α precipitates in fully recrystallized material after aging 8 h at 540 °C

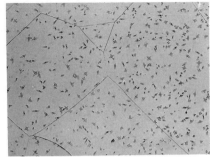

Fig. 1b - Homogeneous distribution of α precipitates in fully recrystallized material after aging 4 h at 440 °C

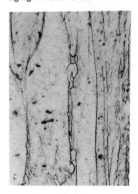

Fig. 2a - Elongated β grains after hot rolling Beta-C.

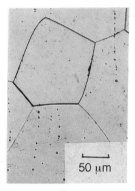

Fig. 2b - Equiaxed β grains after recrystallizing material in Fig. 2a.

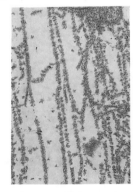

Fig. 2c - α precipitate nucleation at prior β grain boundaries after 4 h 455 °C.

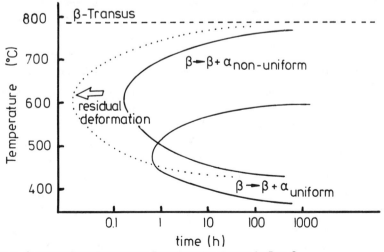

Fig. 3 - Schematic Time-Temperature-Transformation diagram for Beta-C

Mechanical Behavior

Material and Experimental Procedures

Two different charges of this alloy were received in plate form from the RMI Titanium Company, Niles, Ohio through Robert Zapp Werkstofftechnik in Düsseldorf. Chemical analyses are shown in Table I. Both charges were hot rolled to roughly 13 mm: Charge A 70 % at 927 °C and Charge B 63 % at 850 °C. Solution heat treatments (SHT) were performed in air on specimen blanks. Aging treatments were carried out in flowing argon utilizing zirconium foil as an oxygen getter. Tensile tests were performed in the L direction on (German) standard cylindrical specimens with a gage length five times the specimen diameter. Fracture toughness was evaluated according to ASTM E399 on C(T) specimens in the L-T orientation with W=60 mm and B=13 mm. Smooth specimen fatigue results are for rotating bending at a stress ratio of R=-1 at 50 Hz in air. Fatigue specimens were electropolished to eliminate machining effects.

Table I Chemical Composition of the Charges Investigated

Charge	Al	V	Cr	Mo	Zr	Si	Nb	Y*	Fe	O	N	C
A	3.0	8.0	6.0	3.9	3.8	.36	.10	<50	.06	.093	.018	.01
B	3.4	8.1	5.9	4.4	4.3	.38	.07	<50	-	.083	.014	.02

* numbers represent weight percent except for Y content, which is given in ppm.

Microstructure, Tensile Properties and S-N Behavior

Intermediate Strength Level (1050 MPa<UTS<1250 MPa) An underaged condition, two fully aged conditions and one duplex aged (also fully aged) condition were developed by SHT and aging to exhibit comparable tensile properties (Table II). SHT at 927 °C produced completely recrystallized grains with an average diameter of roughly 160 μm. The as-SHT condition exhibits a $\sigma_{0.2}$ and UTS of 850 MPa, el 25 % amd RA 62 %. The underaged condition (16 h 500 °C, Fig. 4a) is characterized by an inhomogeneous, sparse distribution of α-precipitates (dark areas) and a lower ductility than the fully aged conditions. Microstructures after aging 16 h 530 °C (Fig. 4b) and 540 °C (Fig. 4c) are virtually indistinguishable by optical microscopy, and both possess precipitate-free regions approximately 25 × 100 μm. The slightly higher strength in material aged at the lower temperature is accompanied by a slightly lower ductility. Duplex aging resulted in a more homogeneous distribution of the α phase (Fig 4d). The final aging tempera-ture for this duplex cycle, 560 °C, was chosen to obtain $\sigma_{0.2}$ and UTS values similar to those for the simplex treatment of 16 h 540 °C. Qualitatively, there appears to be a correlation between the volume fraction of the precipitate-free regions and the proportional limit (as assessed by $\sigma_{0.02}$), i.e., the more regions remaining, the lower the proportional limit. Microhardness measurements (9) indicate that the local strength in these regions is comparable to that of the as SHT condition.

Table II Tensile Properties for Ti-3Al-8V-6Cr-4Mo-4Zr (intermediate strength level)*

Heat Treatment (Charge)	Microstructure	$\sigma_{0.02}$ MPa	$\sigma_{0.2}$ MPa	UTS MPa	el %	RA %
927 °C/AC 16 h 500 °C (B)	underaged, inhomogeneous	960	1065	1130	9	21
927 °C/AC 16 h 530 °C (A)	fully aged, inhomogeneous	1035	1140	1220	12	21
927 °C/AC 16 h 540 °C (A)	fully aged, inhomogeneous	1005	1085	1165	13	23
927 °C/AC 4 h 440 °C + 16 h 560 °C (A)	fully aged, homogeneous	1035	1085	1140	12	24

* Data from Ref. 9

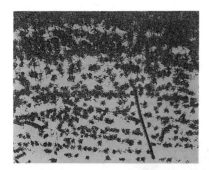

Fig. 4a - Underaged condition after 16 h 500 °C   Fig. 4b - Fully aged condition after 16 h 530 °C

50 μm

Fig. 4c - Fully aged condition after 16 h 540 °C

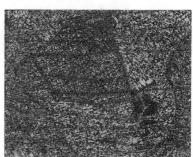

Fig. 4d - Duplex- aged condition after 4 h 440 °C + 16 h 560 °C

S-N curves are shown for the four intermediate strength conditions in Fig. 5. The endurance limit, which for titanium alloys is an appropriate measure of the resistance to crack initiation, scales more reliably with the values of $\sigma_{0.02}$ than with $\sigma_{0.2}$. Previous work (9) had confirmed that cracks tend to nucleate in the precipitate-free regions. An example is shown in Fig. 6. Since these regions tend to be larger in those microstructures which exhibit lower values of $\sigma_{0.02}$, both monotonic and cyclic plastic deformation (where the former controls the proportional limit and the latter ultimately leads to the formation of cracks) are controlled by the size of these locally weak regions. Hence, duplex aging is effective in improving endurance limits.

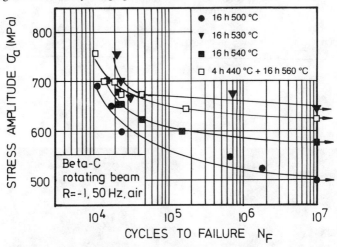

Fig. 5 - S-N curves for the intermediate strength conditions, rotating beam, R=-1, 50 Hz, air

Fig. 6 - Crack nucleation in a precipitate-free region; aged 16 h 530 °C, tested at a stress amplitude of 660 MPa

High Strength Level (UTS>1300 MPa) High strength levels can be developed by fully aging at 500 °C with either simplex (Fig. 7a) or duplex cycles (Fig. 7b) for fully recrystallized material. Higher strengths can also be achieved by lowering the SHT temperature to avoid recrystallization during solutionizing, thus retaining stored energy in the form of dislocations. An SHT at 800 °C was found to yield microstructures with a significant fraction of unrecrystallized grains (2). Both

Table III Tensile Properties for Ti-3Al-8V-6Cr-4Mo-4Zr (high strength level)

Heat Treatment (Charge)	Microstructure	$\sigma_{0.02}$ MPa	$\sigma_{0.2}$ MPa	UTS MPa	el %	RA %
927 °C/AC 30 h 500 °C (B)	fully aged, inhomogeneous	1120	1230	1320	8	14
927 °C/AC 72 h 440 °C + 16 h 500 °C (B)	fully aged, homogeneous	1230	1325	1410	5	11
800 °C/AC 8 h 535 °C (A)*	partially recrystallized, fully aged, inhomogeneous	1150	1225	1320	8	15
800 °C/AC 4 h 425 °C + 8 h 560 °C (A)*	partially recrystallized, fully aged, inhomogeneous	1150	1220	1300	10	13

* Data from Ref 2

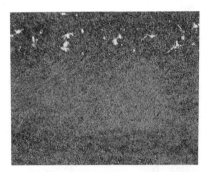

Fig. 7a - Fully aged condition after 30 h 500 °C

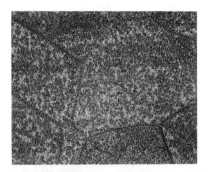

Fig. 7b - Fully aged condition after 72 h 440 °C + 16 h 500 °C

50 μm

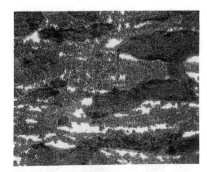

Fig. 7c - Fully aged, partially unrecrystallized condition after 8 h 535 °C

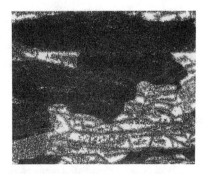

Fig. 7d - Duplex- aged, partially unrecrystallized condition after 4 h 425 °C + 8 h 560 °C

simplex aging (Fig. 7c) and duplex aging (Fig. 7d) treatments were carried out to vary the distribution of the α phase. Tensile properties are given in Table III. Whereas in fully recrystallized material, final aging temperatures of 530-560 °C resulted in strength levels of 1100-1250 MPa (see Table II), the microstructure generated by SHT at 800 °C has retained residual deformation from hot rolling (unrecrystallized grains are visible in Figs. 7c and 7d as the dark areas), resulting in strengths of 1300 MPa or greater, even at shorter aging times. Although duplex aging significantly improved the α-phase distribution in partially recrystallized material, it was not as successful as in material SHT at 927 °C. With the exception of the duplex treatment with a final age at 500 °C, strength levels were essentially identical. The pre-age at 440 °C causes significantly higher strengths to be reached upon a final age at 500 °C than the simplex aging treatment of 30 h.

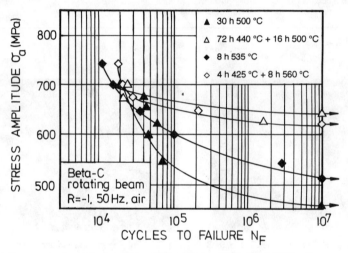

Fig. 8 - S-N curves for the high strength conditions, rotating beam, R=-1, 50 Hz, air

Fatigue behavior for the four high strength conditions is shown in Fig. 8. Again, a correlation is observed between the endurance limit and $\sigma_{0.02}$. However, when compared with the curves in Fig. 5, these fatigue results are somewhat disappointing. The highest fully reversed endurance limit for both strength levels (without resorting to surface treatments) is 650 MPa, which can be obtained at a yield strength of 1140 MPa. Further increases in strength apparently do not bring about a corresponding increase in endurance limit.

Fracture Toughness

Values for K_{Ic} are shown graphically in Figs. 9 and 10 vs. yield stress and elongation, respectively. For the purpose of obtaining a rough estimate of K_{Ic} from tensile properties, there appears to be a fair correlation with strength or ductility. No significant, systematic effect of duplex aging on fracture toughness is observed. However, somewhat poorer toughness values are observed in material with a partially recrystallized microstructure, irrespective of the aging treatment. When compared on the basis of yield stress, (Fig. 9), the underaged condition (16 h 500 °C), exhibits a significantly lower K_{Ic} value than would be expected. This is not surprising in view of the low ductility of this condition.

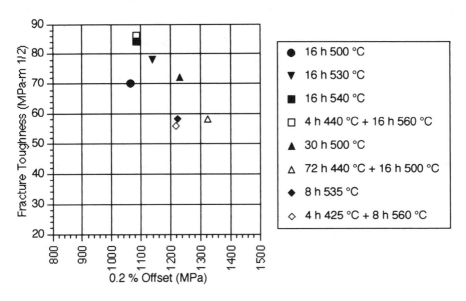

Fig.9 - Fracture toughness K_{IC} vs yield strength (open symbols are for duplex-aged material)

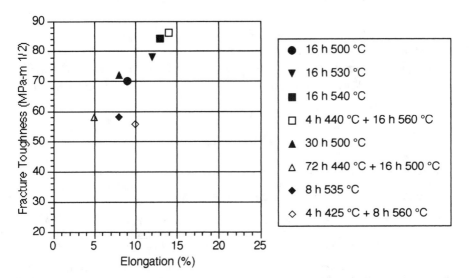

Fig. 10 - Fracture toughness K_{IC} vs elongation to failure (open symbols are for duplex-aged material)

Summary and Conclusions

The success of a duplex treatment depends primarily on the pre-age, since aging response in the range of 420-460 °C is extraordinarily sensitive to slight temperature variations as well as processing history. A duplex treatment for which the pre-age generates a homogeneous distribution of α precipitates (such as that shown in Fig. 1b) evidently results in a smaller mean spacing between α precipitates than the simplex treatment at the same final aging temperature. Hence, higher strengths can be obtained by duplex than with simplex aging. This is consistent with results from similar work on tensile behavior of Ti-15V-3Al-3Sn-3Cr by Okada (6). For a given final aging temperature, a successful pre-age results in strength levels after duplex aging which are roughly 100 MPa higher than after the corresponding simplex aging treatment. Equivalently, when developing duplex aging treatments for Beta-C to attain a desired strength level in the range of 1000-1300 MPa, final temperature cycles should be about 20 °C higher than the simplex treatment which results in the same strength level. This appears to be valid for both fully recrystallized and partially recrystallized microstructures.

Aside from the refinement of the α precipitates, duplex aging can in some cases eliminate precipitate-free regions which contribute to poor endurance limits. Even in cases where duplex aging does not produce an ideally homogeneous microstructure (as in Fig. 7d), the size of the precipitate-free regions is at least significantly reduced, thus improving S-N behavior. This is considered to be the primary benefit to be gained by duplex as opposed to simplex aging.

No systematic effect of duplex aging on ductility or fracture toughness is observed.

Acknowledgments
This work was performed under the GKSS Research and Development program. The authors are grateful to H.-J. Mann (GKSS) for technical assistance and to W.-V Schmitz for the optical microscopy.

References

1. W.M. Parris and H.W. Rosenberg, "Producing Ti-13V-11Cr- 3Al Mill Product at TMCA Historical Note II," Beta Titanium Alloys in the 1980s, ed. R.R. Boyer and H.W. Rosenberg (Warrendale, PA: The Metallurgical Society, 1984), 9-15.

2. H.-E. Krugmann and J.K. Gregory, "Microstructure and Crack Propagation in Ti-3Al-8V-6Cr-4Mo-4Zr," Microstrcture/Property Relationships in Titanium Aluminides and Alloys, ed. Y.-W. Kim and R.R. Boyer (Warrendale, PA: The Metallurgical Society, 1991), 549-561.

3. H.W. Rosenberg, "Property Scatter in Beta Titanium: Some Problems and Solutions," Beta Titanium Alloys in the 1980s, ed. R.R. Boyer and H.W. Rosenberg (Warrendale, PA: The Metallurgical Society, 1984), 145-160.

4. D.E. Thomas, "Microstructure and Mechanical Properties of Investment Cast Beta-C Titanium," Sixth World Conf. on Titanium, ed. P. Lacombe, R. Tricot and G. Béranger (les édition de physique) 1989, 147-152.

5. D.E. Thomas, "Structure and Properties of a Welded Beta Titanium Alloy," Sixth World Conf. on Titanium, ed. P. Lacombe, R. Tricot and G. Béranger (les édition de physique) 1989, 1421-1426.

6. M. Okada, "Strengthening of Ti-15V-3Al-3Sn-3Cr by 2 Step Aging," Sixth World Conf. on Titanium, ed. P. Lacombe, R. Tricot and G. Béranger (les édition de physique) 1989, 1625-1628.

7. S. Ankem and S.R. Seagle, "Heat Treatment of Metastable Beta Titanium Alloys," Beta Titanium Alloys in the 1980s, ed. R.R. Boyer and H.W. Rosenberg (Warrendale, PA: The Metallurgical Society, 1984), 107-126.

8. J.C. Williams and F.H. Froes, "Microstructure and Properties of the Alloy Ti-11.5Mo-6Zr-4.5Sn (Beta III)," Titanium and Titanium Alloys, ed. J.C. WIlliams and A.F. Below, Plenum Press, New York, 1982, 1421-1436.

9. J.K. Gregory and L. Wagner, "Heat Treatment and Mechanical Behavior in Beta-C™," (Paper presented at the VII International Meeting on Titanium organized by Ginatta Torino Titanium, Turin, Italy, November 15, 1991) also GKSS Report GKSS 92/E/7, Geesthacht, Germany.

MICROSTRUCTURE/MECHANICAL PROPERTY RELATIONSHIPS

IN BAR PRODUCTS OF BETA-C™ (Ti-3Al-8V-6Cr-4Mo-4Zr)

J. G. Ferrero, J. R. Wood, P. A. Russo

RMI Titanium Company
1000 Warren Avenue
Niles, OH 44446

Abstract

Beta-C™ is a metastable beta titanium alloy which can be heat treated to achieve a good combination of properties such as tensile strength, fatigue strength, ductility and toughness, such as needed in aircraft fasteners. Variations in hot and cold processing parameters and solution heat treatment can affect the resultant grain structure which in turn affects subsequent aging response and mechanical properties. Several processing/heat treating routes on round bar are explored in this study to determine the effects on microstructure and various mechanical properties.

Beta Titanium Alloys in the 1990's
Edited by D. Eylon, R.R. Boyer and D.A. Koss
The Minerals, Metals & Materials Society, 1993

Introduction

Beta-C™ (Ti-3Al-8V-6Cr-4Mo-4Zr) is a metastable beta titanium alloy which can be heat treated to achieve a good combination of properties such as tensile strength, fatigue strength, toughness and ductility. Beta-C™ is capable of achieving yield strengths from 120 to 210 ksi depending on process route and heat treatment. The alloy achieves this high strength by the precipitation of alpha phase during aging. Because Beta-C™ has a wide range of strength capabilities, it has been used in a large number of applications. Its combination of light weight, good strength, and excellent corrosion resistance has made it an excellent replacement for steel components in various applications, such as oil field tooling, fasteners, springs, etc.

The thermomechanical processing route chosen to produce bar products can have dramatic effects on the final microstructure. The thermomechanical processing route not only affects the final structure but also subsequently affects the recrystallization behavior and aging kinetics[1]. The alloy maintains the beta structure upon cooling allowing bar products 1" (25 mm) and less to be cold drawn. This gives it an advantage over alpha/beta alloys.

This study will evaluate the effects of processing, structure, and heat treatment on the mechanical properties of several sizes of Beta-C™ bar.

Procedure

The bar material used for this study came from production size ingots, 30" diameter 14,000 lbs. The bar sizes studied ranged from 3" (76 mm) to 0.433" (11 mm) diameter and the chemical composition for each of the sizes is shown in Table I. Three of the bar sizes were produced by hot rolling and one by the hot rolling and cold drawing process. A diagram of the typical mill processing sequence is shown in Figure 1.

Table I. Beta-C™ Composition

Bar Size, in. (mm)	Average Chemical Composition, wt%									
	Al	V	Cr	Mo	Zr	C	N	Fe	O	Si
0.433 (11)	3.30	7.70	6.30	4.20	4.20	.01	.02	.10	.09	.02
0.625 (16)	3.28	8.06	5.77	4.24	4.18	.02	.01	.05	.09	.02
1.25 (32)	3.07	7.66	5.74	4.08	3.82	.01	.01	.06	.08	.04
3.0 (76)	3.23	8.13	5.65	4.10	4.15	.02	.01	.08	.08	.03

Figure 1. Typical Mill Process for Beta-C™ Bar Products

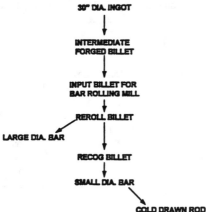

The study was broken down into four phases: I) effect of solution treatment temperature, II) microstructure characterization, III) mechanical properties, and IV) fatigue fracture surface evaluation.

Phase I was designed to evaluate the effect of solution treatment temperature on recrystallization and aging kinetics and thus determine the optimum temperature for both. This was performed using the 3" (76 mm) diameter bar material.

Phase II was designed to characterize the microstructure in both the solution treated and solution treated and aged conditions for the four bar sizes.

Phase III was designed to evaluate the mechanical properties (tensile, toughness, shear, and fatigue) obtained with these various structures. The tensile strength, toughness and shear strengths were evaluated over a wide range of strength levels. The fatigue strength was evaluated at only one strength level.

Tensile tests were performed in accordance with ASTM-E8 using a standard, 0.250" diameter sample with a 1" gauge length. The fracture toughness tests were performed on the 3" (76 mm) and 1.25" (32 mm) diameter bar material. W-2 compact tension specimens machined from the 3" bar in the L-R orientation were tested in accordance with ASTM-E399. Duplicate samples were run for each heat treat condition. On the 1.25" diameter, slow bend pre-cracked charpy's were tested in the L-R orientation according to ASTM-E812. Double shear tests were performed on several bar sizes and run in accordance with MIL-STD-1312. Fatigue tests were performed on all four bar sizes and run in accordance with ASTM-E466.

Phase IV evaluated the fracture surfaces of the fatigue samples only.

Discussion/Results

I) Effect of Solution Treatment Temperature

The solution treatment temperature has been shown to have a dramatic effect on both the recrystallization and aging kinetics of beta alloys. The typical solution treatment temperatures for Beta-C™ are above the beta transus, which for Beta-C™ is approximately 1350°F (732°C). The three temperatures evaluated are as follows: 1450°F (788°C), 1500°F (815°C), and 1550°F (843°C). All treatments were run for 1 hour at temperature followed by an air cool. Higher temperatures were not explored due to the rapid grain growth that occurs at temperatures greater than 1550°F (843°C)[1].

Figure 2 shows the structures and tensile properties obtained with the three solution heat treatments discussed above. The samples were given a decoration age of 925°F(496°C)-30Min-AC after solution annealing[2]. This technique darkens unrecrystallized grains while leaving recrystallized grains light. As can be seen in Figure 2, the 1450°F (788°C) and 1500°F (815°C) treatments show worked grain boundaries and a moderate amount of residual work in the grain interiors indicating incomplete recovery and/or recrystallization. The 1550°F (843°C) treatment shows no grain boundary alpha precipitation and a more uniform equiaxed structure typical of recovery and recrystallization. Based on the similar tensile properties obtained with these three treatments, it appears that the solution treatment had very little effect on tensile properties.

Figure 2. Effect of Solution Treatment Temperature on the
Recrystallization of 3" Ø Beta-C™ Bar After
Decoration Aging ksi (MPa)

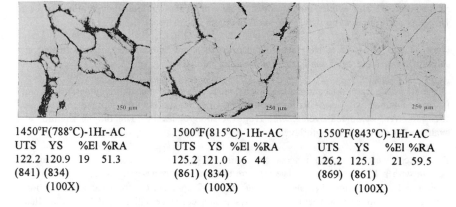

1450°F(788°C)-1Hr-AC	1500°F(815°C)-1Hr-AC	1550°F(843°C)-1Hr-AC
UTS YS %El %RA	UTS YS %El %RA	UTS YS %El %RA
122.2 120.9 19 51.3	125.2 121.0 16 44	126.2 125.1 21 59.5
(841) (834)	(861) (834)	(869) (861)
(100X)	(100X)	(100X)

Figure 3 shows the effect of solution treatment temperatures on the aging kinetics at 900°F (482°C). There is a definite difference in the amount and distribution of alpha as higher solution treatment temperatures are used which is borne out in the tensile properties as shown.

Figure 3. Effect of Solution Treatment Temperature
on the Aging Behavior of 3" ∅ Beta-C™ Bar

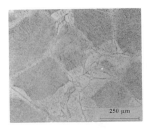  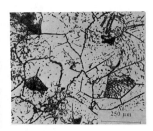

1450°F(788°C)-1Hr-AC +900°F(482°C)-24Hr-AC	1500°F(815°C)-1Hr-AC +900°F(482°C)-24Hr-AC	1550°F(843°C)-1Hr-AC +900°F(482°C)-24Hr-AC
UTS YS %El %RA	UTS YS %El %RA	UTS YS %El %RA
199.8 191.5 8.0 15.0	198.0 189.0 8.3 14	161.0 150.5 8.5 18.7
(1379) (1324)	(1365) (1303)	(1110) (1034)
(100X)	(100X)	(100X)

This difference in aging response and properties can be explained by the following scenario. As discussed above, higher solution treatment temperatures result in more recovery and recrystallization thus decreasing the stored energy and dislocation density in the samples. Unrecrystallized grains have a larger driving force for precipitation and more sites for the alpha to precipitate during aging. The samples solution treated at 1550°F (843°C), which exhibit a fully recovered/ recrystallized structure, have a lower driving force for precipitation of alpha. Precipitation is shown to occur mainly at the grain boundaries, which are high energy locations, whereas precipitation in the grain interiors is minimal and the alpha is rather coarse. This is also shown by the decrease in tensile properties as the solution treatment is increased for the same aging temperature.

There is a delicate balance between the amount of recovery, recrystallization and aging response. Although low solution annealing temperatures result in good aging response, directional properties can occur due to crystallographic texture. Completely recrystallized structures, on the other hand, generally inhibit precipitation of alpha during aging. It is necessary to balance these competing metallurgical processes by judicious selection of the solution temperature. Based on the information shown, a solution treatment temperature of 1500°F (815°C) was chosen because it resulted in the best combination of recrystallization and aging response.

II) Microstructure Characterization

Four different Beta-C™ bar diameters, ranging from 3.0" (76 mm) to 0.433" (11 mm) were used for this study. The solution treated microstructures can be seen in Figure 4. The 0.433" (11 mm) diameter rod was cold drawn to size and the remaining bar sizes were produced by hot rolling. The grain size for each of the sizes was determined by the intercept method outlined in ASTM E-112. The two smaller diameter bars, 0.433" (11 mm) and 0.625"(16 mm), have a fine equiaxed beta grain structure. As the diameter is increased to 1.25" (32 mm) a relatively fine duplex microstructure is developed. The 3" (76 mm) diameter, also exhibits the duplex structure, but has a larger overall grain size and the prior beta grain boundaries, decorated by silicides, are clearly visible.

Figure 4. Typical Beta-C™ Bar Microstructures

Solution Treated (ST) Condition
1500°F(815°C)-1Hr-AC

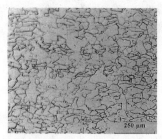

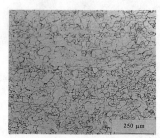

a. 0.433" (11 mm) Ø   Long.   100X
ASTM G.S. #6.0

b. 0.625" (16 mm) Ø   Long.   100X
ASTM G.S. #6.5

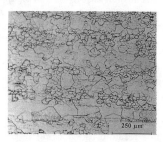

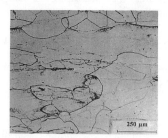

c. 1.25" (32 mm) Ø   Long. 100X
ASTM G.S. #4.3

d. 3.0" (76 mm) Ø   Long.   100X
ASTM G.S. #2.7

It has been proposed that, during thermomechanical processing, silicides, primarily $(Ti,Zr)_5$ Si_3, precipitate on the beta grain boundaries during cooling from the high hot working temperatures[3]. During solution treating, or reheating for further hot working, the structure will begin to recrystallize leaving behind evidence of the prior beta grain boundaries in the form of silicide particles. During aging, the silicides, which have a hexagonal crystal structure, act as nucleation sites for alpha precipitates[3].

The duplex nature of the bar in the larger diameters is believed to be due to nonhomogeneous working across the billet diameter during the forging and rolling operations. As smaller diameter bar is produced, deformation is imparted more uniformly across the diameter. This allows the eventual recrystallization of the large nonuniform grains resulting in a finer, more equiaxed structure. The mechanism of producing the duplex structure and the proposed theories to prevent it are discussed in several papers on the process window theory[4,5,6].

Figure 5 shows the solution treated and aged microstructures of the bar tested in this study after an age of 925°F(496°C)-24Hr-AC. In all cases a fine alpha precipitate is achieved. The precipitate distribution is uniform in Figure 5a, b, and c. However, there are a few areas in Figure 5d, 3" (76 mm) diameter, which show areas where the precipitation is not uniform. As was discussed earlier, this is most likely due to the effects of the recovery/recrystallization process that occurs during annealing.

Figure 5. Typical Beta-C™ Bar Microstructures

Solution Treated + Aged (STA) Condition
1500°F(815°C)-1Hr-AC + 925°F(496°C)-20Hrs-AC

a. 0.433" (11 mm) Ø   Long.   100X
   ASTM G.S. #6.0

b. 0.625" (16 mm) Ø   Long.   100X
   ASTM G.S. #6.5

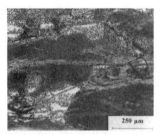

c. 1.25" (32 mm) Ø   Long.   100X
   ASTM G.S. #4.3

d. 3.0" (76 mm) Ø   Long.   100X
   ASTM G.S. #2.7

III) Mechanical Properties

Tensile Properties. The tensile properties of the bar after solution treating at 1500°F (815°C) and aging at 925°F(496°C)-24Hr-AC are shown in Table II. A constant solution treat and age cycle was applied to all four bar sizes and a fairly consistent strength level was achieved.

Table II. Average Tensile Properties of Beta-C™ Bar

Size, in. (mm)	Condition, °F (°C)	UTS, ksi (MPa)	YS, ksi (MPa)	%El	%RA
0.433 (11)	1500(815)-1Hr-AC + 925(496)-20Hr-AC	194 (1337)	182 (1255)	13.0	34.0
0.625 (16)	1500(815)-1Hr-AC + 925(496)-20Hr-AC	192 (1324)	172 (1188)	11.0	27.8
1.25 (32)	1500(815)-1Hr-AC + 925(496)-20Hr-AC	193 (1331)	177 (1220)	9.0	18.8
3.0 (76)	1500(815)-1Hr-AC + 925(496)-20Hr-AC	206 (1424)	195 (1344)	7.5	14.3

Figure 6 shows a plot of yield strength versus aging temperature for a wide range of bar sizes. The data used to generate this graph included data from this study, as well as other production heats. The yield strength was quite consistent over the entire aging range regardless of the bar diameter. This graph suggests that there is no direct dependence of yield strength on grain size in aged Beta-C™. Unlike the yield strength, the tensile ductility, both %elongation and %RA, showed a dependence on grain size, increasing as the grain size decreased. This is shown graphically in Figure 7.

Figure 6. Beta-C™ Properties Strength vs Aging Temp

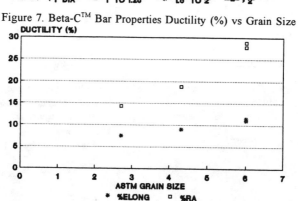

Figure 7. Beta-C™ Bar Properties Ductility (%) vs Grain Size

218

Fracture Toughness. Fracture toughness tests were performed on the 3" (76 mm) and 1.25" (32 mm) diameter bar material at several strength levels. Duplicate samples were run for each heat treated condition. The results of the tests, plotted in Figure 8, show the expected trend of increased toughness with lower yield strengths.

Figure 8. Beta-C™ Bar Properties Fracture Toughness vs Yield Strength

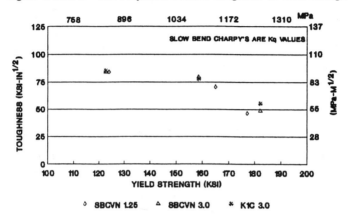

There is good agreement between the compact tension and slow bend pre-cracked charpy results. This is an indication that the pre-cracked charpy test is a valid means of testing the toughness of Beta-C™ when sample size restricts the use of standard compact tension samples. The data also shows that the effect of grain size on fracture toughness is minimal. (See Figure 4 for a comparison of grain size between the 1.25" (32 mm) and 3" (76 mm) diameter bar.)

Double Shear. Because Beta-C™ is being considered as a potential fastener material, double shear tests were performed on several bar sizes in the range of 0.433" (11 mm) to 1.25" (32 mm) diameter. In order for Beta-C™ to be considered, the shear strength must be similar to those of the steel alloys it is aiming to replace. Required double shear values can range from 108 (745) to 125 (862) ksi (MPa) depending on the type of fastener and the application.

Three test conditions were evaluated in this study: 1) as cold drawn, 2) cold drawn and direct aged, and 3) hot rolled solution treated and aged (STA). The data generated is plotted in Figure 9 and shows double shear strength vs yield strength. The shear strength to yield strength ratio for Beta-C™ averaged approximately 60%. As would be expected, the shear strength increases as the yield strength increases. The highest shear strengths were achieved on bar material that was cold drawn and direct aged. This processing option is limited to sizes up to approximately 0.750" diameter. Over that size, the ability to successfully produce cold worked material is reduced and the only other option is the hot rolled STA product.

Figure 9. Beta-C™ Bar Properties Double Shear Strength vs Yield Strength

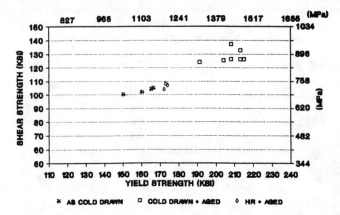

Fatigue Strength. Fatigue strength is a very important material property from the design standpoint. A review of the literature showed that there was limited data on axial load fatigue on Beta-C™ and other beta titanium alloy bar products. One source of data showed some definite trends of improved fatigue strength with decreasing grain size[7]. Typically, one expects the fatigue strength/endurance limit to increase with increasing strength level. However, a review of the current literature indicates lower than expected endurance limits for similar strength levels[8].

In an effort to evaluate the effect of grain size on the fatigue behavior of Beta-C™, axial load fatigue tests were performed on the four bar sizes shown in Figure 4. The bars were aged to a strength level of approximately 195 ksi (1344 MPa) and tested according to ASTM-E466. The test parameters for the tension-tension test were as follows: R = 0.1, frequency = 30 Hz, and tests were terminated at 5 million cycles. In addition to the aged samples, two sizes of bar were tested in the solution treated or low strength condition. These tests were run primarily as a screening method to determine if the precipitation of alpha in conjunction with grain size had any effect on fatigue behavior.

Figure 10a, b, and c, show the plots of maximum stress versus cycles to failure for all the bar material tested. Figure 10a shows two sizes of cold drawn bar both with similar grain size. The only difference between the two is that one was direct aged (DA) and the other was solution treated and aged (STA). The material that was solution treated and aged shows improved fatigue characteristics at the higher stress levels. However, both conditions show run-out at the same level. This indicates a minimal effect of recrystallization on fatigue behavior.

Figure 10a. Stress vs Number of Cycles Cold Drawn Beta-C™ Bar Various Sizes

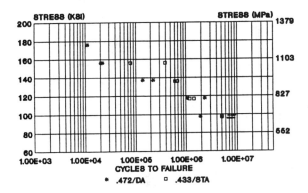

Figure 10b is a plot of three bar sizes in the hot rolled and STA condition, which shows that the 0.625" (16 mm) and 1.25" (32 mm) diameter bar have similar fatigue behavior. However, the 3" (76 mm) diameter bar exhibits a large scatter in the test data. It is felt that this high degree of variability is due to the large duplex grain size of the 3" (76 mm) diameter bar material.

Figure 10b. Stress vs Number of Cycles Hot Rolled Beta-C™ Bar Various Sizes

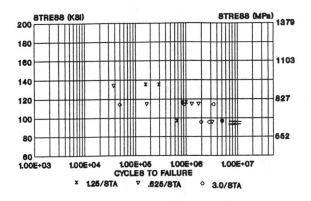

Figure 10c is a plot of the 0.625" (16 mm) and 1.25" (32 mm) diameter bar in the solution treated condition. The data shows fatigue run-out for both bar sizes to be 45 to 50 percent of the ultimate strength. This is similar to the data obtained for the STA material indicating that the fine precipitated alpha had very little impact on the fatigue ratio.

Figure 10c. Stress vs Number of Cycles Hot Rolled Beta-C™ Bar Various Sizes

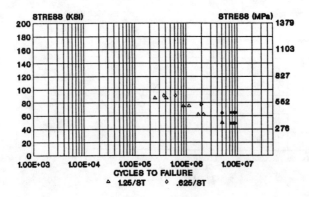

Another method for analyzing fatigue data is to evaluate the ratio of the endurance limit to ultimate strength. Because the data for the 3" (76 mm) bar had so much scatter, and several samples in the other size range failed below the 5 million cycle run-out, an endurance limit of 1 million cycles was chosen to develop this graph. Figure 11 shows that the ratio of endurance limit to ultimate decreases with larger bar diameter and grain size. This trend is not only true for the STA bar but also for the solution treated material indicating that grain size and not alpha precipitate is the controlling factor for determining fatigue life. The fatigue data generated in this study seems to be in rather good agreement with the published data on beta alloy fatigue[8]. In the majority of cases, the endurance limit ranges from 50 to 60% of the ultimate strength.

Figure 11. Ratio of Endurance Limit to Ultimate Strength Beta-C™ Bar

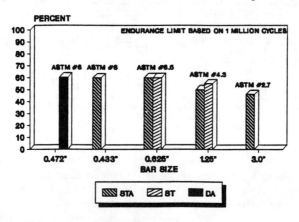

IV) Fatigue Fracture Surface Evaluation

Since this study evaluated such a wide range of grain sizes, which appears to play a major role in fatigue, an evaluation of the fatigue fracture surfaces was warranted. Figures 12 through 15 show SEM fractographs of the initiation sites on the four bar sizes tested. As can be seen, the initiation site in all cases was subsurface and not surface related. An internal initiation site is an indication of a microstructure related phenomenon. Based on the fact that the fatigue fractures had the same characteristics over the entire size range, grain size could not be the only issue. An examination of the lower magnification fractographs clearly shows the effect of grain size on fracture morphology. Another interesting feature of the fractures was that at the initiation site in all the failures there was a flat facet presumably at a grain boundary interface.

Figure 12. Beta-C^{TM} Bar Fatigue Fracture Surfaces

.433" (11 mm) Ø - STA Condition, Axial Load Fatigue R = .1

Figure 13. Beta-C^{TM} Bar Fatigue Fracture Surfaces

.625" (16 mm) Ø - STA Condition, Axial Load Fatigue R = .1

Figure 14. Beta-C™ Bar Fatigue Fracture Surfaces

1.25" (32 mm) ∅ - STA Condition, Axial Load Fatigue R = .1

Figure 15. Beta-C™ Bar Fatigue Fracture Surfaces

3.0" (76 mm) ∅ - STA Condition, Axial Load Fatigue R = .1

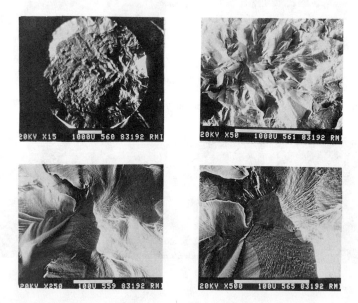

One theory for this type of fracture appearance is related to a thin layer of grain boundary alpha[9]. It has been shown that a layer of grain boundary alpha can influence the tensile ductility[10]. The grain boundary alpha layer is essentially a long soft zone relative to the higher strength matrix. This difference in strength can result in preferential deformation and strain localization which could lead to intergranular fracture. It has also been shown that smaller grain sizes can increase the ductility by decreasing the length of the grain boundary alpha layer, thus reducing the magnitude of the localized strains[10]. This seems to be a plausible theory, however, more work is needed in order to completely verify this mode of failure.

Another possible failure mechanism could be related to the partially aged, or unaged areas in the microstructure which are controlled by the balance between the solution treatment temperature and aging kinetics. The partially/unaged areas would have a lower strength than the surrounding matrix and could be potential sites for crack nucleation based on a similar argument to that given for grain boundary alpha. This phenomenon has been shown in fatigue work performed on aged Beta-C^{TM} plate[11]. As with the grain boundary alpha theory, this failure mechanism has merit. However, more work is needed to totally understand the mechanisms of fatigue failure in Beta-C^{TM}.

Summary/Conclusions

1. A solution treatment temperature of 1500°F (815°C) results in the optimum combination of recrystallized grain structure and aging kinetics on hot rolled bar products.

2. The yield strength of Beta-C^{TM} does not appear to be influenced by grain size, however, the ductility does tend to increase with finer grain structures.

3. Fracture toughness and shear strength vary directly with yield strength as expected, and there appears to be no direct correlation between grain size and toughness or shear strength.

4. The fatigue strength of Beta-C^{TM} showed a definite correlation with grain size; increasing as the grain size decreased.

5. Examination of the fatigue fracture surfaces showed an internal fracture initiation for all sizes tested. The exact nature of this internal initiation is not known. However, it is surmised that grain boundary alpha could be a possible cause. Further work needs to be performed in this area.

References

1. G. A. Bella et al., Microstructure/Property Relationships in Titanium Aluminides and Alloys, edited by Young-Won (Y-W.) Kim and Rodney R. Boyer, TMS, 1991, pp. 493-510.

2. F. H. Froes, J. M. Capenos, and C. F. Yolton, Metallography, 1976, vol.9, pp. 535-537.

3. S. Ankem et al., "Silicide Formation in Ti-3Al-8V-6Cr-4Zr-4Mo", Met. Trans. A, 1987, vol.18A, p. 2015-2025.

4. F. H. Froes et al., Beta Titanium Alloys in the 80's, edited by R. R. Boyer and H. W. Rosenberg, AIME, 1984, pp. 161-184.

5. I. Weiss, R. Srinivasan, and F. H. Froes, Recrystallization '90; Proceedings of the 1st International Conference on Recrystallization in Metallic Materials, edited by T. Chandra, TMS, 1990, pp. 609-616.

6. I. Weiss and F. H. Froes, Titanium Science and Technology, edited by G. Lütjering, U. Zwicker, and W. Bunk, 1985, pp. 499-506.

7. R. Chait and T. S. DeSisto, "The Influence of Grain Size on the High Cycle Fatigue Crack Initiation of a Metastable Beta Ti Alloy", Met. Trans. A, 1977, vol.8A, p. 1017.

8. Internal RMI Titanium Company, Tech Memo 92-17, 1992.

9. R. R. Boyer et al., Microstructure/Property Relationships in Titanium Aluminides and Alloys, edited by Young-Won (Y-W.) Kim and Rodney R. Boyer, TMS, 1991, pp. 511-520.

10. G. T. Terlinde, T. W. Duerig, and J. C. Williams, "Microstructure, Tensile Deformation, and Fracture in Aged Ti-10V-2Fe-3Al", Met. Trans. A, 1983, vol.14A, pp. 2101-2115.

11. J. K. Gregory and L. Wagner, "Heat Treatment and Mechanical Behavior in Beta-C^{TM}", paper presented at the VII. International Meeting on Titanium, November 15, 1991, in Torino, Italy, organized by GTT, Ginatta Torino Titanium S.p.A., Torino.

EFFECT OF THE MORPHOLOGY OF THE PRIMARY ALPHA PHASE ON THE

MECHANICAL PROPERTIES OF BETA-CEZ ALLOY

D. Grandemange, Y. Combres, and D. Eylon*

CEZUS, Centre de Recherches d'Ugine, 73403 Ugine Cedex, France
* The University of Dayton, 300 College Park, Dayton, OH 45469, USA

ABSTRACT

A new metastable beta titanium alloy, Beta-CEZ, has been developed by CEZUS. This alloy was designed for use at moderate temperatures up to 450°C (840°F). Alloy requirements also included high mechanical strength, high fracture toughness and improved creep resistance as compared to existing beta alloys. The understanding of the relationships between processing / microstructure / mechanical properties in the alloy may lead to applications of such promising material in current and future airplanes. In this study, a combination of several forging routes, which included through-transus, hot-die and warm-die forging, with subsequent heat treatments were selected in order to evaluate the microstructure-mechanical property relationships. It was found that the recrystallized prior beta grain boundary alpha, characteristic of the through-transus forging process, lead to an improved tensile-fracture toughness balance. The size of the primary alpha structure observed in through-transus forged specimens also enhances fracture toughness, fatigue crack propagation and creep resistance. The resulting mechanical properties were compared with similarly processed beta titanium alloys such as Ti-17 and Ti-10-2-3. The results show that Beta-CEZ displays tensile, fracture toughness, and creep properties above those obtained with Ti-17 and Ti-10-2-3. Fatigue crack propagation properties are similar for the three alloys.

The authors acknowledge Pechiney Group Research and Development Directorate for funding this work, which was performed at the University of Dayton. In addition, the authors are thankful for the useful discussions and comments from Dr. H. G. Suzuki and W. J. Porter.

Beta Titanium Alloys in the 1990's
Edited by D. Eylon, R.R. Boyer and D.A. Koss
The Minerals, Metals & Materials Society, 1993

INTRODUCTION

In order to meet the demands of future gas turbine engines, there is a need to introduce alloys with high strength and fracture resistance. Since these properties frequently control the size of a fracture-critical component, improvements in one or several of them, allow to design lighter rotating components resulting in higher engine performance. In the case of titanium alloys, mechanical properties can be very dependent upon the microstructural constituents and their morphologies, which in turn are specific to the applied processing and heat treatment.

The present study concerns a high strength metastable beta titanium alloy, Beta-CEZ, which potentialy may produce a combination of strength and fracture resistance superior to titanium alloys currently used in the fan and low pressure sections of the compressor[1]. A combination of several forging routes which included through-transus, hot-die forging and warm-die forging with subsequent heat treatments were selected in order to evaluate the effect of the microstructure on the mechanical properties of the alloy. Among other microstructural features, the morphology of the primary alpha phase at prior beta grain boundaries and inside prior beta grains is of significant importance. Therefore its effects on the room temperature tensile, fracture toughness, and fatigue crack propagation properties as well as medium temperature creep and tensile properties of the alloy were studied. The mechanical properties obtained with Beta-CEZ alloy were also compared with other similarly processed metastable beta titanium alloys.

EXPERIMENTAL PROCEDURE

The chemical composition of the material used in this study is given in Table I. The beta transus temperature was determined using both differentiel thermal analysis (DTA) and metallographic observations and was found to be 890°C (1635°F).

Table I. Chemical composition (w%) of Beta-CEZ alloy used in this study.

Al	Sn	Mo	Zr	Cr	Fe	O	C	N	H	Ti
4.91	2.01	4.01	4.35	2.02	0.93	0.09	0.06	0.06	-*	Bal.

* below 100 ppm

T_β = 890°C (1635°F)

Medium size pancakes (270 mm [10.6"] diameter x 60 mm [2.35"] thick) were forged using three processing routes: through-transus forging, hot-die forging with 830°C (1525°F) die temperature and warm-die forging with 700°C (1290°F) die temperature. They are discribed in the next section along with the resulting microstructures. Prior final heat treatment, the specimens were machined in such a way that the loading axis of the tension and creep specimens was parallel to the loading axis of the fracture toughness and fatigue crack propagation specimens and were all in the tangential orientation.

Tensile testing was performed at room temperature on cylindrical specimens with a diameter of 6.35 mm (0.25") and a 25.4 mm (1") gage length. The tests were conducted using a 0.005 min⁻¹ strain rate through 0.2% yield and a 1.25 mm.min⁻¹ mean crosshead rate thence to failure.
Fracture toughness tests were performed on compact-tension specimens (12.7 mm [0.5"] thick) which were machined, precracked and tested at room temperature per ASTM E399-90 specifications with the loading axis in the tangential orientation of the pancakes.

Fatigue crack propagation tests were run per ASTM E647 on 12.7 mm (0.5") thick compact-tension specimens. An MTS servo-hydraulic testing system was used both for the precracking and testing which were done in laboratory air at room temperature. An R value (the minimum load divided by the maximum load) of 0.1 was used as well as sinusoidal loading at a frequency of 20 Hz. The crack length was measured using a clip gage and a traveling microscope.

Creep testing was performed at 400°C (750°F) at a stress level of 600 MPa (87 ksi) per ASTM E139 on specimen similar to the tensile specimens. Times to 0.1% and 0.2% creep were measured.

RESULTS AND DISCUSSION

Description of the resulting microstructures

Two distinct microstructures resulted from the three forging routes. In all cases, prior to forging, the material was heated above the beta transus temperature to 920°C (1690°F) to produce an homogeneous, equiaxed beta structure. During through-transus forging, a succession of forging steps at decreasing temperatures was performed in order to ensure that the primary alpha phase which precipitates at beta grain boundaries becomes fully recrystallized and that all beta grain boundaries transform into that morphology[2]. Finally, the material was air cooled down to room temperature. The resulting microstructure exhibited elongated prior beta grains with recrystallized primary alpha phase at prior beta grain bounbaries (Figure 1a). Inside the prior beta grains, a lamellar primary alpha microstructure was observed (Figure 1a). After heat treatment, typical primary alpha plates length was around 40 μm.

In the case of the hot-die or warm-die forging, deformation was applied at a constant ram speed for a duration which allowed to obtain the same amount of deformation as in the through-transus forging process. After forging, the material was air cooled. The resulting microstructure displayed large and equiaxed prior beta grains (around 0,25 mm) which were outlined with continuous or semi-continuous primary alpha phase (Figure 1b). After heat treatment, the length of the alpha plates within the prior beta grains was around 5 μm.

In summary, the through-transus forged structure had discontinuous grain boundary alpha phase and coarser primary alpha plates, while the hot-die or warm-die forged material had continuous grain boundary alpha with finer primary alpha plates.

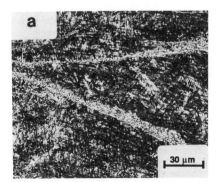 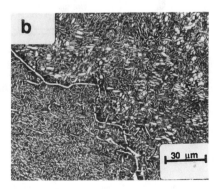

Figure 1 - As forged microstructures of: a)the through-transus forged material. b)the hot-die or warm-die forged material.

Tensile tests

A wide range of strength versus elongation was studied for the hot-die and warm-die forged material while only usuable conditions (El>6%) were investigated for the through-transus forged pancake. The variations of the elongation as a function of the ultimate tensile strength are plotted in Figure 2. It is observed that elongation is inversely proportional to strength which is common to all titanium alloys. The best combination of strength and elongation was obtained with the through-transus forging. This is attributed to the recrystallized and discontinuous nature of the grain boundary alpha obtained by forging the material through the transus. Indeed, it has been reported in several studies that continuous grain boundary alpha is detrimental to ductility[2][3][4]. Since cracks tend to propagate along, or near interfaces, microstructures with continuous or semi-continuous grain boundary alpha are more sensitive to crack initiation and propagation, thereby leading to premature failure.

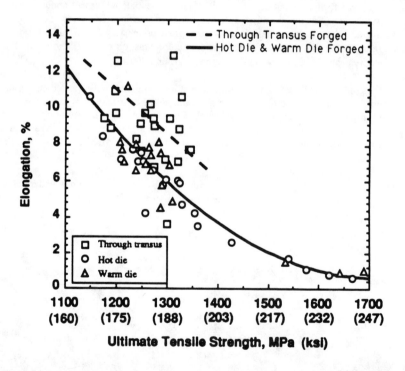

Figure 2 - Relationship between tensile elongation and strength for the three forging conditions of Beta-CEZ alloy.

Fracture toughness tests

The relationship between fracture toughness, ultimate tensile strength and yield strength is plotted in Figure 3. A trend which is common to all titanium alloys is obseved: higher strength levels are associated with lower fracture toughness values.

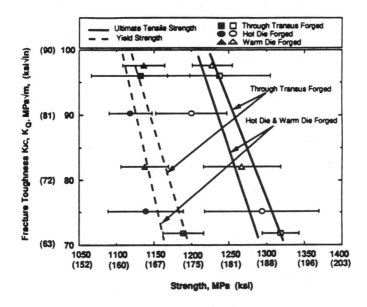

Figure 3 - Relationship between fracture toughness, yield strength and ultimate tensile strength for three forging conditions of Beta-CEZ alloy.

However, through-transus forged samples exhibit a slightly better combination of tensile and fracture toughness properties. This result can be explained by observing the crack path morphology in the direction parallel to the crack propagation. The alpha plate length has been compared to the plastic zone size at the tip of the crack. Typical plane strain plastic zone size ranged from 200 μm to 380 μm[5]. Since the alpha plates of the hot-die or warm-die forged specimens are usually at least forty times smaller than the plastic zone size at the crack tip, they are not effective in defecting the crack (Figure 4b). However, in the case of the through-transus forged material, it is occasionally observed that the primary alpha plates or colonies size are in the order of the plastic zone size. At these locations, the alpha plates become effective in deflecting the crack (Figure 4a). Therefore, tortuosity of the crack path is increased thus contributing to improve the fracture toughness values[6]. It is also observed that, when favorably oriented, with respect to the crack path, the continuous prior beta grain boundary alpha characteristic of the hot-die or warm-die forged microstructures, promotes an intergranular crack path (Figure 5). This path is entirely along alpha/beta interfaces and therefore is a low energy path, which is detrimental to fracture toughness. Intergranular fracture modes can be beneficial if they lead the main crack in a direction away from the main path or if they create crack branching[7]. These last two cases were not observed in our study.

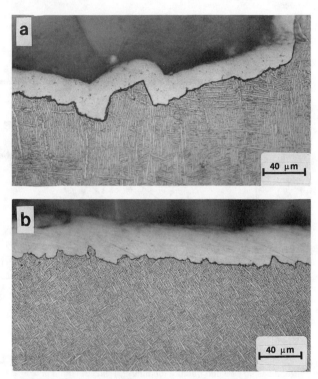

Figure 4 - Effect of the primary alpha plate length on the crack path morphology in a direction parrallel to the crack propagation: a) through-transus forged material, b) hot-die or warm-die forged material.

Figure 5 - Effect of a continuous prior beta grain boundary alpha on the crack path morphology in a direction parallel of the crack propagation direction.

Fatigue crack propagation tests

The results of the fatigue crack propagation tests are reported in Figure 6. For all tested ΔK levels, the scatter band in fatigue crack propagation rate is relatively narrow. Specimens exhibiting the lower yield strength values, therefore the larger plastic zone size at the crack tip for a given ΔK, are expected to display the slowest fatigue crack propagation rates. This phenomena is observed in Figure 6: the specimen exhibiting the higher yield strength value had the higher fatigue crack propagation rate.

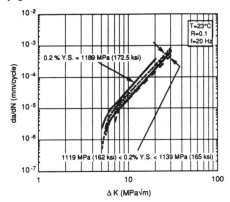

Figure 6 - Comparison of fatigue crack propagation rate curves for the three forging conditions (——— through-transus forged, ------ hot-die or warm-die forged).

Although it does not appear that microstructure plays an important role on fatigue crack propagation properties, it was observed that the microstructural differences between through-transus forged and hot-die or warm-die forged specimens lead to different fracture modes. In the planes parallel to the crack propagation direction, secondary cracks are observed along the fracture path of the through-transus forged specimens at all ΔK values (Figure 7a). This secondary cracking can be attributed to the larger primary alpha plate colonies, which become effective in dividing the crack energy. On the other hand, in the hot-die or warm-die forged specimens, although the fracture is mostly transgranular, fast crack propagation along continuous prior beta grain boundary alpha can be observed (Figure 7b).

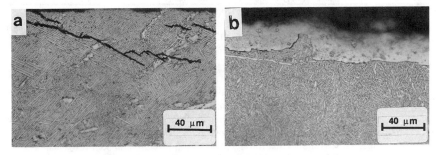

Figure 7 - a) Secondary cracking observed in through-transus forged specimens
b) Crack propagation along continuous prior beta grain boundary alpha observed in hot-die or warm-die forged specimens.

Creep tests

The creep tests results are reported in Table II. The through-transus forged specimens exhibit a higher creep resistance than the hot-die or warm-die forged specimens. It is considered that the coarser primary alpha structure inside the prior beta grains is responsible for the higher creep results obtained in the case of the through-transus forged material. The same result was obtained on Beta-CEZ in a recent study[8].

Table II - Beta-CEZ creep tests results.*
Temperature = 400°C (750°F), Stress = 600 MPa (87 ksi)

Forging	Time (hrs) to 0.1% Creep	Time (hrs) to 0.2% Creep
Through-Transus	27 - 31	95 - 126
Hot-Die or Warm-Die	21 - 25	70 - 86

*Average value of two tests

Comparison with other titanium alloys

A comparison of the room temperature tensile, fracture toughness and fatigue crack propagation properties along with creep properties of Beta-CEZ and two other beta metastable titanium alloys (Ti-17 and Ti-10-2-3) is reported in Table III. This table does not only reflect the higher mechanical property level obtained with through-transus forged Beta-CEZ over hot-die or warm-die forged Beta-CEZ but also the superior mechanical properties, in general, of Beta-CEZ as compared to other high strength structural alloys such as Ti-17 or Ti-10-2-3. This can be attributed both to the composition of Beta-CEZ alloy and its microstructure. The advantage is especially significant for the balance of room temperature tensile strength, tensile ductility and room temperature toughness as well as for creep resistance.

Table III - Comparison of the mechanical properties of Beta-CEZ and two other beta Ti-alloys

Material	UTS (RT) MPa (ksi)		El (%)	K_Q MPa$\sqrt{m}$ (ksi$\sqrt{in}$)		da/dN for ΔK=10 MPa$\sqrt{m}$ (mm/cycle)	Creep* (h)
Beta-CEZ Through-Transus	1250	(181)	10	93	(84)	3.10^{-4}	110
Beta-CEZ Hot-Die/Warm-die	1250	(181)	7.5	85	(77)	3.10^{-4}	75
Ti-17	1250	(181)	8[9]	60	(54)[11]	3.10^{-4}	35[1]
Ti-10-2-3	1250	(181)	6[10]	60	(54)[11]	3.10^{-4}	-NA-

* Time to 0.2% plastic strain under 600 MPa (87 ksi) at 400°C (750°F)

CONCLUSION

Although all processing routes studied in this work for Beta-CEZ alloy lead to a combination of high tensile strength and fracture toughness, the best combination was obtained with the through-transus forged material. The recrystallized prior beta grain boundary alpha observed in this microstructure particularly enhances tensile properties. The coarser primary alpha plate structure observed in this microstructure promotes fracture toughness because of an increased crack path tortuosity, fatigue crack propagation by introducing crack branching and creep resistance. Finally it was shown that for a given strength level, the fracture toughness of through-transus forged material is increased by more than 50% over Ti-17 or Ti-10-2-3. The three alloys exhibit very similar fatigue crack propagation rates. At 400°C (750°F), under 600 MPa (87 ksi), the creep resistance of Beta-CEZ is 2 to 3 times higher than that of Ti-17 in the same microstructural and experimental conditions.

References

1. B. Prandi, J. F. Wadier, F. Schwartz, P. E. Mosser, A. Vassel. " The Beta-CEZ, a High Performance Titanium Alloy for Aerospace Engines, " Proceedings of the 1990 International Conference on Titanium and Applications. Titanium Development Association, Dayton, OH, Vol. 1, 1990, pp. 150-159.
2. Y. Combres, B. Champin. Matériaux et Techniques, May-June 1991, pp. 31-41.
3. T. W. Duerig, J. C. Williams. " Overview: Microstructure and Properties of Beta Titanium Alloys, " Beta Titanium in the 1980's, R. R. Boyer, H. W. Rosenberg, eds., TMS-AIME, Warrendale, PA, 1984, pp. 107-126.
4. M. J. Donachie. "Relationships of Properties and Processes, " Titanum a technical Guide, M. J. Donachie, ed., 1988, p. 167.
5. R. W. Hertzberg. Deformation and Fracture Mechanics of Engineering Materials, Eds. Wiley, Third Edition, 1989, p. 292.
6. D. Eylon, J. A. Hall, C. M. Pierce, and D. L. Ruckle. " Microstructure and Mechanical Properties Relationships in Ti-1100 Alloy at Room and Elevated Temperatures, " Metallurgical transactions A, Vol. 7A (1976), pp. 1817-1826.
7. G. R. Yoder, F. H. Froes, and D. Eylon. "Effect of Microstructure, Strength, and Oxygene Content on Fatigue Crack Growth Rate of Ti-4.5Al-5.0Mo-1.5Cr (Corona 5), " Metallurgical Transactions A, Vol.15A, January 1984, pp; 183-197.
8. P. E. Mosser, N. Marnier, and Y. Honnorat. " Thermomecanical Beta-Processing of High Strength Medium Temperature, " Proc. of the Seventh World Conf. on Titanium, 1992, To be published.
9. H. W. Rosenberg. " Ti-17 Properties, " Beta Titanium in the 1980's, R. R. Boyer, H. W. Rosenberg, eds., TMS-AIME, Warrendale, PA, 1984, pp. 433-439.
10. J. L. Shannon. " Ti-10V-2Fe-3Al Data Sheet, " Aerospace Structural Metals Handbook, " Cindas/Purdue University, West Lafayette, IN, 1991, pp. 39-45.
11. Y. Kawabe, and S. Mineki. " Strengthening and Toughning of Titanium Alloys, " ISIJ, 1991, pp. 785-791.

EFFECT OF STEP-AGING ON THE FRACTURE TOUGHNESS

OF TI-15V-3Cr-3Sn-3Al ALLOY

Naotake NIWA[1] and Hideo TAKATORI[2]

[1] Faculty of Engineering, The University of Tokyo
7-3-1 Hongo Bunkyo-ku, Tokyo, 113 Japan

[2] NIKKO ANCO-TECH INC.
2525 Beech Daly Road Dearborn Heights, Michigan 48125 U.S.A.

ABSTRACT

Development and an application of a new high-low step-aging to improve the fracture toughness-strength balance of a Ti-15V-3Cr-3Sn-3Al alloy are studied. The high-low step-aging of aging at higher temperatures followed by aging at lower temperatures produces bi-modal microstructure composed of coarse and fine alpha precipitates in beta matrix. It greatly improves fracture toughness-strength balance compared with aging at a single temperature. Homogeneous distribution of coarse alpha precipitates produced by adding pre-aging at 573K before the high-low step-aging tends to reduce the superiority of the bi-modal microstructure in fracture toughness. The improvement is provided by the formation of microcracks and voids in the coarse alpha precipitates and rugged crack propagation due to the uneven microstructure. The high-low step-aging is applied to a TIG weldment of the alloy to improve the mechanical properties of the weldment. In the TIG weldment, strength of a fusion zone becomes much higher than that of a base metal after aging at a single temperature because of different aging response. In the first high temperature aging of the high-low step-aging, coarse alpha particles that strengthen little and suppress strengthening by fine alpha precipitation in low temperature re-aging, precipitate more in fusion zone than in base metal because of the enhancement of aging in fusion zone. Therefore, strengthening of fusion zone in re-aging is less than in the base metal, resulting in comparable strength between the fusion zone and the base metal after re-aging. The bi-modal microstructure produced by the step-aging also improves the fracture toughness of the fusion zone of the weldment.

Beta Titanium Alloys in the 1990's
Edited by D. Eylon, R.R. Boyer and D.A. Koss
The Minerals, Metals & Materials Society, 1993

INTRODUCTION

An alloy Ti-15V-3Cr-3Sn-3Al(Ti-15-3) was developed as an age-hardenable metastable beta titanium alloy and the interrelationship of microstructures and mechanical properties has been studied recently[1)-10)]. Detail investigation on fracture toughness, however, has not been performed. The alloy is age-hardened through the precipitation of alpha phase during aging after solution treatment. Mechanical properties of metastable beta titanium alloys depend very much on aged microstructure composed of alpha precipitates and beta matrix. With lowering an aging temperature, size of precipitated alpha phase becomes finer and distribution of it becomes more homogeneous. It is known that aging at a higher temperature produces coarse alpha precipitates which cause little increase of strength and loss of ductility and it is detrimental for the improvement of mechanical properties of metastable beta titanium alloys.

In the former part of this study, with a view to achieving the better balance of fracture toughness and strength of the alloy Ti-15-3, the effect of bi-modal microstructure produced by combination of agings at high temperatures and low temperatures after solution treatment on mechanical properties was investigated. In the latter part, the results of an application of the high-low step-aging developed in the former part to the TIG weldment of Ti-15-3 were discussed.

EXPERIMENTAL PROCEDURES

The alloys were supplied in the form of hot-rolled plate with 12 mm thick from Nippon Mining & Metals Co., Ltd. (Alloy*) and NKK Corp. (Alloy**). Table 1 shows the chemical compositions of the alloys used.

Solution-treatment was carried out as heated at 1123K for 3.6ks in inert gas atmosphere and air-cooled. For the specimens of TIG weldment, solution-treatment was carried out as heated at 1061K for 1.2ks in vacuum and cooled with inert gas. Aging conditions adopted in the former part of this article are listed in Table 2. Aging conditions for specimens of TIG weldments are listed in Table 4.

In this study, three kinds of aging treatments following solution treatment were planned. One is ordinary aging (designated as STA hereafter) in which the effects of aging temperatures of 873K, 773K, 723K and 673K and aging time of up to 360ks were investigated. The second one is high-low step-aging i.e. aging at higher temperatures and re-aging at lower temperature (designated as STDA) in which the effects of aging temperatures and time and especially those of volume fraction of coarse alpha precipitates produced by the first aging at a high temperature on fracture toughness were investigated. The third one is to add pre-aging at 573K to high-low step-aging in order to investigate the effect of distribution of coarse alpha precipitates on fracture toughness (designated as STTA). Shape and size of tensile and compact tension specimens are shown in Fig. 1. Tensile testing and fracture toughness testing were carried out at a room temperature and at crosshead speed of 0.017 mm per second. Tensile properties and fracture toughness were measured in the LT direction and plane strain fracture toughness was determined according to ASTM E399. Solution treatment and TIG welding for specimens of TIG weldments were performed in the Aerospace Division of Nissan Motor Co., Ltd..

Table 1 Chemical compositions of the alloys(mass%).

	V	Cr	Sn	Al	O	N	C	H	Fe	Ti
Alloy*	15.1	3.36	3.04	3.37	0.14	0.008	0.004	0.0061	0.17	bal.
Alloy**	15.3	3.18	3.24	2.87	0.09	0.014	0.010	0.0248	0.03	bal.

Alloy** was used for TIG weldment.

Table 2 Outline and names (A1,...,S3) of agings after solution treatment of heating at 1123K for 3.6ks and air cooling.

	T_1 /K	t_1 /ks	T_2 /K	t_2 /ks	T_3 /K	t_3 /ks	
STA	673	270 ∫ 1Ms					A1
	723	65 ∫ 252					A2
	773	30 ∫ 100					A3
	873	100					A4
STDA	773	22	673	22 ∫ 360			D1
	873	11		43 ∫ 360			D2
		50		180			D3
		100		360			D4
	773	100		360			D5
STTA	573	36	873	11	673	360	S1
				11	723	360	S2
				22			S3

STA :Aging at T_1 for t_1
STDA:Re-aging at T_2 for t_2 after
    aging at T_1 for t_1
STTA:Pre-aging at 573K for 36ks
    before STDA

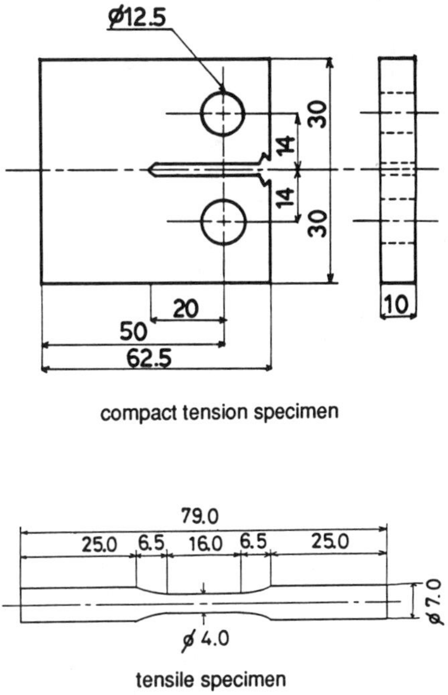

compact tension specimen

tensile specimen

Fig. 1 Configuration of specimens

RESULTS AND DISCUSSION

Effect of Bi-modal Microstructure on Strength-Fracture Toughness Balance

Fig.2 shows Vickers hardness variation with aging time and temperatures in STA. There is the tendency that the lower the aging temperature is, the slower the hardness rises and the higher the attained hardness is. Aging at 873K brought about little hardening.

Within the range of STA, the higher the aging temperature was, the coarser the size of alpha phase became and the more dominant the preferential alpha precipitation at grain boundary became. Hence, the lower the aging temperature was, the more homogeneous the aged microstructure became and the more the strength increased. The conditions of STA, STDA and STTA were chosen as follows. In STA, aging times that give complete hardening and allow half the value of complete hardening were chosen for each temperature. In STDA, the conditions of the first aging at higher temperatures were planned to allow adequate amount of alpha phase precipitated at a higher temperature and at the same time leave margin for further hardening by re-aging. 673K and 723K were chosen as re-aging temperatures to attain sufficient hardening. In STDA, time of aging at 873K was varied from 11ks to 100ks in order to investigate the effect of volume fraction of coarse alpha precipitates on fracture toughness. STTA was planned to get more homogeneous distribution of coarse alpha precipitates by adding pre-aging at 573K to STDA and to investigate the effects of nonuniformity of distribution of coarse alpha precipitates on fracture toughness-strength balance.

Fig. 3 shows hardness variation with re-aging at 673K in STDA. The specimens aged at 773K for 22ks in the first aging (D1) shows the highest initial hardness, attains complete hardening in 180ks and shows the highest value of hardness within the range of this experiment. In the series of specimens aged at 873K, the specimen aged at 873K for 11ks with the lowest initial hardness (D2) attains almost the same hardness level as that of the specimen aged at 773K and re-aged at 673K (D1). In the series of specimens aged at 873K (D2, D3 and D4), prolonged first aging reduces hardening response and attainable hardness in re-aging.

Fig. 4 shows the variation of alpha precipitates with agings. In Fig. 4 (a), coarse alpha precipitates and preferential precipitation of alpha phase at grain-boundaries are observed. Fig. 4 (b) shows microstructure after adding re-aging at 673K for 90ks to the microstructure shown in Fig. 4 (a). Fig. 4 (c) shows the high magnification aspect of the alpha precipitates shown in Fig. 4 (b). It is notable that fine alpha particles precipitate around coarse alpha precipitates produced in the first aging. Although the condition of aging at 873K was the same in both specimens of Fig. 4 (a) and (d), Fig. 4 (d) shows more homogeneous distribution of coarse alpha precipitates compared with that shown in Fig. 4 (a).

The variation of relations between ultimate tensile strength and reduction of area in these treatments is presented in Fig. 5. Prolonged aging in all aging conditions results in increase of strength and corresponding decrease of ductility. There is a tendency that aging at higher temperature produces better ductility when compared at the same

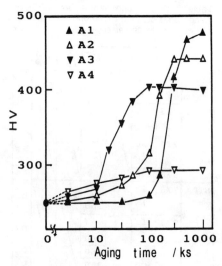

Fig. 2 Variation of hardness with aging time in STA. Aging conditions of A1,...,A4 are shown in Table 2.

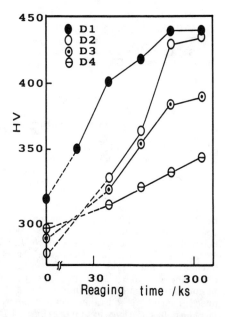

Fig. 3 Variation of hardness with re-aging time in STDA. Aging conditions of D1,...,D4 are shown in Table 2.

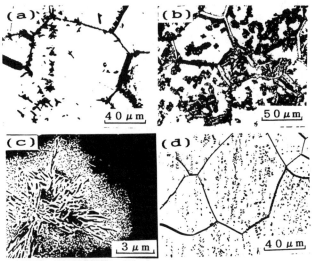

(a)STA:aging at 873K for 11ks
(b)STDA:re-aging at 673K for 90ks after aging at 873K for 11ks
(c)STDA:SEM micrographs of fine alpha precipitates around coarse
 alpha precipitates of the specimen shown in (b)
(d)Homogeneously distributed coarse alpha precipitates obtained by
 adding pre-aging at 573K for 36ks before aging at 873K for 11ks in
 STTA.

Fig. 4 Microstructures obtained by STA, STDA and STTA.

strength level in STA. Strength ductility balance obtained by STDA is close to that of STA of 773K in lower strength level and becomes close to that of STA of 723K and 673K in higher strength level.

Fig. 6 illustrates the effect of aging conditions on the relation between fracture toughness and ultimate tensile strength. All the data of fracture toughness except A4 are valid as plane strain fracture toughness according to E399 of ASTM. In STA (A1, A2 and A3), it is indicated that aging at higher temperature produces better fracture toughness and strength relationship, however, to get higher strength can be difficult. This figure also shows that D1 and D2 of STDA yields more preferable combination of fracture toughness and strength than STA (A1, A2 and A3) does. It is of great interest that adding aging at 873K or 773K prior to aging at 673K increases

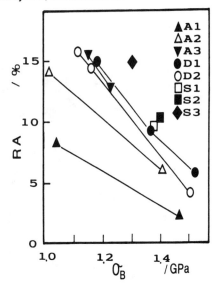

Fig. 5 Relation between ultimate tensile strength (σ_B) and reduction of area (RA) as influenced by the aging conditions shown in Table 2.

fracture toughness greatly in comparison with aging at 673K alone. The improvement of fracture toughness-strength balance would be owing to the bi-modal microstructure produced by the high-low step-aging (Fig. 4 (b) and (c)). However, the STTA specimen (S1) with the homogeneous distribution of coarse alpha precipitates (shown in Fig. 4 (d)) does not give better fracture toughness-strength balance compared with the result of STDA.

The effects of quantity of coarse alpha precipitates on the fracture toughness-strength balance are also indicated in this figure. With increase of volume fraction of coarse alpha phase i.e. increase of aging time in the first aging at 873K (D1, D3 and D4), fracture toughness increases greatly at the same strength level of about 1.2 GPa.

These results suggest that the high-low step-aging is very effective in getting better fracture toughness-strength balance.

Fig. 7 shows fracture path profiles parallel to crack propagation and fracture surfaces

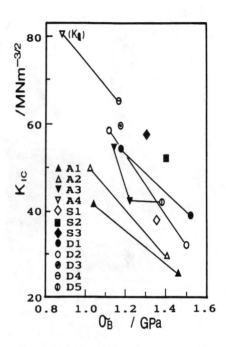

Fig. 6 Relation between ultimate tensile strength (σ_B) and plane strain fracture toughness (K_{IC}) as influenced by the aging conditions shown in Table 2.

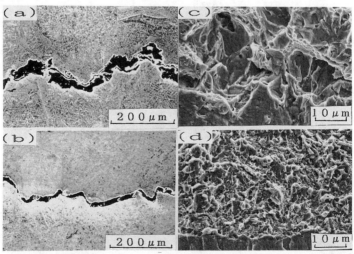

(a) and (c):STDA(D4)   (b) and (d):STA(A3 : t_1=100ks)

Fig. 7 Fracture path profiles ((a) and (b)) and fracture surfaces ((c) and (d)) ahead of fatigue precrack of compact tension specimens.

ahead of fatigue precrack of compact tension specimens. Specimens were unloaded after crack propagation occurred during fracture toughness testing and sectioned vertical to crack plane at the center of thickness of specimens. Although transgranular fracture path is predominant, crack profile of the D4 specimen of STDA (Fig. 7 (a)) is more tortuous than that of the A3 specimen of STA (Fig. 7 (b)). The two aging conditions produced nearly the same strength level, however, the STDA yielded much higher fracture toughness of 65.6 $MNm^{-3/2}$ than 42.5 $MNm^{-3/2}$ of STA as shown in Fig. 6. The tortuosity of the former was composed not only of steeper macroscopic turnabout of crack propagation but also of more rugged microscopic crack propagation on every step of turnabout. Both fracture surfaces are composed of dimples and present ductile aspects, however, the specimen of STDA (Fig. 7 (c)) has much larger and deeper dimples than that of STA (Fig 7 (d)). This means that larger plastic deformation took place on crack propagation from fatigue precrack in the STDA specimen than in the STA specimen.

Fig. 8 shows microcracks in the vicinity of the main crack that was presented in Fig. 7 (a). A large number of microvoids at coarse alpha phases precipitated in grain such as shown in Fig. 8 (a) were observed along the main crack. Fig. 8 (b) shows a microcrack through coarse alpha precipitates ranged along grain boundary.

Although it was reported that coarse alpha precipitates produced by aging at higher temperatures are detrimental to obtaining good combination of strength and ductility/fracture toughness[11], these results demonstrate that microcracks at coarse alpha precipitates produce extensive tortuosity of fracture path and ductile aspect of fracture surface as shown in Fig. 7, and consequently yield high fracture toughness. The fact that the specimen (S1) with homogeneous distribution of coarse alpha precipitates tends to reduce the superiority of the bi-modal microstructure in fracture toughness as shown in Fig. 6 would support the mechanism above.

The bi-modal microstructure of coarse and fine alpha precipitates in beta matrix that was produced by aging at a higher temperature followed by aging at a lower temperature yielded preferable combination of fracture toughness and strength.

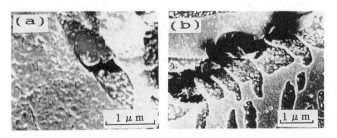

Fig. 8 Void (a) and microcrack (b) in coarse alpha precipitates of the specimen D4.

Application of High-Low Step-Aging on a TIG Weldment

TIG welding is useful for titanium alloys because a vacuum chamber is not necessary for welding, however, it is known that there are two problems. One is that there are different aging responses in a weldment. The other is low fracture toughness of a fusion zone.

The idea of the application of the high-low step-aging to solve these two problems is outlined in Table 3. During the first aging at a high temperature, in a base metal, a little amount of alpha phase precipitates,

Table 3. Outline of the application of the newly developed high-low two-step aging to a TIG weldment of Ti-15-3. In order to attain even strength between a fusion zone and a base metal and improve the fracture toughness of the fusion zone.

	First aging at high temperature	Second aging at low temperature	After high-low two-step aging
Strengthening	Fusion zone>Base metal	Fusion zone<Base metal	Fusion zone=Base metal
Size of alpha precipitates	Strengthening effect Coarse alpha ≪ Fine alpha		Bi-modal microstructure ↓↓
Amount of alpha precipitates	Fusion zone>Base metal	Fusion zone<Base metal	Improvement of fracture toughness

Table 4 Outline and names (AW1,...,DW6) of agings for TIG weldments after welding.

	T_1/K	t_1/ks	T_2/K	t_2/ks	
STA	723	36			AW1
		108			AW2
		360			AW3
STDA	873	10.8	723	108	DW1
		10.8		360	DW2
	923	10.8		108	DW3
		10.8		360	DW4
	873	3.6		360	DW5
	923	3.6		360	DW6

STA : Aging at T_1 for t_1
STDA: Re-aging at T_2 for t_2 after aging at T_1 for t_1

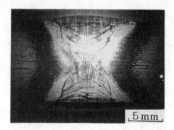

Fig. 9 Macroscopic aspect of a Ti-15-3 TIG weldment after aging at 873K for 3.6ks.

on the other hand, in a fusion zone, larger amount of alpha phase precipitates because there is aging enhancement in the fusion zone, however, increase of hardness is not so much because alpha precipitates are coarse. During the second aging at a low temperature, in the base metal, large hardening can be obtained and, in the fusion zone, aging response is suppressed because certain amount of coarse alpha phase already precipitated in the first aging as shown in Fig. 3, therefore, if we choose proper combination of aging conditions, difference of strength between a base metal and a fusion zone after aging can be reduced. Concerning microstructure, bi-modal microstructure is obtained by this heat-treatment, therefore, the improvement of fracture toughness of the fusion zone can be expected according to the result of the former part of this study. Aging conditions for TIG weldments are listed in Table 4.

Fig. 9 shows the macroscopic aspect of a Ti-15-3 TIG weldment after decoration.

Fig. 10 shows the schematic diagram of the weldment. On the basis of microstructure observation on the weldment as shown in Fig. 9, the weldment is divided into four zones where Z1 corresponds to a fusion zone and Z4 corresponds to a base metal.

Fig. 11 shows the variations of hardness with aging time of STA in a

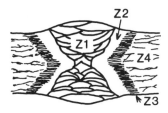

Fig. 10 Schematic diagram of four zones (Z1, Z2, Z3 and Z4) in the TIG weldment on the basis of the macroscopic observation of the weldment shown in Fig. 9.

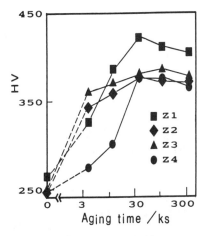

Fig. 11 Variation of hardness in the four zones of the weldment shown in Fig. 10 with aging time in STA at 773K.

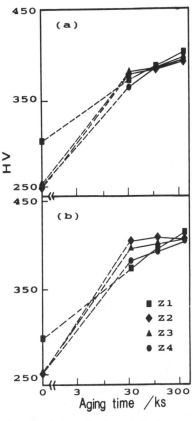

Fig. 12 Variation of hardness in the four zones of the weldment shown in Fig. 10 with aging time in re-aging at 723K of STDA. The specimens were aged at 873K for 10.8ks (a) and 923K for 10.8ks (b) before re-aging respectively.

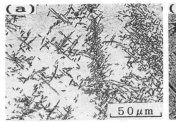
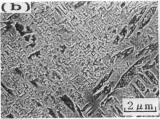

Fig. 13 Microstructures of the fusion zone of a weldment aged in STA and STDA.
(a)Coarse alpha precipitates produced by aging at 873K for 10.8ks.
(b)Bi-modal microstructure of coarse and fine alpha precipitates produced by re-aging at 723K for 360ks after aging at 873K for 10.8ks.

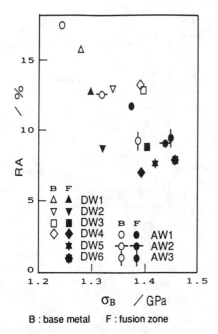

B : base metal   F : fusion zone

Fig. 14 Relation between ultimate tensile
strength (σB) and reduction of area (RA) as
influenced by the aging conditions shown in
Table 4.

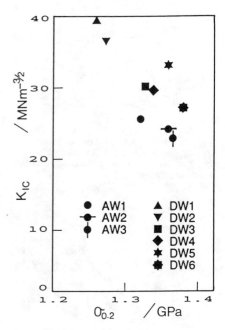

Fig. 15 Relation between 0.2% proof stress
(σ0.2) and plane strain fracture toughness
(KIC) as influenced by the aging conditions
shown in Table 4.

TIG weldment. There are differences of aging response and attained hardness
especially between a base metal and a fusion zone. It indicates that
strength between the fusion zone and the base metal is different in STA.

Fig. 12 shows the variation of hardness with re-aging time in STDA.
After the first aging, the fusion zone has a little higher hardness than
other areas, however, aging response in the second aging is suppressed. So,
after re-aging, hardness values consolidate rather well among 4 areas(Z1,
Z2, Z3 and Z4).

Fig. 13 shows the microstructures of the fusion zone of a weldment
after the first aging at 873K and the second aging at 723K in STDA. The aged
microstructure of the fusion zone of STA specimen shows nonuniform
distribution of coarse alpha precipitates (Fig. 13 (a)) and that of STDA
specimen is composed of coarse and fine alpha precipitates (Fig. 13 (b)).

Fig. 14 shows the relation between tensile strength and reduction of
area. In this figure, blank marks indicate the data of base metal, solid
marks indicate those of fusion zones, circles indicate the data obtained by
STA and others by STDA. Comparing strength in each aging condition, there is
big difference between the base metal and the fusion zone in STA, on the
other hand, strength level is almost the same in STDA. Thus, the first
problem of TIG welding of Ti-15-3 was improved by the high-low two step-
aging.

Fig. 15 shows the relation between tensile strength and fracture
toughness of the fusion zone. Circles indicate the results obtained by STA
and other marks indicate the data obtained by STDA. It is clear that STDA
gives higher fracture toughness compared with STA. The second problem of TIG
welding of Ti-15-3 was also improved by the high-low two step-aging.

CONCLUSIONS

A newly developed high-low step-aging was evaluated in comparison with an ordinary aging at a single temperature.

Bi-modal microstructure composed of coarse and fine alpha precipitates in beta matrix that is produced by the high-low step-aging improved the fracture toughness-strength balance of a Ti-15V-3Cr-3Sn-3Al alloy. The improvement is provided by the formation of microcracks and voids in the coarse alpha precipitates and rugged crack propagation due to the uneven microstructure.

The high-low step-aging applied to a TIG weldment of the alloy produced more even strength between fusion zone and base metal of the weldment and improved the fracture toughness of the fusion zone.

ACKNOWLEDGEMENTS

The authors are grateful for preparing the alloys to Nippon Mining & Metals Co.,Ltd., and Dr. C. Ouchi and Mr. H. Suenaga of NKK Corp., for TIG welding to Mr. H. Sato and Mr. S. Ishimoto of the Aerospace Division of Nissan Motor Co.,Ltd. and for the help of Mr. H. Yagishita, Mr. Y. Fukasawa, Mr. T. Sato, Mr. S. Oshiumi and Mr. K. Yano in the experimental work.

REFERENCES

1. H.W. Rosenberg, "Ti-15-3 : A New Cold-Formable Sheet Alloy," J.Metals,35(11), (1983)30-34
2. M. Okada, D. Banerjee and J.C. Williams, "Tensile Properties of Ti-15V-3Al-3Cr-3Sn Alloy," Titanium Science and Technology, ed. by G. Lutjering, U. Zwicker and W. Bunk, DGM(1985)1835-1842
3. P.J. Bania, G.A. Lenning and J.A. Hall, "Development and Properties of Ti-15V-3Cr-3Sn-3Al(Ti-15-3)," Beta Titanium Alloys in the 1980's, ed. by R.R. Boyer and H.W. Rosenberg, (1984)209-229
4. M. Okada and T. Nishikawa, "The Effect of Cold Rolling and Heat Treatment Conditions on Microstructures and Mechanical Properties of Ti-15V-3Cr-3Sn-3Al Alloy," J. Japan Inst. Metals, 50(1986),552-562
5. Y. Tsumori, T. Matsumoto and Y. Koyama, "Workability and Heat Treatment Characteristics of Beta Titanium Alloy, Ti-15V-3Cr-3Sn-3Al," Tetsu-to-Hagane, 72(1986),603-609
6. C. Ouchi, H. Suenaga and Y. Kohsaka, "Strengthening Mechanism of Ultra-High Strength Achieved by New Processing in Ti-15%V-3%Cr-3%Sn-3%Al Alloy," Proc. of Sixth World Conference on Titanium ed. by P. Lacombe, R. Tricot and G. Beranger, (1988)819-824
7. N. Niwa, T. Demura and K. Ito, "Effects of Chemical Composition on the Heat-treatment Response of Ti-15V-3Cr-3Sn-3Al Based Beta Titanium Alloys," ISIJ Int. 30(1990),773-779
8. N. Niwa et al., "Mechanical Properties of Cold-worked and High-Low Temperature Duplex-aged Ti-15V-3Cr-3Sn-3Al Alloy," ISIJ Int. 31(1991),856-862
9. N. Niwa, "Effects of Bi-modal Microstructure on Fracture Toughness of Ti-15V-3Cr-3Sn-3Al Alloy," Tetsu-to-Hagane, 78(1992),493-499
10. N. Niwa and T. Umeda, "Improvement of Mechanical Properties of Ti-15V-3Cr-3Sn-3Al TIG Weldment by High-Low Two-step Aging," Tetsu-to-Hagane, 78(1992),941-946
11. F.H. FROES and H.B. BOMBERGER, "The Beta Titanium Alloys," J.Metals,37(7), (1985)28-37

EFFECT OF SOLUTION-TREATMENT CONDITIONS ON AGING RESPONSE

IN Ti-15V-3Cr-3Sn-3Al

Hideki Fujii and Hirowo G. Suzuki

Stainless Steel & Titanium, Steel Research Labs.,
Nippon Steel Corporation
20-1 Shintomi Futtsu, 299-12 JAPAN

ABSTRACT

In a beta-type titanium alloy, Ti-15V-3Cr-3Sn-3Al, solution-treatment conditions have a significant influence on the age-hardening and the aged microstructures. In the specimens which are believed to contain a high density of quenched-in vacancies, such as the specimens quenched from high temperatures in the beta region or the ones having a low density of the vacancy sinks (grain boundaries, dislocation, etc.), the aging response is greatly accelerated and the precipitation is relatively homogeneous. In contrast, in the specimens which are believed to content a low density of excess vacancies, such as the specimens quenched from the temperature just above the beta transus, the aging response is delayed and the precipitation is inhomogeneous. The acceleration of the precipitation is probably due to the promotion of the nucleation sites formation by excess vacancies.

The phenomenon is also seen in the real products having thicker sections such as forgings, thick plates, etc., especially in the larger beta grains. The cooling rates also affect the concentration of quenched-in vacancies, resulting in the morphological difference between the surface area and the inside in the thick products. The phenomenon can be applied to the surface hardening method, in which the surface layer of the specimen having the balanced mechanical properties(HV350) can be hardened to HV470.

Beta Titanium Alloys in the 1990's
Edited by D. Eylon, R.R. Boyer and D.A. Koss
The Minerals, Metals & Materials Society, 1993

INTRODUCTION

Ti-15V-3Cr-3Sn-3Al is one of the most popular beta-type titanium alloys in which high strength can be obtained by the so-called solution-treatment and aging(STA)(1). Like other beta-type alloys, the solution-treatment is conducted usually at the temperature just above the beta transus. However, there have been some reports suggesting that the conditions of the solution-treatments influence the aging response in beta-type alloys(2)-(5). A systematic study on this issue has been conducted by the authors(6)(7); those results show that the aging response in Ti-15V-3Cr-3Sn-3Al is strongly affected by the solution-treatment conditions and the phenomenon is believed to be related to the concentration of quenched-in vacancies. Since the details of the phenomenon are described in the reference (6) and (7), some of the major experimental results and discussions are briefly shown here in the present report. After that, some practical examples related to the phenomenon and the applications are described.

EFFECT OF SOLUTION-TREATMENT CONDITIONS ON AGING RESPONSE

All of the beta-phase can be retained in beta-type alloys by quenching from temperatures in the beta-phase region. However, the concentration of pre-existing excess vacancies in the quenched specimens is extremely different depending on the solution-treatment conditions. Specifically the quenching temperatures (equilibrium concentration of vacancies), the density of the vacancy sinks (dislocations and grain boundaries), and the cooling rates after the solution treatment (annihilation time of vacancies) can be controlled by heat treatment. Some of the phenomena investigated by the authors using a hot-rolled Ti-15V-3Cr-3Sn-3Al plate of 10mm in thickness are shown below in this section.

Fig.1 shows the age-hardening behavior at the aging temperatures of $200 \sim 300\,°C$ after the solution-treatment at $800 \sim 1300\,°C$ for 30min and water quenching. Hardness of the as solution-treated specimens is equally HV255

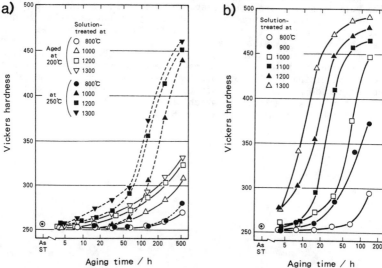

Figure 1   Age-hardening behavior at (a)200 and 250 °C, and (b)300 °C. Solution-treatment is conducted at 800 ~1300 °C for 30min followed by water quenching (1-step solution-treatment).

regardless of the solution-treatment temperatures. However, the age-hardening behaviors is quite different. Fig.1 shows that the higher solution-treatment temperature leads to the faster age-hardening. In the specimen solution-treated at 1300 °C for 30min and water quenched, which is referred as the "1300 °C 1-step solution-treated specimen" from now on, the age-hardening begins only in 16h at 200 °C(Fig.1-(a)). On the other hand, in the 800 °C 1-step solution-treated specimen, i.e. the specimen solution-treated at 800 °C which is just above the beta transus(760 °C), it takes 256h to start the age-hardening at the same aging temperature(Fig.1-(a)). This phenomenon can be more clearly seen at the aging temperature of 300 °C(Fig.1-(b)) and the precipitation is faster in the specimen solution-treated at the higher temperature(Fig.2), which corresponds to the age-hardening behavior.

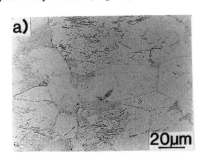 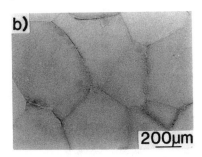

Figure 2   Microstructures of the specimens aged at 300 °C (a)for 128h after the 1-step solution-treatment at 800 °C, and (b)for 64h after the 1-step solution-treatment at 1200 °C.

The above observations can be explained by considering the differences in the initial concentrations of excess vacancies. The specimens solution-treated at higher temperatures should have a larger concentration of vacancies at the quenching temperatures as well as a smaller density of the vacancy sinks ; a lower density of the dislocations and the beta grain boundaries (larger beta grain size). Hence, a larger concentration of excess vacancies are retained upon subsequent aging. On the other hand, in the specimens solution-treated at lower temperatures, such as the 800 °C 1-step solution-treatment, almost all of excess vacancies migrate to the beta grain boundaries during quenching and annihilate, resulting in a smaller concentration of excess vacancies in the beta matrix.

Fig.3 shows a pattern of the "2-step solution-treatment" (1200 °C,30min, air cooling to 800 °C + 800 °C,30min,water quenching). It gives an example in which the beta grain boundaries act as the vacancy sinks. Fig.4 shows the microstructures of the as 2-step solution-treated specimen and the subsequently aged specimens at 300 °C. Inside the beta grains, a relatively rapid and homogeneous precipitation occurs. However, the precipitation process is delayed considerably near the grain boundaries, resulting in the denuded zones along the beta grain boundaries.

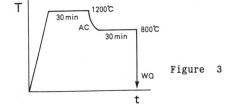

Figure 3   A pattern of the "2-step solution-treatment" (1200 °C,30min,air cooling to 800 °C + 800 °C,30min,water quenching).

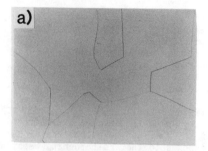

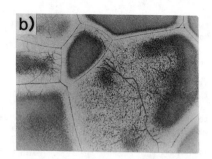

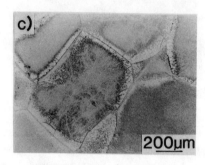

200µm

Figure 4   Microstructures of the 2-step solution-treated specimens. (a)As solution-treated, (b)aged at 300 °C for 32h, and (c)aged at 300 °C for 64h.

The difference between the 2-step solution-treated specimen and the 1200 °C 1-step solution-treated one (which shows the homogeneous and rapid precipitation in the extensive area in the specimen(Fig.2-(b))) indicates an effect of the quenching temperatures (800 and 1200 °C) on the aging response. Both specimens are considered to have the same density of the vacancy sinks, i.e. same grain size (about 500µm) and dislocation density. Although the grain boundaries must act as the vacancy sinks in both specimens, the equilibrium concentration of vacancies at the quenching temperature of 1200 °C is extremely high compared with 800 °C. Thus, a sufficient concentration of vacancies is present to suppress denuded zone formation in the vicinity of the beta grain boundaries in the 1200 °C 1-step solution-treated specimen. In contrast, the vacancy supply is insufficient to suppress the formation of the denuded zones in the 2-step solution-treated specimen.

Dislocations may also act as the vacancy sinks. Fig.5 is a pattern of the "3-step solution-treatment" (1200 °C,30min,water quenching + 500 °C,96h, air cooling + 800 °C,30min,water quenching). The grain size of the 3-step solution-treated specimen is about 500µm, which is almost equal to those of the 1200 °C 1-step and the 2-step solution-treated specimens. A feature of the 3-step solution-treated specimen is that it has a relatively high density of the dislocations, many of which form networks or arrays, although the density is not as high as that in the deformed specimens(6). As in the 800 °C 1-step solution-treated specimen, the 3-step solution-treated specimen shows the sluggish aging response as seen in Fig.6

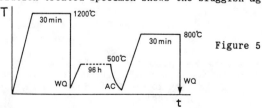

Figure 5   A pattern of the "3-step solution-treatment" (1200 °C,30min,water quenching + 500 °C, 96h, air cooling + 800 °C,30min, water quenching).

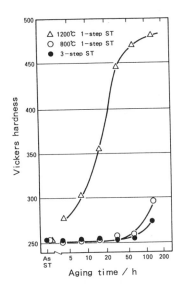

Figure 6   Age-hardening behavior at 300 °C of the 3-step solution-treated specimen (●), the 800 °C 1-step solution-treated one (○), and the 1200 °C 1-step solution-treated one (△).

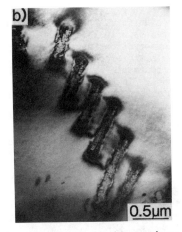

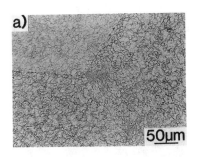

Figure 7   (a)Optical, and (b)TEM microstructures of the specimen aged at 300 °C for 128h after the 3-step solution-treatment.

Figure 8   An effect of the plastic deformation after the 1200 °C 1-step solution-treatment on the alpha precipitation (aged at 400 °C for 4h).

253

(aging temperature is 300 °C). This is believed to be caused by a decreased concentration of excess vacancies as a result of the dislocations acting as the vacancy sinks. It appears to be a result of the influence of dislocations as the vacancy sinks exceeding their effect as the precipitation sites(Fig.7).

In Fig.8, a microstructure of the specimen which is 1-step solution-treated at 1200 °C, plastically deformed by being struck on the surface with a punch, and aged at 400 °C for 4h. Near the surface and inside of the specimen, the precipitation is rapid because of high levels of plastic strain or quenched-in vacancies. However, the precipitation delays in the area between them, in which a small amount of plastic strain added, and near the isolatedly formed twins, because the dislocations and the twin boundaries act as the vacancy sinks rather than the precipitation sites.

As described above, the concentration of excess vacancies significantly influences the aging response. Then, why are the precipitation and the age hardening responses accelerated by quenched-in vacancies? The answer seems to be related to the aged microstructures. In specimens or areas in which a small amount of excess vacancies exist (such as in Fig.2-(a), the vicinity of the grain boundaries in Fig.4, Fig.7, and the intermediate area in Fig.8), the precipitation occurs heterogeneously on the dislocations or the grain boundaries. Meanwhile, in the specimens which are believed to contain a large concentration of excess vacancies (Fig.2-(b), the interior of the beta grains in Fig.4, the inside of the specimen in Fig.8), the precipitation is uniform. These results indicate that a large number of quenched-in vacancies have an strong influence on the precipitation sites formation.

TEM observations support our hypothesis. Fig.9 shows the omega precipitation in the specimen aged at 250 °C for 64h after the 1200 °C 1-step solution-treatment. It is known that the omega phase precipitates quite homogeneously in the beta matrix. However, in Fig.9, the omega phase precipitates in the form of aggregates although the aggregates are distributed relatively homogeneously in the beta matrix. It suggests that something promoting the precipitation is created here at the aggregates by the early stage of the aging. At the aging temperature of 300 °C, the equiaxed aggregates composed of the small alpha particles also precipitate relatively homogeneously in the beta matrix(Fig.10). In addition to that,

 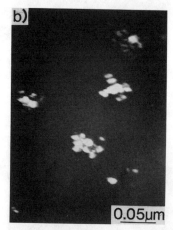

Figure 9   An early stage of the omega precipitation in the aged specimen at 250 °C for 64h after the 1-step solution-treatment at 1200 °C (dark field images).

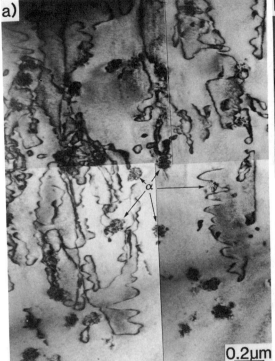

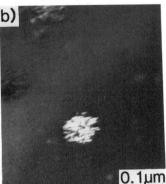

Figure 10 Aged microstructures at 300 °C for 2h after the 1-step solution-treatment at 1200 °C. (a)Relatively homogeneous precipitation of the alpha phase and dislocations with helical and loop configuration. (b)Dark field image of an aggregate of the small particles.

there are many dislocations with the specific configuration such as loops and herixes, which are probably introduced at the time of annihilation of excess vacancies. It is noticeable that the alpha precipitation is independent of those dislocations, which also strongly suggests that the nucleation sites are formed here at the very early stage of the aging. Since almost all of excess vacancies are probably annihilated at this moment, it is very difficult to think that the accelerated aging response is due to the acceleration of solute atoms by excess vacancies.

As described above, the accelerated aging response are considered to be due to the promotion of the precipitate nucleation by excess vacancies. However, the nature of the nucleation sites has not been clarified yet. One of the likeliest candidates as the nucleation sites is a complex cluster composed of the excess vacancies and the solute atoms. Interstitial impurities such as oxygen may also be a constituent of the complex cluster. The detection of that or other possible candidates of the nucleation sites and the elucidation of the relationship with the beta zone (beta prime) have to be examined in the future.

The accelerated age-hardening phenomenon by quenched-in vacancies is observed especially at the aging temperatures below 350 °C. It is not pronounced at the aging temperatures above 400 °C(Fig.11). That is believed to be related to the morphological features of the alpha phase (so called Type I and II). The detail discussion can be seen in the reference (7).

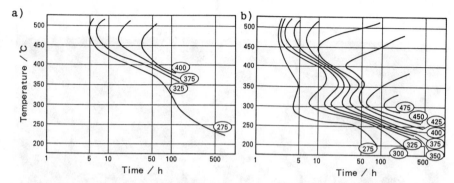

Figure 11 Change of Vickers hardness during the aging in the specimens which are 1-step solution-treated at (a)800°C and (b)1200°C. The data are derived from ref.(6).

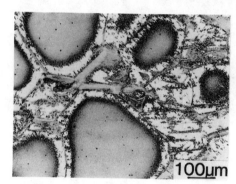
100μm

Figure 12 "Eyeball" structures in a forged block with 25mm in maximum thickness. The specimen is aged at 480°C for 10min after solution-treated at 800 °C for 60min followed by water quenching.

PRACTICAL EXAMPLES AND APPLICATIONS

The accelerated aging response described above also can be seen often in the real products. Fig.12 shows the microstructure of a forged Ti-15V-3Cr-3Sn-3Al block with 25mm in maximum thickness. The specimen is solution-treated at 800°C for 60min, water quenched, and aged at 480°C for 10min. Like the 2-step solution-treated specimen (Fig.4), a strange microstructure is observed which is composed of the homogeneously distributed alpha phase in the center of the beta grains and the denuded zones near the grain boundaries. Such a microstructure is frequently observed in the larger beta grains and is not often seen in the smaller grains. The formation mechanism of this type of microstructure, which is referred as an "eyeball", is similar to that of the 2-step solution-treated specimen. Namely, in the large grains, excess vacancies contained inside of the beta grains cannot reach the vacancy sinks (grain boundaries) because of the distance for vacancies to migrate, and consequently a large amount of excess vacancies are retained except near the grain boundaries.

Eyeballs are often observed not only in forged specimens but also in the materials having large grain size such as welded materials, cast materials, and thick plates. Fig.13-(a) is a microstructure of a hot-rolled thick plate of 10mm in thickness which is solution-treated at 800°C for 30min, water quenched, and aged at 480°C for 60min. Eyeballs are extensively formed near the surfaces, which indicates that the concentration of quenched-in vacancies is also affected by the cooling rates. In the thick

256

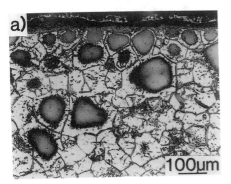
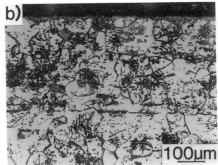

Figure 13   Microstructures near the surface of a thick plate of 10mm in thickness. The specimens are aged at 480 °C for 60min after solution-treated at 800 °C for 30min followed by (a)water quenching or (b)air cooling.

plates, the cooling rates during water quenching are considerably different depending on the distance from the surfaces, while the grain size and the quenching temperature are equal at any position in thickness. In locations far from the surfaces, the cooling rate is slow enough for the excess vacancies in the beta grains to reach the grain boundaries, resulting in the "normal" microstructure after the aging. On the other hand, if the excess vacancies within the beta grains cannot reach the grain boundaries, such as near the surfaces because of the higher cooling rate, the eyeball microstructure results.

The eyeball structure is considered to be unfavorable because plastic strain is probably accumulated preferentially in the soft denuded zones. It can be avoided by the cooling rate control, that is, the proper cooling rates which are low enough for all of excess vacancies in the beta grains to reach the grain boundaries but not so low that the alpha phase precipitates during cooling. An example of that is shown in Fig.13-(b), which is a microstructure of the specimen which is solution treated at 800 °C for 30min, air cooled, and aged at 480 °C for 60min. The eyeballs near the surface observed in the water quenched specimen (Fig.13-(a)) completely disappear by air cooling after the solution treatment.

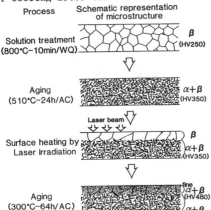

Figure 14   Schematic representation of the surface hardening method using the accelerated age-hardening phenomenon in Ti-15V-3Cr-3Sn -3Al.

257

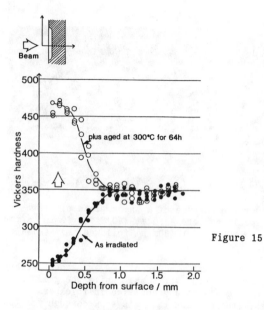

Figure 15 Depth profile of hardness in the as laser irradiated specimen (●) and the specimen after the final procedure (○).

Finally, an application of the accelerated age-hardening phenomenon to the surface hardening is briefly described(7). The concept of the surface hardening method is illustrated in Fig.14. First, a conventionally heat treated material(e.g. solution-treated at about 800 °C and aged around 500 °C for 8~32h) is prepared ; this has the excellent strength-ductility relationship. Second, only the surface layer is rapidly heated to high temperature in the beta region, followed by rapid cooling. This procedure can be attained, for example, by laser irradiation under the suitable irradiating conditions. Rapid heating is needed in order to minimize the heat transport to the base material. Heating to high temperature and rapid cooling are needed in order to retain a large number of quenched-in vacancies in the beta phase, which is formed by the inverse transformation. Finally, the material is aged again at 300~350 °C. During the second aging, the fine and relatively homogeneous precipitation of the alpha phase occurs in the surface layer in a relatively short time (32~64h), resulting in the hardened surface layer. This is due to the large concentration of quenched-in vacancies. Meanwhile, the base material is not affected because of the low aging temperature compared with the first aging temperature of about 500 °C(1).

The experimental results are shown in Fig.15. The material used in the experiment is a cold-rolled thin plate (5mm in thickness) with 1190MPa tensile strength, HV350 Vickers hardness, and 12% elongation after solution-treated at 800 °C for 10min, water quenched and aged at 510 °C for 24h. After removing the oxidized layer (4mm in thickness), CO_2 laser beam is used to irradiate the specimen under the following condition : output power - 5kW, beam size - 10mm x 10mm, scanning velocity - 33m/s. The second aging is conducted at 300 °C for 64h. As seen in Fig.15, the surface layer is initially softened by the inverse transformation after the irradiation while the second aging hardens it again to HV470. The base material hardness (HV350) is kept unchanged throughout the process. Although HV470 may not be enough in the industrial application, a harder surface may be obtained by optimizing the conditions of the heat treatments and irradiation. Moreover, the present experimental result suggests that there is some possibility that the accelerated age-hardening phenomenon might be applied to the real process in the future.

SUMMARY

The effects of the solution-treatment conditions on the subsequent aging response of Ti-15V-3Cr-3Sn-3Al is significant. The observed responses can be understood in terms of the concentration of quenched-in vacancies, which is a function of the quenching temperature, the density of vacancy sinks (dislocations, grain boundaries, twin boundaries, etc.), and the cooling rates. It is believed that excess vacancies play an important role in the nucleation site formation of the precipitates, although some metallurgical issues remain to be solved. Since the phenomenon affects the precipitate microstructures and the resulting mechanical properties, quantitative and kinematic analyses should be conducted in addition to the application developments like the surface hardening methods mentioned above.

REFERENCES

1. P.J.Bania, G.A.Lenning, and J.A.Hall, Beta Titanium Alloys in the 1980's, ed. by R.R.Boyer and H.W.Rosenberg (AIME, New york, 1984),209.

2. Chen Hai-Shan, Titanium'80 Science and Technology, ed.by H.Kimura and O. Izumi(AIME, New York, 1980),1591.

3. M.Fujita, Y.Kawabe, and H.Irie, Tetsu to Hagane (in Japanese),73(1987),S. 700.

4. Y.Shirosuna, A.Nozue, T.Okubo, K.Kuribayashi, R.Horiuchi, S.Ishimoto, and H.Satoh, Tetsu to Hagane(in Japanese),77(1991),1489.

5. T.Inaba, K.Ameyama, and M.Tokizane, J.Japan Inst. Metals, 54(1990),853.

6. H.Fujii and H.G.Suzuki, Mater.Trans.JIM, 34(1993),373.

7. Idem : Ibid,382

8. H.Fujii, I.Takayama, and H.G.Suzuki, Proc. of the 1st Japan International SAMPE Symposium and Exhibition, ed. by N.Igata, I.Kimpara, T.Kishi, E. Nakata, A.Okura, and T.Uryu(SAMPE,JAPAN,1989),93.

MODIFICATION OF ALPHA PHASE PRECIPITATION
BY COLD WORK OF THE Ti-15-3 ALLOY

*M. A. Imam, +P. K. Poulose, and *B. B. Rath

*Naval Research Laboratory
Washington, DC 20375-5320 USA

+University of the District of Columbia
Washington, DC USA

ABSTRACT

The purpose of the present investigation was to examine the mechanical property changes in the Ti-15-3 alloy resulting from different degrees of cold work (10% to 50% cold reduction) and heat treatments at various temperatures (480°C to 565°C), and to study the underlying microstructures in selected cases, so that method of developing suitable microstructures for structural applications can be established. The dislocation structures and vacancies produced by cold work results in faster and uniform nucleation and growth of precipitates. Higher strength can be obtained in a shorter time without appreciable loss of ductility. A cold work of 10% was found to be too small to produce uniformity in deformation.

Beta Titanium Alloys in the 1990's
Edited by D. Eylon, R.R. Boyer and D.A. Koss
The Minerals, Metals & Materials Society, 1993

INTRODUCTION

Beta titanium alloys have received considerable attention in aircraft industry because of their cold-formability and age hardening characteristics. The Ti-15-3 alloy is one of the most promising alloys in this group.

Several studies have been conducted to determine its cold formability [1-5] and age-hardening properties [2,6,7]. A few attempts have also been made to study the effect of cold work on age-hardening [3,8]. In beta titanium alloys beta phase is stable at high temperature, whereas at room temperature both alpha and beta phases co-exist. Because of the slow rate of transformation of beta to alpha on cooling, beta can be retained at room temperature as the meta stable phase even after slow cooling. Alpha is separated by aging at the higher end of the alpha-beta region. The alpha precipitated directly from beta at high aging temperature is heterogeneous, and the precipitation is mainly along grain boundaries and dislocations. Low temperature aging results in intense and homogeneous precipitation of intermediate phases, contributing to the very high strength of the alloy, while rendering it very brittle. In solute rich alloys the intermediate phase is beta prime (bcc), and in solute lean alloys it is omega (hcp) [1,9]. A low temperature aging followed by high temperature aging resulting dissolution of the intermediate phase in beta and in-situ precipitation of alpha. The alpha thus produced is more homogeneous than that produced by direct precipitation. Such duplex aging treatment is effective in reducing the aging time for required strength. Direct aging and duplex aging have been found to result in comparable tensile ductility at comparable yield and tensile strength values, while duplex aging appears to have better fracture toughness [6,10]. A high temperature- aging followed by low-temperature aging also yields similar results. However, the properties are strongly influenced by the processing history of the material, and hence, results from different studies show large differences in strength and ductility values.

As cold working increases the density of dislocations and vacancies it also enhances homogeneity of alpha precipitation. Hence, cold working and aging would be a viable alternative to duplex aging. Cold working in itself increases strength, and aging directly at high temperature avoids formation of the brittle phase. Some aspects of this approach have been investigated [6,11,12]. Preliminary results of a study on Ti-15-3 alloy by the authors have shown [8] that cold work not only reduces the tendency for formation of grain boundary alpha but also produces finer and homogeneous precipitation of alpha.

This paper presents additional results from the study of the effect of cold work on precipitation of alpha and tensile properties. The range of cold work has been extended and tensile test results after different treatments yielding moderate strength of the alloy are discussed.

Experimental Procedure

The material for this study was furnished by Timet Corporation in the form of 1.5 inch thick hot rolled plates. The chemical composition of the alloy in weight percent was: V-14.79, Al-2.91, Cr-3.06, Sn-3.03, Fe-0.128 and N-0.009. Heat treatments were performed on the material in (a) as received (hot rolled), (b) 10% cold worked, (c) 25% cold worked and (d) 50% cold worked conditions. Aging treatments were done is a salt bath furnace at 480°C, 540°C and 565°C. Progress of aging was determined by Vickers and Knoop hardness measurement. Compared to the earlier study where the knoop hardness was measured, the Vickers hardness results showed less scatter and higher reproducibility. In the earlier study where tensile tests were performed on cylindrical specimen, a large number of failures occurred at the threaded grips. Hence, in the present study, flat tensile specimens of gauge length 1 inch, width 0.25 inch, and thickness 0.125 inch were used. The tensile tests were confined to those specimen aged at 540°C as they were found to yield moderate strength and ductility, while those aged at 480C were brittle and those aged at 565°C were soft. The tests were performed on a hydraulic Instron machine. Light, scanning and transmission electron microscopic studies were also conducted. Thin films were produced by electrolytic thinning using a solution composed of 12.5% methanol, 31% butanol and 6.5% of 70% perchloric acid at ~14 volts and -40 to -50°C.

RESULTS AND DISCUSSION

The hardness curves for specimens aged at 480°C, 540°C and 565°C after different percentages of cold work are shown in Fig. 1A. Although the initial hardness is increased significantly by cold work, peak aged hardness shows only a small increase due to cold work. Cold work increased the age hardening rate, as seen by the earlier peak in cold worked conditions. The rate of hardening increased with cold work from 0 to 25%; the increase in rate due to additional deformation (50%) was not significant. In cold worked specimens aged at 540 and 565°C rapid hardening occurred in the first two hours of aging, and after reaching the peak, the drop was gradual.

In specimens tested after aging at 540°C, no specimen had an aging time less than 3 hours. This range was avoided due to the expected large scatter in the data. The results of the tension tests of specimens aged at 540°C are given in Table 1 and shown in Fig. 1B. The tension test results show a more significant increase in yield strength and tensile strength due to cold work for any given aging period than that shown by the hardness values. Cold work is expected to have two effects, increase in strength due to strain hardening and, in addition, that due to increased density of alpha nuclei formed. However, they do not seem to be additive. Although strength increase is seen in both cold worked and undeformed materials, an enhancement of strength due to additional nucleation is not seen in cold worked specimens. Actually increase in strength due to precipitation seem to be somewhat less in cold worked condition than in undeformed specimens, Table 2. The increase in strength is less pronounced in 50% cold worked than in 25% cold worked alloy. Nevertheless, increased cold work consistently showed higher strength before and after aging. Such differences were not observed in the hardness results. It shows that the hardness values are not sensitive enough to indicate small changes in tensile properties. When cold work increased from 25 to 50% the increase in strength in aged condition occurred without appreciable drop in ductility.

The lack of enhancement of nucleation due to cold work may be only apparent. All tensile tests on the cold worked specimens were done in overaged condition. The enhancement of hardening rate due to cold work is obvious from the hardness data as seen from the earlier peak in hardness. The more numerous and small particles that caused the earlier peak also coalesce faster than the larger particles in the undeformed condition. Hence overaging also can be expected to be faster in cold worked specimens than in undeformed specimens. It is also likely that deformation beyond a certain amount of cold work does substantially increase nucleation rate. Another factor that could contribute a lower increase in strength is recovery. Small amount of plastic deformation does not cause softening due to recovery, but the rate of recovery increases with increase in plastic deformation.

The tensile properties of unaged specimens is noteworthy. An increase in yield strength due to strain hardening is well recognized. A substantial change in tensile strength is unlikely due to strain hardening. Although the shape change in rolling is different from that in tension test, the mechanism of plastic deformation is essentially the same. The decrease in ductility due to cold work is in line with this expectation. Hence a large difference is tensile strength observed in the present study may be attributed to specimen thickness. The nominal thickness of all tensile specimens were the same. This is equivalent to having the cold worked specimens started at a higher initial thickness. The peak load where tensile strength is determined is reached when the effect of decrease in cross sectional area supersedes the effect of strain hardening. This occurs faster in a thinner specimen than in a thicker specimen. This effect will be more pronounced in alloys with low strain hardening coefficient (n), as is the case with the present alloy. A better assessment of strength can be made only by comparing true stress values.

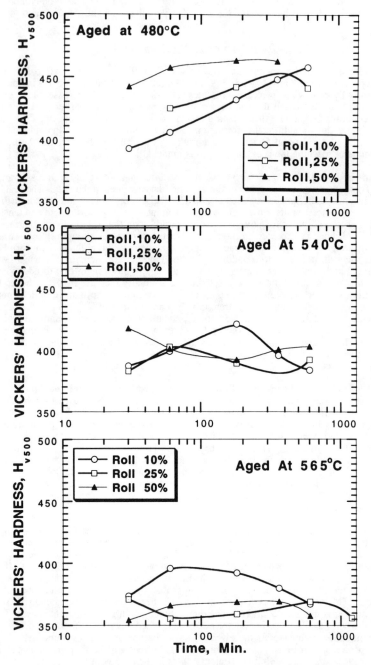

Figure 1A. Variation of microhardness with cold work and aging
time at temperatures 480°C, 540°C and 565°C.

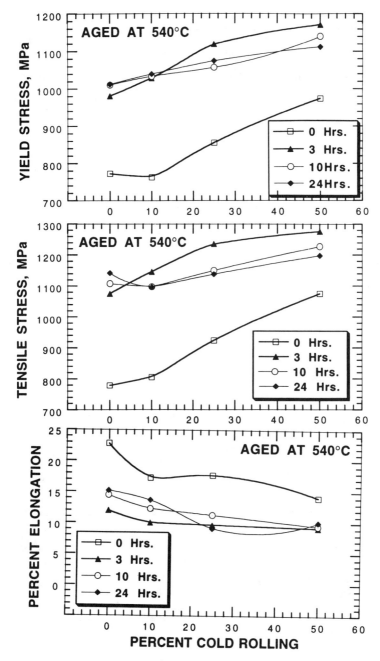

Figure 1B. Tensile test results showing yield stress, tensile stress and percentage elongation as a function of percent cold rolling.

Light micrographs of undeformed and cold worked specimens are shown in Fig. 2. The undeformed specimen reveals completely recrystallized condition after hot working. The microstructures also reveal the change of the grains from equiaxed shape to elongated shape with increasing cold work. The nonuniformity of deformation in the 10% cold worked condition is indicated by the microstructure. Persistent dark streaks were observed even after repeated polishing and etching (Fig. 2B), which is attributed to bands of deformed regions separated by undeformed regions. Irregular changes in strength and ductility observed in 10% cold worked specimens may be due to the nonuniformity of deformation during cold work. Such banded deformation was not observed in 25% and 50% cold worked specimens.

Transmission electron micrographs of aged specimens are shown in Fig. 3. In the undeformed specimens the precipitates are coarse, and the number of precipitates increases as aging time is increased from 3 to 24 hours (peak strength). In the cold worked specimens all the micrographs are of those in the near-peak or overaged condition. The size of the precipitates decreases and the density increases with increasing amount of cold work increasing the strength of the material. The grain boundaries in the cold worked specimens were blurred, and no preferred precipitation along the grain boundaries was observed. The dislocation structure produced by cold work was retained in all aged specimens, indicating absence of recrystallization. The nonuniformity of deformation in the 10% cold worked specimens is revealed in the electron micrograph as well (Fig. 3C). The deformation bands are subdivided into smaller bands seen in the electron micrograph. The strength and ductility results correspond to the microstructures observed.

The present study was a continuation of an earlier study [8] where the effect of cold work on subsequent precipitation characteristics and mechanical properties of Ti-15-3 alloy was investigated. The selection of a wider range of cold work and tensile tests after several aging periods enhanced the understanding of the structure-property relationships in this alloy in the intermediate strength levels. The dislocation structures and vacancies produced by cold work results in faster and uniform nucleation and growth of precipitates. Higher strength can be obtained in a shorter time without appreciable loss of ductility. A cold work of 10% was found to be too small to produce uniformity in deformation. Duplex aged alloys have shown higher fracture toughness than single-stage aged alloys at comparable strength values, even though improvement in ductility was not observed [6,7]. Possible similar effects in cold worked alloys are being investigated.

CONCLUSIONS

The strengthening of Ti-15-3 alloy by direct aging and after cold work was investigated. The results show that:

1. Cold working increases the rate of strengthening, and hence, the aging time for specified strength can be reduced by cold working.

2. Cold working produced consistently higher tensile strength and yield strength values in all aged condition at 540°C, without appreciable drop in ductility.

3. Hardness test results are not very sensitive to changes in tensile properties of Ti-15-3 alloy.

4. Cold working reduces tendency of formation of grain boundary precipitates by providing additional nucleation sites.

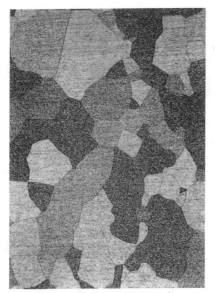

(A) Undeformed (as received)

(B) 10% C.R plus aged at
     540oC for 3 hrs.

___400 μm___

(C) 25% C.R plus aged at
     540oC for 3 hrs.

(D) 50% C.R plus aged at 540oC
     for 3 hrs.

Figure 2. Light micrographs of undeformed and cold worked Ti-15-3.

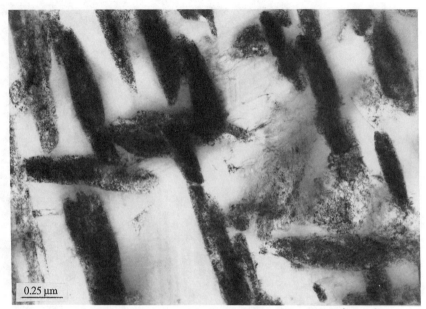

Figure 3(A). TEM of undeformed plus aged specimen (540°C/3 hrs) showing alpha precipitates.

Figure 3(B). TEM of undeformed plus aged specimen (540°C/24 hrs) showing larger number of precipitates as compared to Fig. 3A.

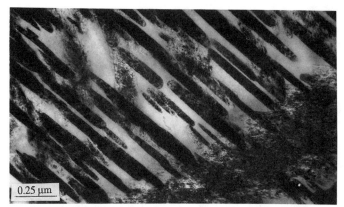

Figure 3(C).   TEM of 10% C·R plus aged (540oC/3 hrs)   Specimen showing alpha precipitates and deformation band.

Figure 3(D).   TEM of 25% C·R plus aged (540oC/3 hrs)   Specimen Showing alpha precipitate.

Figure 3(E).   TEM of 50% C·R plus aged (540oC/3 hrs)   Specimen showing alpha precipitate.

Table 1

Tension Tests Results (Aged at 540°C for Different Times)

Percent Cold Work	Aging Time, hrs.	YS, MPa	TS, MPa	Percent Elongation
0	0	773	779	22.8
10	0	763	806	17.3
25	0	856	924	17.7
50	0	974	1075	14.1
0	3	981	1075	12.0
10	3	1029	1146	10.0
25	3	1120	1236	9.7
50	3	1171	1276	9.2
0	10	1011	1108	14.5
10	10	1034	1098	12.3
25	10	1057	1150	11.3
50	10	1140	1228	9.5
0	24	1013	1142	15.2
10	24	1040	1097	13.7
25	24	1075	1138	9.1
50	24	1112	1198	10.1

Table 2

Increase in Strength Due to Aging

Percent Cold Work	From 0 to 3 hr		From 0 to 10 hr		From 0 to 24 hr	
	YS, MPa	TS, MPa	YS, MPa	TS, MPa	YS, MPa	TS, MPa
0	207	296	241	331	241	365
10	262	62	221	290	276	290
25	262	310	207	228	221	214
50	200	200	138	152	138	124

YS = Yield Strength
TS = Tensile Strength

ACKNOWLEDGMENT

The authors wish to thank Dr. J. Feng for assisting in TEM studies and Mr. K. Robinson for assisting in optical microscopy and microhardness studies.

REFERENCES

1. F. H. Froes and H. B. Bomberger, The Beta Titanium Alloys, Journal of Metals, (July 1985), 28.

2. P. J. Bania, G. A. Yolton and J. A. Hall, Development and Properties of Ti-15v-3Cr-3Sn-3Al (Ti-15-3), (Proc. Sym. on Beta Titanium Alloys in the 80's, AIME, Atlanta, Georgia, March 8, 1983), p. 209.

3. A. G. Hicks and H. W. Rosenberg, Ti-15-3 Foil Properties and Applications, ibid, p. 231.

4. M. E. Rosenblum, A. Shames and W. P. Treppel, Cold Forming of Ti-15v-3Cr-3Al-3Sn, ibid, p. 307.

5. A. E. Leach, Formed Ti-15v-3Cr-3al-3An Tankage, ibid, p. 331.

6. P. K. Poulose and H. Liebowitz, Improvement of Fracture Toughness in High Strength Beta Titanium Alloys, (Final Technical Report, NAVAIR Contract Number N00019-83-C-0177, 1985).

7. N. Niwa, K. Ito, H. Takatori and H. Sukayama, influence of Heat Treatment on Microstructure and Mechanical Properties of Ti-15-3 Alloy, (Proc. of the VI Internat. Conf. on Titanium, France, 1988), p. 1507.

8. M. A. Imam, P. K. Poulose and B. B. Rath, Effect of Cold Work and Heat Treatment in Alpha Region on Mechanical Properties of Ti-15-3 Alloy, (Proceedings of the Seventh World Conference on Titanium, San Diego, California, 1992) (to be published).

9. T. W. Duerig and J. C. Williams, Microstructure and Properties of Beta Titanium Alloys, (Proc. Sym. on Beta Titanium Alloys in the 80's, AIME, Atlanta, Georgia, March 8, 1983), 19.

10. M. Okada, Strengthening of Ti-15-3 by Two-Step Aging, ibid, 1625.

11. M. Okada, The Effect of Cold Work and Heat Treatment Conditions on Microstructure and Mechanical Properties of Ti-15V-3Al-3Sn-3Cr Alloy, ibid, 205.

12. O. Ouchi, H. Suenaga and Y. Kohsaka, Strengthening Mechanism of Ultra-High Strength Achieved by New Processing in Ti-15%V-3%Cr-3%Sn-3%Al Alloy, ibid, 819.

EFFECT OF HIP AND HEAT TREATMENT ON FATIGUE INITIATION AND

TENSILE FAILURE IN Ti-15V-3Cr-3Sn-3Al CASTINGS

W. J. Porter and D. Eylon

University of Dayton
Graduate Materials Engineering
300 College Park Drive
Dayton, OH 45469-0240, USA

ABSTRACT

Castings of the beta alloy, Ti-15V-3Cr-3Sn-3Al, as reported in three works, were compared and evaluated to determine the effects of hot isostatic pressing (HIP) and heat treatment (HT) on fatigue initiation and tensile fracture. The castings were HIP'd and either direct aged (DA) or solution treated and aged (STA) using various schedules to achieve different strength levels. Yield strengths ranging from 150 to 185 ksi (1035 to 1275 MPa) were obtained. Examination of some tensile fracture surfaces revealed transgranular fracture and grain boundary alpha dimpling in lower strength conditions and intergranular fracture in the higher strength conditions. High cycle fatigue test results were compared for the various HIP and heat treatment (HT) conditions. Fatigue initiation sites identified included surface and subsurface porosity, surface flaws such as machining marks and areas where thin, planar grain boundary alpha (GBα) phase was oriented 45° to the applied stress. The different HIP and HT conditions resulted in a wide range of tensile and fatigue properties. Lower cooling rates in the casting stage and thicker grain boundary alpha were associated with lower elongation.

Beta Titanium Alloys in the 1990's
Edited by D. Eylon, R.R. Boyer and D.A. Koss
The Minerals, Metals & Materials Society, 1993

INTRODUCTION

Titanium and its alloys have an excellent combination of properties including low density, good corrosion resistance, and high strengths at low and moderate temperatures. While these properties are attractive for a variety of engineering applications, the high cost of titanium and the subsequent processing required to achieve finished parts have limited its use for broad industrial application. Industries where a material's capabilities outweigh cost concerns, such as aerospace and defense, have traditionally accounted for the bulk of titanium usage.

In an effort to address the cost question, near-net shape manufacturing processes such as powder metallurgy, precision forging, superplastic forming and investment casting have been studied. Investment casting has thus far proven to be an excellent method for producing titanium alloy net-shape products and has seen considerable growth over the past fifteen years [1]. Ti-6Al-4V (Ti-6-4), an alpha+beta ($\alpha+\beta$) alloy, is the most widely used for investment casting. Following HIP and heat treatment, such castings consistently yield mechanical properties equivalent to or exceeding those of wrought Ti-6-4 [1]. The success of Ti-6-4 and other $\alpha+\beta$ alloy castings, such as Ti-6Al-2Sn-4Zr-2Mo, has led to adaptation to casting technology of other titanium alloys including beta alloys such as Ti-3Al-8V-6Cr-4Mo-4Zr (Beta C), Ti-10V-2Fe-3Al, and Ti-15V-3Cr-3Sn-3Al (Ti-15-3) [2].

Ti-15-3 was originally developed as a high strength, strip-producible, and cold-formable alloy for aerospace applications where formability, weldability and deep hardenability are important [3]. Based on the excellent properties of Ti-15-3 in its wrought form [YS=190 ksi (1310 MPa), El=9%], castings of Ti-15-3 were tested and found to provide good mechanical properties after HIP and STA. This paper will focus on comparing the results of three previously reported programs which evaluated Ti-15-3 alloy castings for tensile and fatigue properties [2,4,5]. The prominent mechanisms leading to fatigue crack initiation and tensile failure will also be addressed.

EXPERIMENTAL PROCEDURES

MATERIAL

The chemical compositions of the three casting sources evaluated in this work are shown in Table I. The geometries of the castings are outlined in Table II. The castings are identified by the organization responsible for the machining and heat treating of the samples (**LTV**- LTV Aerospace and Defense[4], **PCC**- Precision Castparts Corporation[2], **UD**- University of Dayton[5]).

Table I: Chemical Compositions of Ti-15V-3Cr-3Al-3Sn Castings (wt%)

Source ID	V	Cr	Al	Sn	Fe	Si	C	O	N	H	Ti
LTV[1]	15.30	3.16	3.04	3.04	0.11	0.03	NA	0.160	0.019	0.011	bal.
PCC[2]	15.40	3.00	2.80	2.90	0.12	NA	0.013	0.110	0.003	0.006	bal.
UD[3]	15.60	3.00	3.00	3.01	0.11	NA	0.009	0.125	0.013	0.003	bal.

[1]Ref 4, [2]Ref 2, [3]Ref 5

PROCESSING

The HIP and HT schedules used in the various programs are shown in Table II. A direct age treatment was utilized on the LTV castings [4]. In this case the β-solution treatment (β-ST) was accomplished during the 2 hour HIP cycle, which was done at a temperature approximately 300°F (150°C) higher than the beta-transus temperature (β_T) for Ti-15-3 ($\beta_T \approx 1400$°F (760°C) [6]).

MECHANICAL TESTING

All tensile tests were conducted in accordance with ASTM E8.

High cycle fatigue (HCF) testing, in all three works, was done at room temperature using an R-ratio (R=$\sigma_{min}/\sigma_{max}$) of 0.1. The smooth specimens (K_t=1) were tested according to ASTM E466. The frequencies used were 10 Hz for LTV, 60 Hz for PCC, and 5 Hz to 50,000 cycles followed by 10 Hz to failure for UD1 and UD2.

Table II: Processing Conditions and Tensile Properties

Source ID	Cast Geometry	HIP °F/ksi/hr (°C/MPa/hr)	DA or STA °F/hr (°C/hr)		UTS ksi (MPa)	YS ksi (MPa)	El. (%)	R.A. (%)
LTV 1	block-0.5" thick	1700/15/2 (930/105/2)	950/8	(510/8)	180 (1240)	170 (1170)	3.0	NA
LTV 2	block-0.5" thick	1700/15/2 (930/105/2)	900/16	(485/16)	196 (1350)	184 (1270)	3.0	NA
LTV 3	block-1.0" thick	1700/15/2 (930/105/2)	950/8	(510/8)	180 (1240)	170 (1170)	2.5	NA
LTV 4	block-1.0" thick	1700/15/2 (930/105/2)	900/16	(485/16)	190 (1310)	184 (1270)	2.0	NA
PCC	test bar[1]	1650/15/2 (900/105/2)	1550/0.5/AC+990/16/AC[2] (845/0.5/AC+530/16/AC[2])		163 (1125)	151 (1040)	6.9	13.4
UD 1	0.5" dia. bar	1650/15/2 (900/105/2)	1750/16/GFC[3]+975/12/AC (955/16/GFC[3]+525/12/AC)		181 (1250)	169 (1165)	3.7	9.3
UD 2	0.5" dia. bar	1750/15/2 (955/105/2)	1750/16/GFC[4]+975/8/AC (955/16/GFC[4]+525/8/AC)		174 (1200)	159 (1100)	6.6	15.0

[1]Dia. not available　　[2]AC (Air Cool)　　[3]GFC (Gas Fan Cool) =150°F (65°C)/min　　[4]GFC=350°F (195°C)/min

RESULTS AND DISCUSSION

MICROSTRUCTURE

Representative microstructures for HIP+STA conditions are shown in Figure 1. After HIP above the β_T or following β-ST, the microstructure consisted of recrystallized, equiaxed beta grains. Subsequent aging below the β_T (β_T - 410 to 500°F [210 to 260°C] in this case) resulted in precipitation of α particles within the beta grains and along grain boundaries (Figures 1b and d). Longer aging times result in larger amounts of α precipitation.

TENSILE PROPERTIES

Average room temperature tensile properties for each condition are shown in Table II. Tensile fracture surfaces for selected samples from conditions UD1 and UD2 are shown in Figure 2.

In the LTV series, yield strengths remained the same for material processed using like conditions, regardless of the difference in casting thickness. The tensile elongation, however, was affected by changes in casting thickness. Due to the differences in the thicknesses of the castings used in the LTV material, a small grain size effect is shown. The slower cooling rate and the larger grain size of the 1.0" thick block, when compared to 0.5" thick block, leads to a small decrease in ductility. This is also associated with a thicker grain boundary alpha (GBα) phase in the 1.0" material formed during cooling from the HIP ST. The relationship between decreasing tensile ductility with increasing GBα thickness has been explained elsewhere [5,7]. Although LTV 1 and 2 had only 3% ductility, LTV 3 and 4 had even lower values of 2 and 2.5%, respectively, representing a 17 to 33% drop. Again, grain size and thickness differences of GBα account for this occurrence[5,7].

Subtle processing differences during HIP and HT for UD1 and UD2 material are responsible for the difference in properties between the two groups. The 100°F (55°C) difference in HIP temperature between UD1 and UD2 was chosen to close all residual porosity. The post-HIP solution treatment temperature (1750°F (955°C)) chosen for each group probably canceled any HIP-temperature effects on microstructure. To address the effects of GBα on tensile failure, a faster gas fan cooling (GFC) was chosen for the UD2 material (150°F(85°C)/min for UD1 vs 350°F(195°C)/min for UD2). The faster cooling rate was designed to minimize GBα precipitation (i.e. thickness) upon cooling from the ST temperature. The most significant reason for the differences in tensile properties between UD1 and UD2 is related to the aging times. The 12 hour aging for UD1 material allowed for more precipitation of fine alpha and higher strengths, at the expense of ductility.

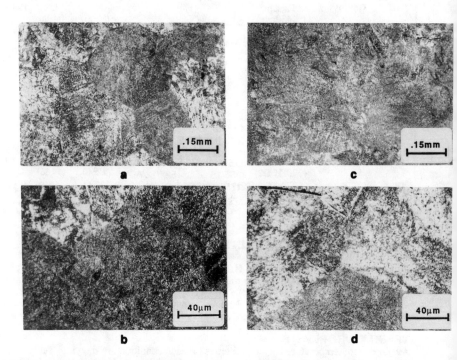

Figure 1: Photomicrographs of the microstructures found in: (a,b) UD1 and (c,d) UD2.

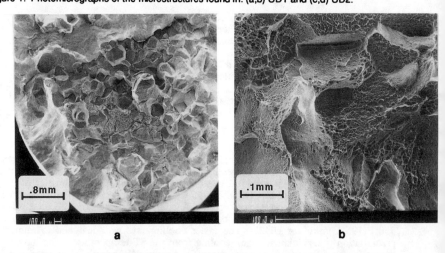

Figure 2: Photographs obtained by SEM of tensile fracture surfaces for conditions (a) UD1 and (b) UD2.

The tensile fracture surfaces of UD1 and UD2 are compared in Figure 2 and exhibit different modes of failure: low ductility, high strength UD1 samples show large amounts of intergranular fracture (Figure 2a) while UD2 tensile samples were mostly transgranular (Figure 2b). The difference in fracture mode can be attributed to the difference of relative strength between the aged beta grains and the surrounding GBα. Where a small mismatch in strength between the GBα and the α+β structure is found, plastic deformation is evenly distributed between the grain and the grain boundary. Tensile voids nucleate within the grains, resuting in transgranular tensile fracture. Intergranular fracture occurs when large localized plastic strains are distributed over the relatively small volume of the GBα material. Due to the planar nature of the GBα-phase and the corresponding long slip path found in the β-STA material, high stress concentration and local strain occur at the grain boundary triple points. The inherent strength of the aged beta grains results in strain localization in the relatively soft GBα. In these triple point regions, fracture occurs at low macroscopic strains even though the local strains which could be accommodated in the GBα are high as evidenced by the dimpled facets (Figure 2a) [5,7].

FATIGUE PROPERTIES

Individual S-N curves for all conditions are shown in Figures 3 a through c. Figure 3d compares the individual results with each other. Table III is a compilation of the HCF data and initiation site identification (where available) for all conditions. Representative examples of each type of initiation site are shown in Figure 4.

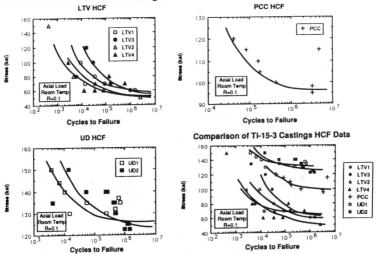

Figure 3: S-N curves for all conditions. (a) LTV [4]; (b) PCC [2]; and (c) UD [5]. 3(d) is a comparison of all S-N data.

The fatigue properties of the three groups of castings vary quite drastically (Figure 3d). Average fatigue ratios (runout fatigue stress/ UTS) of LTV, PCC and UD were 0.33, 0.60 and 0.70, respectively. The relatively poor fatigue strengths of the LTV material can be attributed to a number of factors. Table III indicates that many of the initiation sites of the LTV material were associated with residual porosity from the casting process. Figure 4a is an example of a pore-related initiation site in the LTV material. Pores ranging in size between 50-75 μm in diameter were associated with fatigue initiation. From this information, the castings either had significant numbers of surface connected pores that would not have been healed during the HIP process or the HIP process itself did not attain the pressures required to close non-surface connected pores. Also, many of the pores related to fatigue initiation were not spherical in shape and, as reported by the author, were planar in occurrence [4] and is an indication of an incomplete HIP. It should be noted that the HIP cycle cited for this group, 1700°F/15ksi/2hr (925°C/105MPa/2hr), is commonly used with Ti-15-3 castings and other alloy castings and is typically sufficient to close and heal all porosity.

Fractographic analysis of the UD series revealed two dominant sites for fatigue initiation. The most common site for both UD conditions was related to GBα. GBα–related features include long, planar and continuous GBα phase inclined about 45° to the tensile or load direction which coincides with the angle of maximum shear. An example of a GBα initiation site is shown in Figure 4c. GBα initiated failure has been seen in Ti-6Al-4V and in the beta alloy, Ti-8Mo-8V-2Fe-3Al [8,9].

Sub-surface microporosity was also identified as a common initiation site in the UD samples. These pores were commonly 10-15µm in diameter and surprisingly were associated with longer life tests (N>620,000 cycles). Also, the pores were found to be present on grain boundaries. However, these areas of GBα were usually normal to the applied loads and therefore not strong candidates for GBα initiation. It is possible that the samples exhibiting micropore-initiated failure had only thin or non-planar regions of GBα so that initiation was not possible at such locations and therefore shifted to the next weakest link in the chain - the areas of stress concentration surrounding the micropores. This indicates strong microstructural integrity within the other regions of the material, namely the α-precipitated β matrix [5].

Figure 5 compares the fatigue properties of all of the Ti-15-3 castings with those of Ti-6-4. While the fatigue ratios of the UD and PCC material (0.70 and 0.60, respectively) are nearly the same as those for Ti-6-4 (0.65 to 0.71), because of the higher tensile strength of Ti-15-3, the runout stress is considerably higher by as much as 30ksi (205 MPa). Ti-15-3 castings can therefore be used in high stress, fatigue sensitive applications.

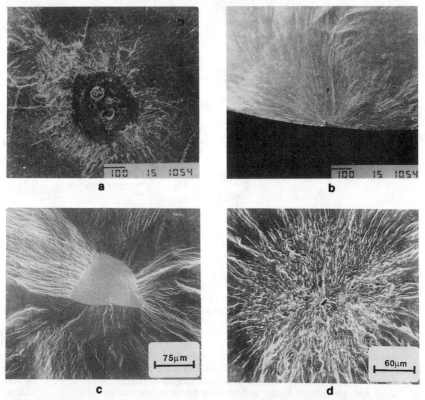

Figure 4: Typical fatigue crack initiation sites. (a) sub-surface porosity (LTV) [4]; (b) surface (LTV) [4]; (c) grain boundary alpha (GBα)(UD) [5]; and (d) sub-surface microporosity (UD) [5].

Table III: Fatigue Test Data for Ti-15-3 Castings

Condition	Maximum Stress ksi	(MPa)	Cycles to Failure	Initiation Site
LTV 1[1]	100	(690)	9,147	surface
	80	(550)	34,283	surface
	60	(415)	1,503,360	(runout)
	70	(485)	89,110	surface porosity
	70	(485)	14,057	casting defect
LTV 3[1]	120	(830)	15,665	surface
	100	(690)	29,548	surface
	60	(415)	1,001,541	(runout)
	80	(550)	6,144	surface
	70	(485)	159,852	sub-surface porosity
LTV 2[1]	70	(485)	36,797	surface
	50	(345)	2,485,232	(runout)
	150	(1035)	435	test malfunction
	80	(550)	4,660	sub-surface porosity
	60	(415)	80,553	sub-surface porosity
LTV 4[1]	80	(550)	131,175	surface
	100	(690)	2,673	surface porosity
	70	(485)	577,836	surface
	60	(415)	34,430	sub-surface porosity
	60	(415)	12,354	sub-surface porosity
PCC[2]	95	(655)	3,243,000	(runout)
	98	(675)	3,210,000	(runout)
	100	(690)	432,500	NA
	105	(725)	175,500	NA
	110	(760)	159,000	NA
	115	(795)	4,700,000	(runout)
	115	(795)	81,800	NA
	120	(830)	42,000	NA
	120	(830)	40,500	NA
UD1[3]	125	(860)	1,315,433	sub-surf. micropore
	130	(895)	242,750	GBα
	130	(895)	15,652	GBα
	132	(910)	478,074	GBα
	135	(930)	57,535	GBα
	135	(895)	430,477	grip
	135	(895)	697,160	sub-surf. micropore
	137	(945)	621,756	sub-surf. micropore
	140	(965)	10,502	GBα
	144	(995)	6,796	GBα
	150	(1035)	3,883	GBα
UD2[3]	123	(850)	984,237	GBα
	123	(850)	1,459,937	sub-surf. micropore
	125	(860)	1,234,985	sub-surf. micropore
	130	(895)	487,220	grip
	130	(895)	514,153	GBα
	135	(895)	4,196	GBα
	135	(895)	454,753	GBα
	140	(965)	55,430	GBα
	140	(965)	234,040	grip
	140	(965)	252,920	grip
	140	(965)	263,780	grip
	150	(1035)	14,380	GBα

[1]Ref 4   [2]Ref 2   [3]Ref 5

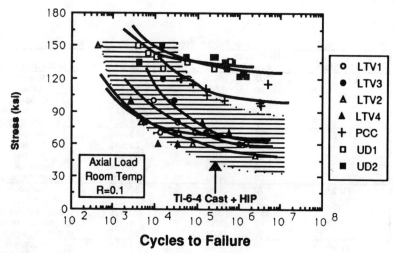

Cycles to Failure

Figure 5: Comparison of S-N data for castings of Ti-15-3 and Ti-6-4.

SUMMARY AND CONCLUSIONS

Three groups of Ti-15-3 investment castings (PCC, LTV, and UD) were compared to determine the effects of HIP and HT on tensile and fatigue properties. While the tensile properties were consistent, large differences in fatigue behavior were identified.

1. Ultimate tensile strengths ranging from 163 to 196 ksi (1125 to 1350 MPa) were obtained using various HIP and HT schedules.

2. Ductility ranged from 2.0% for high strength condition material (UTS=190 ksi (1310 MPa)) to 6.9% for lower strength material (UTS=163 ksi (1125 MPa)).

3. A drop in tensile elongations was found when comparing the 1" block castings (LTV3 and LTV4) to the 0.5" block castings (LTV1 and LTV2). This was due to the larger grain size found in the slower cooled 1" block material.

4. The large strength mismatch between the higher strength, finely precipitated α+β structure and the lower strength grain boundary alpha (GBα) phase in UD1 casting, resulted in localized ductile intergranular tensile fracture surfaces yet with low macrodeformation. The smaller strength mismatch in UD2 allowed for more transgranular fracture and therefore higher elongation.

5. Fatigue data for the three groups of castings was significantly different. The highest fatigue strength (UD) was 120 ksi (825 MPa), while the lowest (LTV) was 45 ksi (310 MPa). Average fatigue ratios for LTV, PCC and UD were 0.33, 0.60 and 0.70, respectively.

6. Some LTV specimens exhibited fatigue initiation sites related to residual casting porosity. The appearance of these pores suggests significant surface connections or a HIP cycle that may not have attained the pressures cited.

7. Fatigue crack initiation for UD1 and UD2 castings occurred at GBα phase and was sometimes associated with occasional microporosity. GBα fatigue crack initiation was due to the presence of a thin, continuous, planar GBα phase oriented at about 45° to the direction of

the applied stresses. GBα initiation resulted in samples with occasional short fatigue life for each condition.

8. Fatigue cracks which initiated at micropores in UD1 and UD2 samples were found only in specimens exhibiting longer than average fatigue lives. These were always associated with GBα regions normal to the tensile direction.

9. Ti-15-3 castings have tensile and fatigue properties that are the same as or much better than Ti-6-4 castings.

ACKNOWLEDGMENTS

Parts of this work were funded by the Boeing Commercial Airplane Group. Rod Boyer of Boeing is acknowledged for his many contributions to this program.

REFERENCES

1. D. Eylon, W. J. Barice, R. R. Boyer, L. S. Steele, and F. H. Froes, "Casting of High Strength Beta Titanium Alloys," Sixth World Conference on Titanium, Part II, P. Lacombe, R. Tricot, and G. Beranger, eds., Les Editions Physique, Les Ulis Cedex, France, 1989, pp. 655-660.

2. S. Soltez, "The Effect of Heat Treatment on the Mechanical Properties of Ti-3Al-8V-6Cr-4Mo-4Zr, Ti-15V-3Al-3Cr-3Sn and Ti-10V-2Fe-3Al Castings," (PCC Internal Report, March, 1987)

3. H. W. Rosenberg, "Ti-15-3 Property Data," Beta Titanium in the 1980's, R. R. Boyer and H. W. Rosenberg, eds., TMS-AIME, Warrendale, PA, 1983, pp. 409-432.

4. J. L. Petty-Galis, R. D. Goolsby, and L. M. Orsborn, "Investigation of Fatigue Behavior of Cast Ti-15V-3Al-3Cr-3Sn," Microstructure/ Property Relationships in Titanium Aluminides and Alloys, Y-W. Kim and R. R. Boyer, eds., TMS, Warrendale, Pa, 1991, pp. 563-578.

5. W. J. Porter, R. R. Boyer and D. Eylon, "Effects of Microstructure on the Mechanical Properties of Ti-15V-3Cr-3Al-3Sn Castings," to be published in the proceedings of the 7th World Conference on Titanium, San Diego, CA, 1992.

6. P. J. Bania, G. A. Lenning, and J. A. Hall, "Development and Properties of Ti-15V-3Cr-3Sn-3Al (Ti-15-3)," Beta Titanium Alloys in the 1980's, R. R. Boyer and H. W. Rosenberg, eds., TMS-AIME, Warrendale, PA, 1983, pp. 209-229.

7. G. T. Terlinde, T. W. Duerig, and J. C. Williams, "Microstructure, Tensile Deformation and Fracture in Aged Ti-10V-2Fe-3Al," Metallurgical Transactions A, Vol. 14A, October 1983, pp. 2101-2115.

8. D. Eylon, "Fatigue Crack Initiation in Hot Isostatically Pressed Ti-6Al-4V Castings," Journal of Materials Science, Vol. 14, 1979, pp. 1914-1922.

9. R. Chait and T. S. Desisto, "The Influence of Grain Size on the High Cycle Fatigue Crack Initiation of a Metastable Beta Ti Alloy," Metallurgical Transactions A, Vol. 8A, June 1977, pp. 1017-1020.

HIGH TEMPERATURE DEFORMATION OF THE NEAR BETA

Ti-15V-3Cr-3Sn-3Al ALLOY

R. Srinivasan and I. Weiss
Department of Mechanical and Materials Engineering
Wright State University, Dayton, Ohio 45435.

Abstract

High temperature deformation of Ti-15V-3Cr-3Sn-3Al was studied in order to better understand the effect of alloy composition, deformation conditions, grain size and grain shape on the hot workability of this alloy. Isothermal constant strain rate compression tests were carried out on as-cast and recrystallized material over a temperature range of 927°C to 1260°C (1700°F to 2300°F) at strain rates of 10^{-3}, 10^{-2}, and 10^{-1} s^{-1}. The as-cast material, with elongated grains with a size of about 25 mm, exhibited a stress peak upon yielding, followed by either flow softening, or strain hardening and then softening, depending on the deformation conditions. The deformed material was annealed to produced samples with smaller equiaxed recrystallized grains of size less than 4 mm. For the recrystallized material similar deformation behavior was observed at yielding, but constant flow stress deformation was observed at higher strains. The yield stress behavior is related to the initial mobile dislocation density and the interaction of these dislocations with atoms of the solute elements, while the flow softening is associated with substructure localization during deformation.

Beta Titanium Alloys in the 1990's
Edited by D. Eylon, R.R. Boyer and D.A. Koss
The Minerals, Metals & Materials Society, 1993

Introduction

Beta and near-beta titanium alloys are formable alloys that allow the fabrication of complex shapes with high strength upon heat treatment. The excellent formability, deep hardenability and good corrosion resistance have made these classes of alloys technologically very important structural materials. The good formability of these alloys in comparison to alpha and alpha+beta titanium alloys is the result of a room temperature microstructure consisting entirely or almost entirely of the beta phase with the bcc crystal structure. Further improvements in room temperature ductility of heat treated near beta alloy can be obtained by grain refinement. Since the hot working conditions during primary processing of the as-cast material play an important role in any attempt to produce mill annealed product with fine equiaxed grains, a study of the high temperature deformation of the near-beta alloy Ti-15V-3Al-3Cr-3Sn (Ti-15333) was undertaken. Though several previous studies on the high temperature deformation of Ti-15333 have been reported, most were carried out mainly to evaluate the superplastic deformation behavior of this alloy [1-4].

During high temperature deformation at constant strain rate of the Ti-15333 alloy a yield stress peak was observed, followed by constant stress deformation [5]. This type of behavior has also been observed in other near-beta titanium alloys such as Ti-8Mn, Ti-15Mo, Ti-13Cr-11V-3Al [6], Beta III, Ti-10V-2Fe-3Al, Ti-3Al-8V-6Cr-4Mo-4Zr [7], Ti-14.8V, and Ti-13Mn alloys [8,9]. Similar behavior has also been observed in beta Zr-Nb alloys [10]. In this work, the effects of temperature, strain rate, grain size and grain shape on the yield stress peak and flow stresses for deformations up to 0.7 true strain are presented, and discussed in terms of alloy chemistry, changes in mobile dislocation density, formation of substructure, and localization of deformation.

Experimental Procedure

Isothermal constant true strain rate compression tests were conducted on cylindrical specimens using a servohydraulic test machine. Tests were conducted in vacuum over a temperature range of 927°C to 1260°C (1700° to 2300°F) at strain rates of 10^{-3}, 10^{-2}, and $10^{-1}s^{-1}$. A graphite based lubricant was used between the compression platens and the specimens. Testing was carried out on large grain as-cast and fine grain recrystallized materials. The starting material was obtained from a 178mm (7 inch) diameter cast ingot that contained large grains, some of which were bigger than 25mm (1 inch) (Figure 1). The as-cast specimens, 50.8mm (2 inch) high and 33.8mm (1.33 inch) diameter, were deformed under the above mentioned conditions to a true strain of 0.7 (50% reduction in height). A second set of samples 19mm (0.75 inch) high and 12.7mm (0.5 inch) diameter were machined by EDM from the deformed first set of specimens. The samples were then recrystallized by vacuum annealing at 1093°C (2000°F) for 1 hour [11] to produce equiaxed grains of about 3 to 4 mm in size. These recrystallized specimens were then deformed under the same strain rate and temperature conditions given above. Figure 2 shows the starting microstructure of the as-cast and recrystallized materials.

Results and Discussion

The flow curves of the Ti-15333 alloy tested at strain rates of $10^{-3}s^{-1}$, $10^{-2}s^{-1}$, and $10^{-1}s^{-1}$, at temperatures between 927°C and 1260°C in both the as-cast and the recrystallized conditions are shown in Figures 3 to 6. Higher flow stress are observed at the higher strain rates and lower temperatures for both as-cast and recrystallized materials. In general, the flow curves display two distinct regions: a peak in the flow stress upon yielding, followed by gradual change in flow stress with strain. The magnitude of the peak stress decreases with increasing

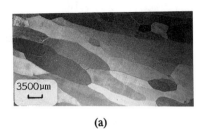

(a)

(b)

Figure 2. Starting microstructure of Ti-15333 compression specimens, (a) As-cast, and (b) Recrystallized materials.

Figure 1. Macrostructure of Ti-15333 cast ingot.

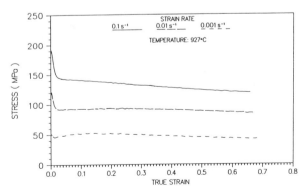

Figure 3.   Flow curves of as-cast Ti-15333 deformed at $\dot\epsilon = 10^{-3}$, 10^{-2}, 10^{-1} s^{-1} at 927°C.

285

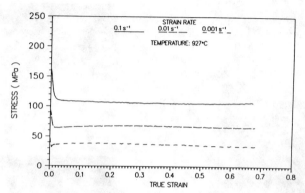

Figure 4.   Flow curves of recrystallized Ti-15333 deformed at $\dot{\epsilon} = 10^{-3}$, 10^{-2}, 10^{-1} s^{-1} at 927°C.

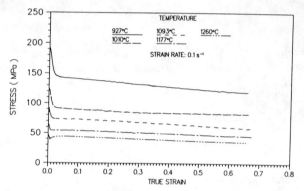

Figure 5.   Flow curves of as-cast Ti-15333 deformed at temperatures in the range of 927° to 1260°C, at $\dot{\epsilon} = 10^{-1}$ s^{-1}.

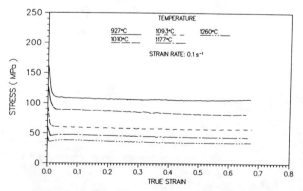

Figure 6.   Flow curves of recrystallized Ti-15333 deformed at temperatures in the range of 927° to 1260°C, at $\dot{\epsilon} = 10^{-1}$ s^{-1}.

286

strain rate. Some of the flow curves exhibit flow softening behavior following the peak stress. Under some deformation conditions, especially at slow strain rates and at high temperatures for the recrystallized material, there is strain hardening followed by a constant flow stress deformation. The variations of the peak stress with temperature and strain rate for the as-cast and the recrystallized conditions of the Ti-15333 alloy are shown in Figures 7 and 8, respectively. At the highest strain rate of $10^{-1}s^{-1}$ and at all temperatures, peaks in the flow stress upon yielding are observed. As the test temperature increased and the strain rate decreased, smaller peak stresses were observed. For the as-cast material yield stress peaks were absent above 1010°C at a strain rate of $10^{-3}s^{-1}$ and above 1093°C at a strain rate of $10^{-2}s^{-1}$. For the recrystallized material, yield stress peaks of about the same magnitude as those for the as-cast material were measured at low deformation temperatures. However, yield stress peaks were also found to exist at higher temperatures and slower strain rates. Thus, the effect of static recrystallization and the formation of smaller grains is to expand the range of test conditions under which yield stress peaks are observed to higher deformation temperatures, as indicated in Figures 7 and 8. The presence of peak stress in both materials indicates that the controlling mechanism of the yield stress peak is not grain size and grain shape dependent, but is associated with the chemical composition of the alloy, and the interaction of crystal defects and solute atoms. Figure 9 shows the flow curves of Ti-15V-3Cr-3Sn-3Al, Ti-6Al-4V and Ti-5Al-2.5Sn alloys deformed at the same temperature and strain rate conditions of 1093°C and $10^{-1}s^{-1}$ respectively. At this deformation temperature all these alloys are in the beta phase. The flow stresses increase with increasing solute content from 7.5% in Ti-5-2.5 to 24% in Ti-15333. The Ti-15333 alloy which is the only one that shows a yield stress peak, is also the only alloy containing Cr.

The flow stress of a material at an imposed strain rate $\dot{\epsilon}$ is related to the mobile dislocation density ρ by the following equation [12]:

$$\dot{\epsilon} = (1/2)\,\rho bv$$

where b is the Burger's vector, ρ is the mobile dislocation density, and v ($= (\sigma/\sigma_o)^n$)is the average velocity of dislocations. σ is the applied stress, σ_o and n are constants. At a constant strain rate $\dot{\epsilon}$, a lower flow stress σ will be obtained when a high density of mobile dislocations is present. A material which shows a rapid increase in mobile dislocations upon yielding will exhibit a yield stress peak. The peak stress (σ_p) for such a material can be expressed as:

$$\sigma_p = \sigma_s + \sigma_i$$

where σ_s is the stress required to operate dislocations sources and σ_i the friction stress needed to move the dislocations [13]. Figure 10 shows TEM micrographs for the as-cast material. Large numbers of dislocation loops can be observed. If the majority of these dislocations interact with solute atoms and are in an immobile configuration, the initial density of mobile dislocations will be low, and peak stress would be observed during deformation at all strain rates. It is also expected that the magnitude of the peak stress would be higher at lower deformation temperature and higher deformation rates. This is observed for the Ti-15333 alloy deformed at 927°C, as shown in Figures 7 and 8, where log σ_p is plotted vs. 1/T for both the as-cast and recrystallized materials. The apparent activation energy obtained from these plots of about 13 kcal/mol. should be related to the activation energy associated with the interaction of solute atoms with dislocations and the dislocation locking process. In an attempt to rationalize the interaction of solute atoms and dislocations, the interaction energy between solute

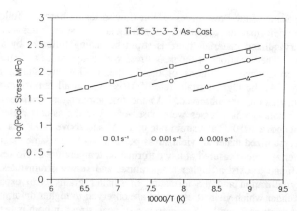

Figure 7.   Variation of peak stress with temperature for as-cast Ti-15333.

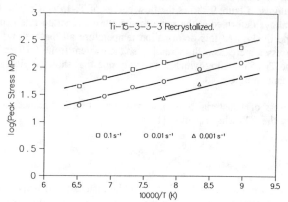

Figure 8.   Variation of peak stress with temperature for recrystallized Ti-15333.

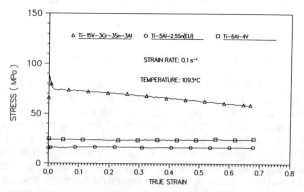

Figure 9.   Flow curves of as-cast Ti-15V-3Cr-3Sn-3Al, Ti-6Al-4V and Ti-5Al-2.5Sn deformed at T = 1093°C and $\dot{\epsilon}$= 10^{-1} s^{-1}.

288

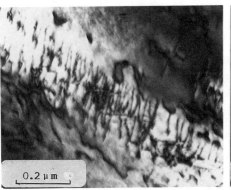 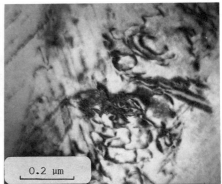

0.2 μm   0.2 μm

Figure 10.   TEM micrograph for as-cast Ti-15333 showing dislocation loops

and edge dislocations was estimated using the following relationship:

$$U_i = 4Gbr_{Ti}^3 \epsilon / R$$

where G is the shear modulus, b is the Burger's vector, $\epsilon = \Delta r / r_{Ti}$ and R is the distance of the solute atom from the dislocation [13]. Table I shows the atomic radii of the various elements found in beta and near beta alloys. Of all the elements listed, the atomic radii of Cr, Fe, and Mn are considerably different than that of titanium and those of the other solute elements. The interaction energies of the different solute atoms with edge dislocations were calculated assuming G = 45 GPa, and R ≈ b. For the Ti-15333 alloy, the interaction energy for Cr atoms is about 10.6 kcal/mol., which is higher than that for Al, V, and Sn atoms, and is approximately equal to the measured activation energy of 13 kcal/mol. from Figures 7 and 8. Based on this type of solute dislocations interaction, the variation in the peak stress with temperature and strain rate in Ti-15333 alloy can now be explained. The

TABLE I: Interaction Energy of Various Alloy-
            ing Element Atoms

	r* (nm)	Δr (nm)	U_i (kcal/mol)
Zr	0.160	+0.013	7.2
Sn	0.158	+0.011	6.1
Ti	0.147		
Al	0.143	−0.004	2.2
Mo	0.140	−0.007	3.9
V	0.136	−0.011	6.1
Cr	0.128	−0.019	10.6
Fe	0.124	−0.023	12.8
Mn	0.118	−0.029	16.2

*   W.F. Smith, Principles of Materials Science
     and Engineering, McGraw-Hill, 1991.

as-cast material contains dislocation loops which are mostly immobile. As the deformation temperature is increased, more dislocations are able to break away from their locked position, and a higher density of mobile dislocations is produced. This results in lower peak stresses. The increase in the mobile dislocation density with increase in temperature also affects the drop in flow stress after yielding, as defined in Figure 11, and shown in Figures 12 and 13, for the

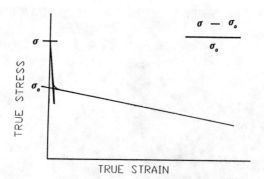

Figure 11. Definition of drop in stress after yielding.

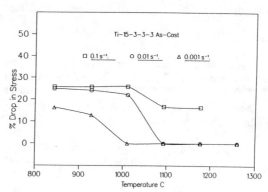

Figure 12. Drop in flow stress after the yield stress peak as a function of temperature and strain rate for as-cast Ti-15333.

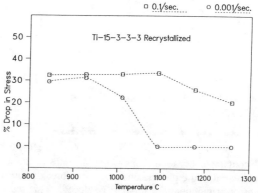

Figure 13. Drop in flow stress after the yield stress peak as a function of temperature and strain rate for recrystallized Ti-15333.

Figure 14.   Substructure in as-cast Ti-15333 deformed at 790°C to a true strain of 1.05.

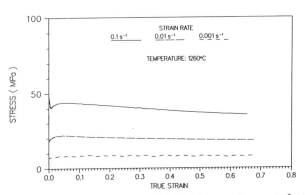

Figure 15.   Flow curves of as-cast Ti-15333 deformed at $\dot{\epsilon} = 10^{-3}, 10^{-2}, 10^{-1}$ s^{-1} at 1260°C.

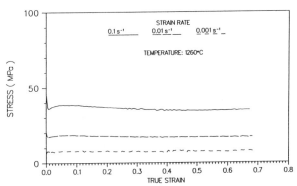

Figure 16.   Flow curves of recrystallized Ti-15333 deformed at $\dot{\epsilon} = 10^{-3}, 10^{-2}, 10^{-1}$ s^{-1} at 1260°C.

291

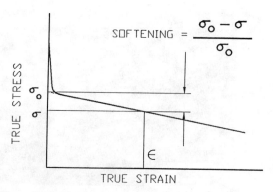

Figure 17.   Definition of flow softening for large strain deformation.

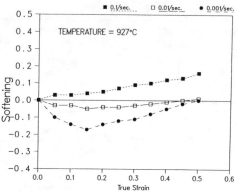

Figure 18.   Flow softening of as-cast Ti-15333 as a function of strain and strain rate at 927°C.

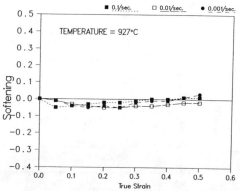

Figure 19.   Flow softening of recrystallized Ti-15333 as a function of strain and strain rate at 927°C.

292

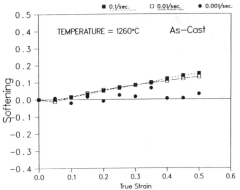

Figure 20.   Flow softening of as-cast Ti-15333 as a function of strain and strain rate at 1260°C.

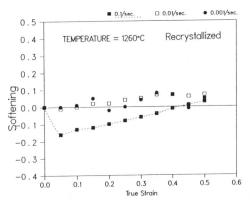

Figure 21.   Flow softening of recrystallized Ti-15333 as a function of strain and strain rate at 1260°C.

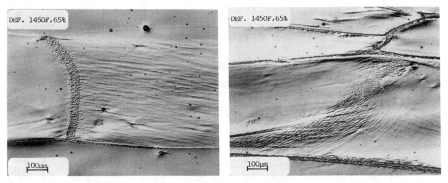

Figure 22.   Non-homogeneous substructure in the vicinity of a grain boundary in as-cast Ti-15333 deformed at 790°C to a true strain of 1.05.

293

cast and recrystallized material, respectively. The percent drop in yield stress decreases as the deformation temperature increases, i.e., as the mobile dislocation density increases. This behavior is analogous to the effect of prestaining observed by Johnston and Gilman [12]. Above the temperature of 1010°C for the as-cast material and 1093°C for the recrystallized material deformed at the strain rate of $10^{-3}s^{-1}$, a sufficient number of mobile dislocations exist and yield stress peaks are not observed. For deformation at the higher strain rate of $10^{-1}s^{-1}$ the density of mobile dislocation is still low, and yield stress peaks are observed even at the highest deformation temperature of 1260°C for the recrystallized material, as shown in Figure 13.

Figures 3 and 4 also show the effect of strain rate on the flow stress of as-cast and recrystallized materials, respectively, deformed at 927°C. For the as-cast material, following the yield stress peak, the flow stresses decrease as strain increases. This flow softening was most pronounced at the highest strain rate of $10^{-1}s^{-1}$ (Figure 3). Deformation at the lower strain rates of $10^{-2}s^{-1}$ and $10^{-1}s^{-1}$ for both as-cast and recrystallized materials results in flow curves typical of a material undergoing dynamic recovery, where the rate of hardening by generation of dislocations is balanced by the rate of softening due to dislocation annihilation, which results in constant flow stress deformation [14,15]. The microstructure produced during dynamic recovery consists of deformed grains containing substructure, as shown in Figure 14 for Ti-15333 deformed at 790°C. Flow curves at the highest deformation temperature of 1260°C (Figures 15 and 16) show that both the as-cast and the recrystallized materials, after the yield stress peak, show an initial period of work hardening followed by flow softening, especially at the highest strain rate of $10^{-1}s^{-1}$.

The percent softening as defined in Figure 17 is shown in Figures 18 to 21 for the different test conditions. The as-cast material deformed at strain rate of $10^{-1}s^{-1}$ at 927°C shows extensive flow softening as a result of substructure localization during deformation. Under these conditions a finer and denser dislocation structure is developed mainly along deformed grain boundaries, and a coarser and less dense substructure is produced in the deformed grain interior, as shown in Figure 22 for Ti-15333 alloy deformed at 790°C. The as-cast material deformed at the lower strain rate of $10^{-3}s^{-1}$ at 927°C, tend to initially work harden followed by slight flow softening until constant flow stress deformation occurs. The recrystallized material containing finer equiaxed grains deform more uniformly at 927°C for all strain rates, as shown in Figure 18. Similar trends were observed for as-cast and recrystallized samples deformed at 1260°C. In general, the as-cast material deforms more uniformly at the higher temperatures and slower strain rates, as demonstrated in Figure 19. The initial work hardening in both the as-cast and recrystallized materials at 1260°C and at the strain rate of $10^{-1}s^{-1}$ (Figures 13 and 14) is probably due to a high density of mobile dislocations generated during yielding which causes the stress to fall to very low levels. The interaction among these dislocations causes work hardening before steady state deformation occurs.

Summary

1.   Ti-15V-3Cr-3Sn-3Al exhibits a yield stress peak during elevated temperature compression. The magnitude of the stress peak decreases with increasing temperature and decreasing strain rate.

2.   The magnitude of the peak stress does not change between as-cast large grained material and recrystallized grain material at low temperature, but stress peaks are observed at higher temperatures for the recrystallized material.

3.   The variations of the magnitudes of the peak stress as a function of strain rate and temperature can be rationalized in terms of the composition of the material and density of mobile dislocations.

4. Flow localization is observed in as-cast material deformed at a strain rate of $10^{-1}s^{-1}$ as a result of substructure localization along deformed grain boundaries.
5. Constant flow stress deformation is observed in recrystallized material under most deformation conditions.
6. Work hardening following the yield stress peak is observed during deformation of as-cast and recrystallized materials at 927°C for $\dot{\varepsilon} = 10^{-3}s^{-1}$ and at 1260°C for $\dot{\varepsilon} = 10^{-1}s^{-1}$.

References

1. N.E. Paton and C.H. Hamilton, Titanium '84, Science and Technology, G. Lutjering, U. Zwicker and W. Bunk, Eds, TMS-AIME, Warrendale, PA, 1984, p. 649.

2. A.K. Ghosh and C.H. Hamilton, Net Shape Technology in Aerospace Structures, Vol. III, National Academy Press, Washington, D.C., 1986, p. 365.

3. C.H. Hamilton, Proc. of the Conf. on Superplastic Forming, S.P. Agrawal, Ed., ASM, Metals Park, Ohio, 1985, p. 122.

4. S.M.L. Sastry, P.S. Pao and K.K. Sankaran, Titanium '80, Science and Technology, H. Kimura and O. Izumi, Eds. TMS-AIME, Warrendale, PA, 1980, p. 874.

5. R. Srinivasan, Scripta Metall. et Materialia, Vol. 27, 1992, p. 935.

6. P. Griffith and C. Hammond, Acta Metall., Vol. 2, 1972, p. 935.

7. G.C. Morgan and C. Hammond, Titanium '84, Science and Technology, G. Lutjering, U. Zwicker and W. Bunk, Eds, TMS-AIME, Warrendale, PA, 1984, p. 717.

8. S. Ankem, J.G. Shyue, M.N. Vijayshankar and R.J. Arsenault, Materials Science and Engineering, April 1989, p. 51.

9. M.N. Vijayshankar and S. Ankem, Recrystallization '90, T. Chandra, Ed, TMS, Warrendale, PA, 1990, p. 673.

10. J.J. Jonas, B. Heritier and M.J. Luton, Metall. Trans., A. Vol. 10, 1979, p. 611.

11. I. Weiss, R. Srinivasan, and F.H. Froes, Recrystallization '90, T. Chandra, Ed, TMS, Warrendale, PA, 1990, p. 609.

12. W.G. Johnston and J.J. Gilman, J. Appl. Phys., Vol. 20, No. 2, 1959, p. 129.

13. G.E. Dieter, Mechanical Metallurgy, McGraw Hill, 1986.

14. D.L. Bourell and H.J. McQueen, J. Appl. Metal Working, 1987, p. 15.

15. H.J. McQueen and J.J. Jonas, Treatise in Materials Science and Technology, Plastic Deformation of Materials, R.J. Arsenault, Ed., Academic Press, New York, NY, Vol. 6, 1975, p. 393.

FATIGUE AND FRACTURE TOUGHNESS PROPERTIES

IN THE BETA-RICH α+β TITANIUM ALLOY SP-700

T. Fujita, M. Ishikawa, S. Hashimoto, K. Minakawa, and C. Ouchi

Materials and Processing Research Center
NKK Corporation
Kawasaki, JAPAN

ABSTRACT

Fatigue and fracture toughness of SP-700, a new β-rich α+β titanium alloy, were investigated. In annealed conditions, excellent fatigue strength and fracture toughness were obtained. In solution treatment and aging conditions, a wide range of tensile properties was provided by selecting aging temperatures. The fatigue strength in this condition was greatly improved with an increase in tensile strength. It is noted that the observed improvement in fatigue and tensile strength was achieved with a reasonable level of fracture toughness reduction. Factors and mechanisms contributing to the observed fatigue and fracture toughness properties are discussed in terms of microstructural features and deformation characteristics.

Beta Titanium Alloys in the 1990's
Edited by D. Eylon, R.R. Boyer and D.A. Koss
The Minerals, Metals & Materials Society, 1993

INTRODUCTION

The newly developed β-rich α+β titanium alloy SP-700 is designed to have improved hot and cold formability over Ti-6Al-4V. The nominal chemical composition is Ti-4.5Al-3V-2Fe-2Mo (wt %) and the β-transus temperature (T_β) is about 900°C. This alloy has several interesting features such as extremely fine microstructure, excellent strength-ductility balance, deep hardenability and most significantly, low temperature superplastic formability at temperatures below 800°C. The microstructural characteristics and tensile properties of SP-700 were reported elsewhere [1, 2].

In addition to tensile properties, fatigue and fracture toughness are an integral part of the design properties for structural materials. For titanium alloys, as seen in other metallic materials, an increase in tensile strength tends to increase the high cycle fatigue strength [3] but to decrease fracture toughness [4]. In recent years, a number of studies have been carried out in order to understand mechanisms and factors responsible for fatigue and fracture toughness of titanium alloys. It is revealed from the previous studies that both the fatigue and fracture toughness properties of titanium alloys are largely controlled by chemical composition, environment and microstructural features [4, 5]. Among the microstructural features, grain size appears to be important for both these properties [6, 7]. Since SP-700 consists of a very fine microstructure, particularly primary α grains, of the order of 2-3 μm, the fatigue and fracture toughness properties of SP-700 are of great interest.

The objective of the present work is to study microstructure-property relationships for the fatigue and fracture toughness of SP-700. Microstructural factors and fracture mechanisms are discussed.

FATIGUE PROPERTIES

Experimental Procedure

The materials were 25 mm diameter bars of SP-700 and Ti-6Al-4V. Ti-6Al-4V was included for a comparison purpose. The oxygen level was 840 ppm in SP-700 and 1410 ppm in Ti-6Al-4V. Various heat treatments listed in Table I were applied to samples prepared from both the materials. Optical micrographs of annealed and solution treated and aged microstructures for both the alloys are shown in Fig. 1. Recrystallization annealed SP-700 shows a very fine microstructure, 60% of which consists of equiaxed primary α grains of about 2 μm size, and 40% retained β phase. Duplex annealed Ti-6Al-4V reveals a much coarser microstructure; the α grain size is much greater than SP-700 grains and the prior β phase mostly transforms to acicular α. Solution treated and aged SP-700 shows much finer primary α and transformed β structure than Ti-6Al-4V. The β phase of SP-700 retains nearly 50% volume fraction and is partially transformed to α″ martensite upon water quenching from the solution temperature. Subsequent aging causes the decomposition of the retained β and α″ martensite resulting in a fine dispersion of α precipitates in the β phase and, as a result, providing very high strength.

Table I Heat treatment conditions

Materials	Symbol	Condition	Heat Treatment
SP-700	MA	Mill annealed	720°C/1h/AC
	RA	Recrystallization annealed	800°C/1h/AC
	STA	Solution treated and aged	825°C/1h/WQ + 510°C/6h/AC
Ti-6Al-4V	MA	Mill annealed	704°C/1h/AC
	DA	Duplex annealed	950°C/1h/AC + 704°C/1h/AC
	STA	Solution treated and aged	950°C/1h/WQ + 538°C/6h/AC

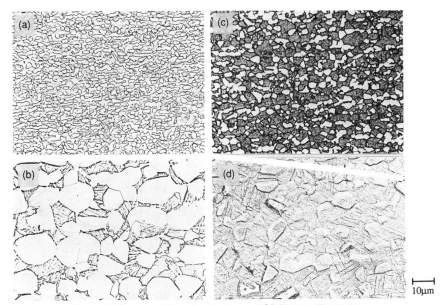

10μm

Figure 1 - Representative microstructures of SP-700 and Ti-6Al-4V.
(a) SP-700 (RA). (b) Ti-6Al-4V (DA). (c) SP-700 (STA).
(d) Ti-6Al-4V (STA).

Tensile testing was conducted on round bar specimens with a 6.25 mm diameter and a 25 mm gage length at a strain rate of 0.006 mm/mm/min at room temperature in accordance with ASTM E8. The resultant tensile properties of both the alloys for all the heat treatment conditions are summarized in Table II. It is noted that annealed SP-700 has excellent strength-ductility combinations; higher strengths and higher ductilities than annealed Ti-6Al-4V.

Axial fatigue testing was performed on round bar specimens with a diameter of 7 mm at room temperature according to ASTM E466. Specimens were carefully machined and polished so as to avoid residual stresses and circumferential tool marks. Specimens were cycled under load control in a servohydraulic testing system. The loading wave form was sinusoidal and all tests were done with a frequency of 10 Hz at a stress ratio of R = 0.

Table II Tensile properties

Materials	Heat Treatment	0.2%PS	TS	EL	RA
		MPa	MPa	%	%
SP-700	MA	1013	1056	20	59
	RA	942	961	22	66
	STA	1360	1495	8	23
Ti-6Al-4V	MA	932	1015	19	40
	DA	846	947	18	37
	STA	1107	1202	11	32

299

Results

Maximum stress versus cycles to failure (S-N) curves for all the heat treated materials are shown in Fig. 2. SP-700 shows superior fatigue behavior to Ti-6Al-4V; the 10^7 cycles fatigue strength of SP-700 is higher than that of Ti-6Al-4V for both the annealed and solution treated and aged conditions. The solution treated and aged SP-700 shows a maximum fatigue strength of 1000 MPa. It is interesting to note that the mill annealed SP-700 exhibits higher fatigue strength than the solution treated and aged Ti-6Al-4V. Figure 3 is the relationship between tensile strength and high cycle (10^7) fatigue strength for both the alloys. The fatigue strength tends to increase with tensile strength. The fatigue to tensile strength ratio of the annealed SP-700 exceeds 80%, about 10% higher than Ti-6Al-4V. For the solution treated and aged SP-700, however, this ratio drops to 70%, the same percentage as Ti-6Al-4V.

A marked difference in the S-N curve shapes can be seen between both the alloys. There is a sudden transition from low cycle to high cycle for SP-700; if the life exceeds 10^5 cycles, the specimen does not break until over 10^7 cycles. However, the fatigue lives of Ti-6Al-4V moderately decrease with a decrease in applied stress; several specimens broke at 10^5 or 10^6 cycle levels, which was not the case for SP-700. It was found from fracture surface observation that all the fatigue cracks initiated at the surface for SP-700 while subsurface cracks nucleated above 10^6 cycles for Ti-6Al-4V in all the heat treatment conditions. This phenomenon is common for Ti-6Al-4V when cyclically stressed at a positive stress ratio (tension-tension) [8, 9, 10].

Figure 4 shows a scanning electron micrograph of a typical subsurface crack initiation site observed in Ti-6Al-4V. Several flat facets existed at the origin. These facets were observed for all the specimens which broke by subsurface crack nucleation. To study more about the cause of subsurface cracking, the sample was sectioned perpendicular to the fracture surface along a line which passes through a facet. Figure 5 is a scanning electron micrograph of the sectioned plane and the arrow indicates the facet. It can be seen that the facet corresponds to the fracture of a primary α grain and that the facet lies at a low angle to the fracture plane.

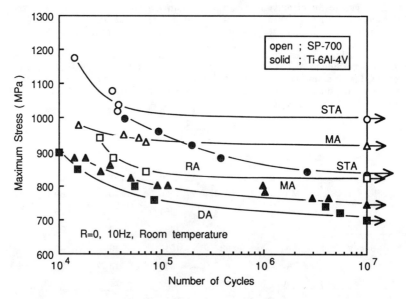

Figure 2 - S-N curves of SP-700 and Ti-6Al-4V.

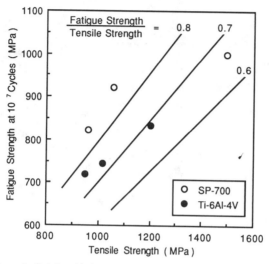

Figure 3 - Relationship between tensile strength and fatigue strength.

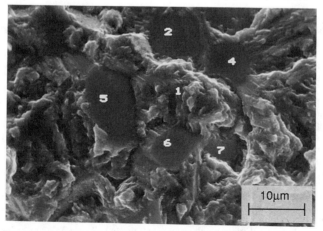

Figure 4 - Scanning electron micrograph of subsurface crack nucleation site.
Regions 2, 4, 5, 6 and 7 correspond to flat facets.

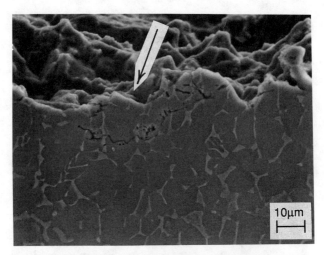

Figure 5 - Scanning electron micrograph of underlying microstructure of the crack initiation. The arrow indicates a flat facet. Loading direction is vertical.

FRACTURE TOUGHNESS

Experimental Procedure

The materials used to evaluate fracture toughness were 15 mm and 26 mm thick SP-700 plates and a 22 mm thick Ti-6Al-4V plate. All the plates were rolled in an $\alpha+\beta$ field and subsequently annealed or solution treated and aged. Heat treatment conditions were the same as Table I except that β annealing (900°C/1h/AC) was included for SP-700. Furthermore, several blanks of 26 mm thick SP-700 plate were solution treated at 825°C for one hour, air cooled and aged for 6 hours at 480°C, 510°C, 540°C, 570°C, 600°C, 650°C and 720°C, respectively, followed by air cooling. The last one (aged at 720°C) corresponds to duplex annealing. Upon air cooling from the solution treatment temperature of 825°C, the microstructure consists of approximately 10% volume fraction of acicular α and 40% volume fraction of retained β. Through the subsequent aging at temperatures lower than 600°C, there were secondary α precipitates formed from the retained β phase, providing higher strength at the cost of fracture toughness and ductility. However, little α precipitation occurred at temperatures higher than 650°C. Hereafter the solution treatment and aging with water quenching is referred to STA-I and the one with air cooling as STA-II.

Fracture toughness testing was performed at room temperature in a universal testing machine in accordance with ASTM E399. Standard 12.5 mm and 25 mm thick compact specimens were prepared from the 15 mm and 26 mm SP-700 plates, respectively, and 20 mm thick specimens from the 22 mm Ti-6Al-4V plate for the two orientations L-T and T-L. In order to investigate the deformation behavior at the crack tip, a small number of the specimens were loaded to levels just below K_{Ic}, unloaded immediately and subsequently sectioned in the mid-thickness .

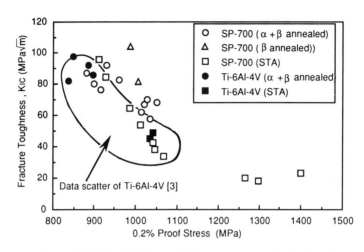

Figure 6 - Relationship between 0.2% proof stress and fracture toughness.

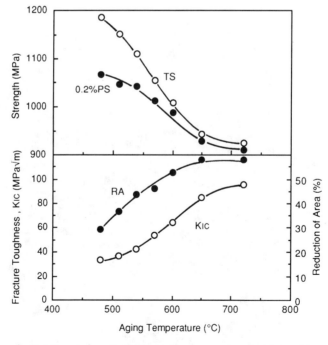

Figure 7 - Variation of strength, ductility and fracture toughness of STA-II SP-700
as a function of aging temperature. Samples were solution treated
at 825°C for one hour followed by air cooling. Aging was done for 6
hours followed by air cooling.

Results

The results of fracture toughness tests are shown as a function of 0.2% proof stress in Fig. 6.. The zone indicated in the Fig. 6 encompasses the data for Ti-6Al-4V which were collected from published data by Judy et al [4]. As can be seen in Fig. 6, K_{Ic} value decreases with increasing strength. The K_{Ic} values of the annealed SP-700 exhibit a somewhat wide range of strength-toughness relations depending on the manufacturing conditions, but are located at the upper limit of the scatter band of Ti-6Al-4V data. The β annealed SP-700 shows a better strength-toughness balance at the cost of ductility as is the case with other alloys. The STA-I presently employed results in very high strength but sacrifices fracture toughness (three data located at lower right).

The STA-II provides a wide range of strength-toughness combinations depending on aging temperature. The variation of strength, ductility and fracture toughness by STA-II is replotted as a function of aging temperature in Fig. 7. As can be seen, the levels of fracture toughness and ductility increase and strength decreases with an increase in aging temperature. A maximum aging temperature of 720°C (duplex annealing) results in the highest toughness of 96 MPa$\sqrt{m}$. Medium strength-toughness combinations can be obtained by varying aging temperature in STA-II, for example, aging at 570°C or 600°C.

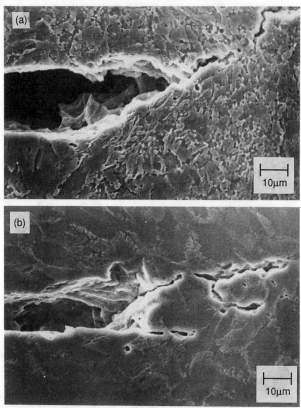

Figure 8 - Scanning electron micrographs of crack tip configurations of compact specimens loaded to levels just below Klc. (a) SP-700 (MA). (b) Ti-6Al-4V (MA).

Scanning electron micrographs of the crack tip configurations of the specimens loaded below K_{Ic} levels for annealed SP-700 and Ti-6Al-4V are compared in Fig. 8. The present observation indicates that there is a striking difference in crack tip deformation process between SP-700 and Ti-6Al-4V. For Ti-6Al-4V, many microvoids were formed at α/β phase boundaries ahead of the advancing main crack. These voids coalesced to form a rather tortuous crack path. For SP-700, on the other hand, such void formation was hardly observed and the advancing crack extended in a quite straight manner. The observation on Ti-6Al-4V coincides with the results obtained by Geyser and Lütjering [6].

DISCUSSION

One of the distinct features of SP-700 is that a very fine microstructure can be obtained in as-rolled and heat treated conditions. The high fatigue-to-tensile strength ratios in excess of 80% for annealed SP-700 (shown in Fig. 3) are partly attributed to its extremely fine microstructure. Lucas and Konieczny investigated the effect of primary α grain size on fatigue strength of Ti-6Al-4V and showed that fatigue strength increased remarkably with a decrease in primary α grain size [11]. Since SP-700 consists of primary α grains of about 2 μm size, it is believed that the observed superior fatigue strength of SP-700 to that of Ti-6Al-4V is primarily due to its fine microstructure.

Furthermore, the low aluminum content (4.5%) of SP-700 appears to make the deformation characteristics of primary α favorable to tensile ductility and fatigue strength. Based on a study on the fracture behavior of binary Ti-Al alloys and Ti-6Al-4V, Williams and Lütjering reported that increasing aluminum and oxygen concentrations and primary α grain size promoted planar slip in the α phase and intense dislocation pile-ups, promoting crack initiation [7]. Figure 9 compares slip characteristics in the primary α grains for the annealed SP-700 and Ti-6Al-4V. The samples were taken from the uniform gage section of fatigue specimens which failed at the 10^4 cycle level. Intense planar slips were observed for Ti-6Al-4V but not for SP-700 even though it was stressed at a much higher level. The previous work on the alloy partitioning of both alloys in the solution treated condition through EDS analysis on thin foil [2] reported that the aluminum concentrations in the α phase of SP-700 and Ti-6Al-4V at T_β - 50°C were 5.0% and 6.8%, respectively. The aluminum concentration of 5.0% alone is not enough to prevent planar slip development but fine grain size together with lower oxygen content helps promote the wavy slip character for SP-700. Accordingly, the present study suggests that the deformation characteristics of α grains contribute to the improvement of fatigue strength of SP-700.

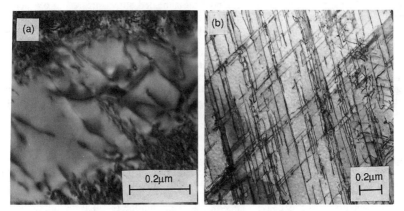

Figure 9 - Transmission electron micrographs of dislocation arrangements in cyclically stressed specimens.
(a) SP-700 (MA); σ_{max} = 941 MPa, N_f = 6.12 x 10^4.
(b) Ti-6Al-4V (MA); σ_{max} = 804 MPa, N_f = 9.10 x 10^4.

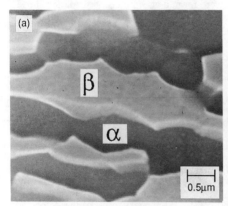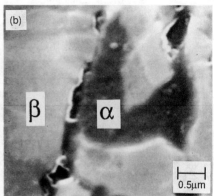

Figure 10 - Scanning electron micrographs of cross sections of tensile specimens stressed to 6% strain. Stress direction is horizontal.
(a) SP-700 (MA). (b) Ti-6Al-4V (MA).

The present observation on subsurface fatigue initiation for Ti-6Al-4V as shown in Figs. 4 and 5 coincides with the result obtained by Neal and Blenkinsop [8]. We did not observe such subsurface fatigue crack initiation for SP-700 in the present experiments. This may be related to the high resistance against crack initiation in the α phase, which resulted from the distinct slip characteristics of the alloy.

As for fracture toughness, different fracture mechanisms were observed between both the alloys. As shown in Fig. 8, the deformation behavior at the crack tip for Ti-6Al-4V was characterized by microvoid formation and their coalescence ahead of the main crack, which consumed energy prior to the onset of the secondary crack. This mechanism was not observed in the annealed SP-700. Figure 10 compares scanning electron micrographs of the longitudinal cross section of the tensile specimens for both the annealed alloys stressed to 6% permanent strains. Microvoids were formed at the α/β phase boundaries for Ti-6Al-4V while such voids were hardly seen for SP-700. The higher aluminum concentrations in the α phases and larger α grain size for Ti-6Al-4V promote planar slip characteristics, resulting in significant dislocation pile-ups at the α/β interfaces, and thus leading to early void nucleation. On the other hand, fine grained SP-700 with low aluminum content and a high volume fraction of retained β phase should have high deformation capability and therefore high resistance to void formation. It is thought from the present observation that in the fracture process of SP-700 more energy is absorbed by crack blunting at the crack tip before unstable fracture occurs as compared with Ti-6Al-4V.

Solution treatment with water quenching and subsequent aging of SP-700 results in a very high fatigue strength and a low level of fracture toughness. A fine dispersion of secondary α precipitates in the β phase contributed to high strength but significantly lowered fracture toughness. Like other α+β titanium alloys, little crack tip blunting was observed prior to unstable fracture. Soft retained β phase should contribute to high fracture toughness in the annealed conditions, but the aged prior β phase promotes easy crack propagation because of high notch sensitivity of this phase. This may also explain the fact that the fatigue to tensile strength ratio as shown in Fig. 3 became lower than 70% although the fatigue strength itself became the higher.

CONCLUSIONS

The β-rich α+β alloy SP-700 has much higher fatigue strengths than Ti-6Al-4V in both the annealed and heat treated conditions. This results from the distinct microstructural characteristics of SP-700: extremely fine microstructure. In the annealed condition, the deformation process of the α grains associated with the low aluminum content appears to be an additional factor for the improvement of fatigue resistance over Ti-6Al-4V. A higher volume of retained β phase together with good formability of the α phase contribute to the high fracture toughness and tensile ductility levels obtained in the annealed and STA-II (air cooling) with high temperature aging conditions. In the STA-I (water quenching) and STA-II with low temperature aging, the deformation and fracture processes of SP-700 are predominantly controlled by the transformed β phase which consists of α precipitates in the β matrix.

ACKNOWLEDGMENT

The authors would like to thank their coworker Mr. H. Iizumi for experimental assistance.

REFERENCES

1. C. Ouchi et al., "Development of β-rich α-β Titanium Alloy: SP-700," NKK Technical Review, No. 65 (1992), 61-67.

2. M. Ishikawa et al., "Microstructure and Mechanical Properties Relationship of β-rich α-β Titanium Alloy; SP-700" (Paper presented at the Seventh World Conference on Titanium, San Diego, California, 29 June 1992), 8.

3. A. W. Bowen and C. A. Stubbington, Titanium and Titanium Alloys, ed. J. C. Williams and A. F. Belov (1982), 1989-2001.

4. R. W. Judy, Jr., B. B. Rath, and R. J. Goode, Titanium, Science and Technology, ed. G. Lütjering, U. Zwicker, and W. Bunk (1985), 1925-1943.

5. G. Lütjering and A. Gysler, Titanium, Science and Technology, ed. G. Lütjering, U. Zwicker, and W. Bunk (1985), 2065-2083.

6. A. Gysler and G. Lütjering, Titanium, Science and Technology, ed. G. Lütjering, U. Zwicker, and W. Bunk (1985), 2001-2008.

7. J. C. Williams and G. Lütjering, Titanium '80, Science and Technology, ed. H. Kimura and O. Izumi (1980), 671-681.

8. D. F. Neal and P. A. Blenkinsop, Acta Met., 24 (1976) 59-63.

9. A. Atrens et al., Scripta Met., 17 (1983), 601-606.

10. S. Adachi, L. Wagner, and G. Lütjering, Titanium, Science and Technology, ed. G. Lütjering, U. Zwicker, and W. Bunk (1985), 2139-2146.

11. J. J. Lucas and P. P. Konieczny, Met. Trans., 2 (1971) 911-912.

AMBIENT TEMPERATURE TENSILE AND CREEP DEFORMATION

BEHAVIOR OF ALPHA AND BETA TITANIUM ALLOYS

C. A. Greene and S. Ankem

Department of Materials & Nuclear Engineering
University of Maryland
College Park, Maryland 20742-2115

Abstract

Systematic studies were undertaken to determine the ambient temperature tensile and creep deformation behavior for α Ti - 0.4 (wt%) Mn, grain size ~30 μm, and β Ti - 13.0 Mn, grain size ~200 μm, alloys and tensile deformation behavior of a β Ti - 14.8 V alloy with a grain size of ~40 μm. The ambient temperature creep mechanisms of α Ti-Mn alloy when tested at 95% Yield Stress (YS) were found to be similar to those of tensile deformation, i.e. fine homogeneous slip, occasional twinning and rare grain boundary sliding. Significant creep deformation was observed in α Ti-Mn alloy followed by creep exhaustion. In the case of β Ti-Mn alloy, no significant creep strain was observed when tested at 95% YS. The tensile deformation mechanisms of β Ti-V alloy were found to be coarse and wavy slip and what appears to be twinning. This is in contrast to the tensile deformation of β Ti-Mn alloy which deformed only by coarse and wavy slip.

This Research is Funded by Office of Naval Research Contract No. 00014-90-J-1907

Beta Titanium Alloys in the 1990's
Edited by D. Eylon, R.R. Boyer and D.A. Koss
The Minerals, Metals & Materials Society, 1993

Introduction

Titanium alloys are technologically important. They have very attractive properties including high strength to weight ratio, high fracture toughness and excellent sea water corrosion resistance. It has been known for sometime that near α and α–β titanium alloys creep at ambient temperatures.[1-4] However, no systematic investigation has been undertaken to determine the effect of alloying element and microstructure on mechanisms responsible for room temperature creep. Most of the earlier investigations were concerned with determining creep properties of specific alloys and little emphasis was placed on fundamental understanding of creep deformation mechanisms. In regards to the ambient temperature creep behavior of near β alloys, no significant information is available.

To understand the creep of two phase, i.e. α–β, titanium alloys we must first understand the creep of the individual phases and how it is altered by alloying element and microstructure. For example, if two alloys with the same microstructure have different alloying elements will the creep deformation mechanism and the creep strain be the same? This question cannot be answered by the current dearth of literature on the ambient temperature creep of specific alloys. The aim of this investigation was to systematically study the effect of microstructure and alloying elements on the room temperature tensile and creep deformation behavior of α, α–β and β Ti-Mn and Ti-V alloys. In this paper, the preliminary results dealing with the ambient temperature tensile and creep behavior of α and β Ti-Mn alloys as well as the ambient temperature tensile behavior of β Ti-V alloys are presented.

Experimental

For this portion of the investigation, two different Ti-Mn alloys and one Ti-V alloy were used. The alloys and their compositions are: α Ti - 0.4 (wt %) Mn, β Ti-13.0Mn and β Ti-14.8V. Oxygen contents were in the range 0.071 to 0.116 wt %.

These alloys were melted as 16.6 kg ingots and processed to 1.74 cm diameter bars at RMI Co. For the final heat treatment, the alloys were sealed in quartz tubes at 10^{-5} to 10^{-6} torr and annealed for 200 hr at 963 K, followed by water quenching. This allowed adequate time for complete recrystallization of the samples which resulted in equiaxed type microstructures as shown in Figure 2.

Tensile and creep specimens were identically machined from the quenched bars with a gauge length of 2.86 cm. Flats were machined on the gauge sections. The flats were mechanically polished, electropolished at -60°C and then etched to reveal grain boundaries. Gold fiducial lines were put on the flats following the method of Starke, et al,[3] Attwood and Hazzledine,[5] and others.[6,7,10] To put the lines on the specimens, a Jeol JXA-840 scanning electron microscope was used to expose a photoresist spun on the polished and etched flats of the tensile and creep specimens. The specimens were sputter coated with gold and when the unexposed photoresist was removed, gold grid lines 1 - 2 μm wide spaced 20 μm apart covering an area 2 mm x 2 mm in the center of the gauge length were obtained. Typical grid lines are also shown in Figure 2. The grid lines served two purposes. First, very fine homogeneous slip was monitored by noting any distortion in the grids. Second, it was possible to keep track of grain boundary sliding where the grid lines became displaced at the grain boundaries.

Scanning electron micrographs using a secondary electron detector were taken of the samples before testing. It was important to take pictures before as well as after deformation since it was difficult in some cases to discern if a twin or other deformation mechanism was present before testing. Typical before and after tensile deformation pictures are shown in Figure 2 (a) and 2 (b), respectively.

Specimens with gold grid lines were tensile tested on a floor model INSTRON machine at room temperature (293 K) and at an initial strain rate of 3.28×10^{-5} per second. For each tensile test two to four specimens were used and the results averaged. Individual test results varied at maximum by 4%. The samples were strained to about 3% total strain. Typical true stress-strain curves are shown in Figure 1. SEM pictures taken after testing are shown in Figures 2 (b) and 3 (b). Optical pictures at 500X were also taken after testing since, for the α Ti-Mn alloy and the β Ti-V alloy, the twins are more easily discerned in the light microscope, see Figure 4.

Specimens with grid lines were creep tested at room temperature in air on an Applied Test Systems (ATS) Lever Arm Tester.[12] The strain was measured by a clip on extensometer and strain module, calibrated in accordance with ASTM E83A. The strain data was recorded on a PC using an Analog Connection™ ACJr™ A/D board and Quicklog PC™ Software, both from Strawberry Tree Incorporated.

The specimens were creep tested at 95% of their respective 0.2% Yield Stresses. The tests were conducted under constant load conditions, however the reduction in cross sectional area was minimal, thus approximating constant stress creep tests. Typical creep curves are shown in Figure 5. SEM pictures of the alloys were taken after testing as shown in Figures 6 (b), (c) and 9 (b). As in the case of tensile tests, optical pictures at 500X were also taken after creep testing and they are shown in Figures 7 and 8.

Results

Tensile Tests

α Ti-Mn. The ambient temperature true stress - true strain curve shown in Figure 1 (a) for α Ti-0.4Mn with grain size ~40 μm has a 0.2% YS of ~235 MPa and shows work hardening. This result agrees well with previous investigations.[8,14] The deformation mechanisms are twinning, fine homogeneous slip, and rare grain boundary sliding.

β Ti-Mn. The ambient temperature true stress - true strain curve shown in Figure 1 (b) for β Ti-13.0Mn with grain size ~200 μm shows very little strain hardening and it has a yield strength of 940 MPa. The deformation mechanism appears to be very coarse and wavy slip as shown in Figure 2 (b). No twinning was observed. Similar results were also reported in previous investigations by Ankem and Margolin.[8,9,10,14]

β Ti-V. The ambient temperature true stress - true strain curve for β Ti-14.8V, at a grain size of ~35 μm is shown in Figure 1 (c). The YS of this alloy is 790 MPa. Note that this is a lower strength than the YS of

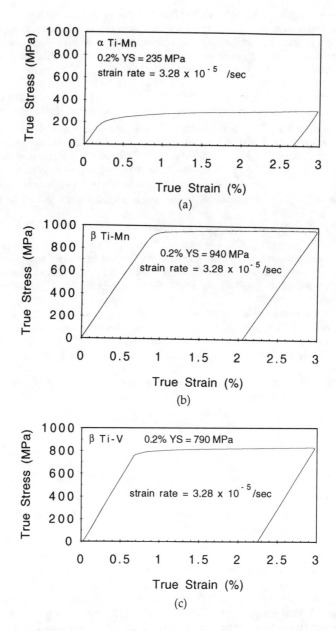

Figure 1 - Ambient temperature true stress-true strain curves of: (a) Ti-0.4Mn (~100% α) alloy, (b) Ti-13.0Mn (~100% β) alloy, and (c) Ti-14.8V (~100% β) alloy. Alloys were heat treated for 200 hr @ 963 K followed by water quench.

β Ti-13.0Mn alloy. As in the β Ti-Mn case, there is very little strain hardening.[18] However, in contrast to the β Ti-Mn alloy, coarse and wavy slip and another mechanism, what appears to be twinning was also observed, see Figures 3 and 4. Profuse twinning in Ti-15.5V deformed between 4-9% by cold rolling has been reported by Oka and Taniguchi.[11]

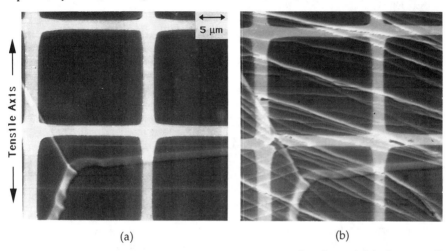

(a)   (b)

Figure 2 - SEM micrographs of Ti-13.0Mn (~100% β) alloy: (a) before tensile deformation and, (b) the same area after tensile deformation, note slip bands. Total plastic strain is 2.1% at a strain rate of 3.28×10^{-5} per second. The grain size is ~200 μm.

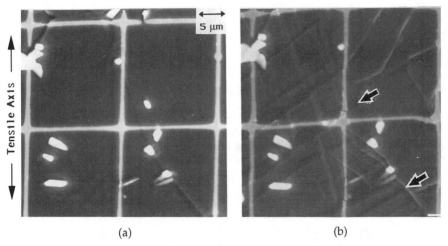

(a)   (b)

Figure 3 - SEM micrographs of Ti-14.8V (~100% β) alloy: (a) before tensile deformation and, (b) the same area after tensile deformation, note slip bands and twinning at the arrows. Total plastic strain is 2.25% at a strain rate of 3.28×10^{-5} per second. The grain size is ~30 μm.

Tensile Axis

10 µm

Figure 4 - Optical micrograph of β Ti-14.8V alloy after tensile deformation to 2.25% plastic strain. Note twinning and slip lines.

Creep Tests

α Ti-Mn. The Ambient temperature creep curve of α Ti-0.4Mn crept at 95% of 0.2% YS is presented in Figure 5 (a). The slope of the curve is decreasing, i.e. the curve is approaching a plateau, indicating an exhaustion type creep mechanism.[3] Comparison of the photomicrographs of before and after creep deformation indicate that the tensile and creep deformation mechanisms are similar in spite of significant differences in strain rates. Figure 6 (b) in comparison with Figure 6 (a) shows elongation of grid lines in the direction of loading indicating very fine slip (not seen in the photomicrograph). In Figure 6 (b) grain boundary sliding is noticeable, but it is to be noted that this is not a common occurrence. Twinning also occurs in this alloy, see Figures 6 (c) and 7, it is noted that for some twins there is an associated grid distortion.

β Ti-Mn. The ambient temperature creep curve of β Ti-13.0Mn crept at 95% YS is presented in Figure 5 (b). This curve shows very little creep strain in contrast to the α phase. The creep strain was only 0.03% in ~400 hr. Despite the fact that there is very little creep strain, coarse slip was observed in a few grains as shown in Figures 8 and 9. No twinning was detected.

Discussion

Observation of Figures 6 (a), (b), and (c) indicates that the α Ti-Mn alloy, with a grain size of ~30 µm crept at ambient temperature at 95% YS, deforms predominately by fine slip, occasional twinning and rare grain boundary sliding. It is to be noted that the fine slip lines are not visible on the specimen, however this conclusion was drawn from a comparison of grid lines before and after deformation. In contrast, the occasional twins can be clearly seen, for example see Figure 6 (c). The twinning in this alloy can be seen even more clearly in the optical micrographs, see Figure 7. It is interesting to note that this α Ti-Mn alloy creeps at a high rate in the beginning followed by creep exhaustion. It is yet to be determined the roles of fine slip and twinning in creep. For example, if twinning plays a significant role in creep of α Ti-Mn alloy, but twinning is limited by grain size, then this would mean significant creep deformation in α Ti-Mn alloys with grain sizes much larger than 30 µm will occur. We are currently in the process of

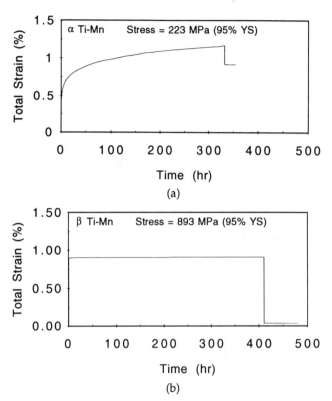

Figure 5 - Ambient temperature creep curves for: (a) Ti-0.4Mn (~100% α) alloy and (b) Ti-13.0Mn (~100% β) alloy, crept at 95% YS. Alloys were heat treated for 200 hr @ 963 K followed by water quench.

studying such effects of grain size on α Ti-Mn alloys.

In contrast to α Ti-Mn alloy, β Ti-Mn alloy exhibits no significant creep strain, see Figure 5 (b). It is not clear at this time why the creep strain is so low but it appears to be related to the low initial mobile dislocation density[17] and absence of twinning. The investigation is underway to understand creep deformation behavior of the β Ti-Mn alloys.

In contrast to β Ti-Mn alloy, the β Ti-V alloy exhibits both coarse slip and what appears to be twinning during tensile deformation, see Figures 3 and 4. In the past there was some controversy over the stress induced plates, whether they are martensite or twins.[11,19] Our results support the contention that these plates are twins, but we are in the process of confirming this. The absence of twinning in the β Ti-Mn alloy and the occurrence of twin like stress induced plates in the β Ti-V alloy may be related to the atomic size differences between the titanium and the solute elements. The atomic size difference between Ti and V is much smaller than the atomic size difference between Ti and Mn. It will be

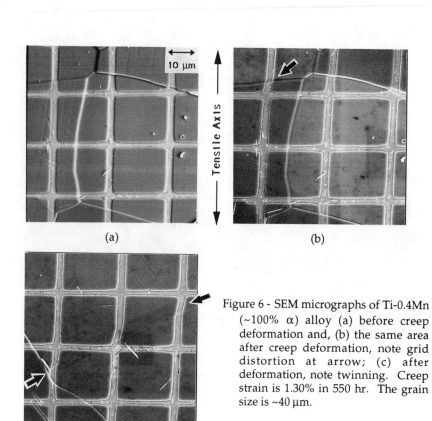

(a)

(b)

(c)

Figure 6 - SEM micrographs of Ti-0.4Mn (~100% α) alloy (a) before creep deformation and, (b) the same area after creep deformation, note grid distortion at arrow; (c) after deformation, note twinning. Creep strain is 1.30% in 550 hr. The grain size is ~40 μm.

Figure 7 - Optical micrograph of α Ti-0.4Mn alloy after creep deformation to 1.30% creep strain. Note twinning at the arrows.

Figure 8 - Optical micrograph of β Ti-13.0 Mn alloy after creep deformation to 0.03% creep strain. Note slip lines.

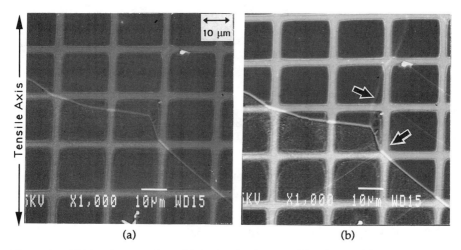

Figure 9 - SEM micrographs of Ti-13.0Mn (~100% β) alloy: (a) before creep deformation and, (b) the same area after creep deformation, note slip bands at the arrows. Total creep strain is 0.03% in 408 hr. The grain size is ~200 μm.

interesting to see whether the creep deformation behavior of β Ti-V is different from that of the β Ti-Mn alloy. Such studies are underway.

Conclusions

1. When the α Ti-Mn alloy was crept at 95% yield stress, significant creep strain was observed in the beginning and this was followed by creep exhaustion. The creep deformation mechanisms appear to be similar to those in tensile deformation, i.e. fine slip, occasional twinning and rare grain boundary sliding.
2. When the β Ti-Mn alloy was crept at 95% yield stress no significant creep strain was observed. This is in contrast to α Ti-Mn alloy where significant creep strain was observed.
3. In β Ti-V, the tensile deformation occurs by coarse and wavy slip and what appears to be twinning. This is in contrast to the tensile deformation of β Ti-Mn alloy which deforms by coarse and wavy slip. This difference appears to be related to the atomic size difference between titanium and the solutes.

Acknowledgements

The Authors wish to thank Dr. George Yoder for his interest, encouragement and helpful discussions throughout the course of this investigation. The Authors also wish to thank Mr. J.R. Wood and Mr. Stan Siegle at RMI Co for supplying the titanium alloys for this investigation and their encouragement.

This Research is funded by Office of Naval Research under contract No. 00014-90-J-1907.

References

1. H. P. Chu, "Room Temperature Creep and Stress Relaxation of a Titanium Alloy," J. of Mater, 5 (3) (1970), 633-642.

2. M. A. Imam and C. M. Gilmore, "Room Temperature Creep of Ti-6Al-4V," Metall. Trans. A, 10A (1979), 419-425.

3. W. H. Miller, Jr., R. J. Chen, and E. A. Starke, Jr., "Microstructure, Creep, and Tensile Deformation in Ti-6Al-2Nb-1Ta-0.8Mo," Metall. Trans. A, 18A (1987), 1451-1467.

4. G. Y. Gao and S. C. Dexter, "Effect of Hydrogen on Creep Behavior of Ti-6Al-4V Alloy at Room Temperature," Metall. Trans. A, 18A (1987), 1125-1130.

5. D. G. Attwood and P. M. Hazzledine, "A Fiducial Grid for High-Resolution Metallography," Metallography, 9 (1976), 483-501.

6. M. Kikukawa, M. Jono, and M. Adachi, "Direct Observation and Microscopic Measurement of Fatigue Damage by Scanning Electron Microscopy", Mechanical Behavior of Materials, JSMS (Proceedings of the 1974 Symposium on Mechanical Behavior of Materials, Volume 1, Kyoto, Japan, August 21-24, 1974), 307-317.

7. A. Karimi, "Plastic Flow Study Using the Microgrid Technique," Materials Science and Engineering, 63 (1984), 267-276.

8. S. Ankem and H. Margolin, "A Rationalization of Stress-Strain Behavior of Two-Ductile Phase Alloys," Metall. Trans. A, 17A (1986), 2209-2226.

9. S. Ankem and H. Margolin, "Finite Element Method (FEM) Calculations of Stress Strain Behavior of Alpha-Beta Ti-Mn Alloys: Part II," Metall. Trans. A, 13A (1982), 603-609.

10. S. Ankem and H. Margolin, "Alpha-Beta Interface Sliding in Ti-Mn Alloys," Metall. Trans. A, 14A (1983), 500-503.

11. M. Oka and Y. Taniguchi, "{332} Deformation Twins in a Ti-15.5 Pct V Alloy," Metall. Trans. A, 10A (1979), 651-653.

12. S. Ankem, C. A. Greene, and X. Liu, "Fundamental Studies On Ambient Temperature Creep Behavior of Alpha, Alpha-Beta, and Beta Titanium Alloys," (Interim Report Submitted to Office of Naval Research/N00014-90-J-190, University of Maryland, November 1991).

13. S. Ankem and H. Margolin, "Beta and Alpha Grain Sizes In Alpha-Beta Ti-Mn Alloys," Metall. Trans. A, 8A (1977), 1320-1321.

14. J. Jinoch, S. Ankem, and H. Margolin, "Calculations of Stress-Strain Curves and Stress and Strain Distributions for an Alpha-Beta Ti-Mn Alloy," Mat. Sci. Eng., 34 (1978), 203-211.

15. J. L. Swedlow et al., "Discussion on 'Calculations of Stress-Strain Curves and Stress and Strain Distributions for an Alpha-Beta Ti-Mn Alloy," Mat. Sci. Eng., 40 (1979), 139-142.

16. S. Ankem and H. Margolin, "The Role of Elastic Interaction Stresses on the Onset of Plastic Flow for Oriented Two Ductile Phase Structures," Metall. Trans. A, 11A (1980), 963-972.

17. S. Ankem et al., "The Effect of Volume Per Cent of Phases on the High Temperature Tensile Deformation of Two-phase Ti-Mn Alloys," Mat. Sci. Eng., A111 (1989), 51-61.

18. T. S. Kuan, R. R. Ahrens, and S. L. Sass, "The Stress-Induced Omega Phase Transformation in Ti-V Alloys," Metall. Trans. A, 6A (1975), 1767-1774.

19. M. K. Koul and J. F. Breedis, "Phase Transformations in Beta Isomorphous Titanium Alloys," Acta Met., 18 (1970), 579-588.

THE TEMPERATURE DEPENDENCE OF THE DEFORMATION AND OXIDATION

BEHAVIOR OF AN AGE-HARDENABLE BETA + ALPHA-TWO TITANIUM ALLOY

D. M. Goto, L. S. Quattrocchi* and D. A. Koss

Department of Materials Science and Engineering
The Pennsylvania State University
University Park, PA 16802

Abstract

The beta Ti alloy Ti-23Nb-11Al (at %) is unique in that it is age hardenable due to the formation of lath-like α_2-phase precipitates based on Ti_3Al. Furthermore, age-hardening occurs at temperatures significantly higher than most conventional beta Ti alloys. This suggests the possibility of the elevated temperature usage of α_2-strengthened beta Ti alloys such as Ti-23Nb-11Al. The study examines the compressive deformation behavior of the Ti-23Nb-11Al alloy in the -196°C to 650°C temperature range and demonstrates that very high strengths are possible even at 600°C by precipitation hardening due to the α_2 phase. The thermogravimetric oxidation behavior in laboratory air from 600°C to 700°C indicates parabolic behavior consistent with oxygen diffusion through a scale.

---

* Currently at Pratt and Whitney, East Hartford, CT.

Beta Titanium Alloys in the 1990's
Edited by D. Eylon, R.R. Boyer and D.A. Koss
The Minerals, Metals & Materials Society, 1993

Introduction

Most Ti alloys which retain a predominantly β-phase microstructure after heat treatment are age hardenable due to the formation of hcp alpha-phase precipitates, which typically occurs in the range of temperatures between 450°C and 500°C [1]. However, use of these "conventional" beta (β) Ti alloys is limited to temperatures well below precipitation heat-treatment temperatures in order to avoid possible microstructural instabilities. Recent research indicates that it is possible to age-harden the β Ti alloy Ti-23Nb-11 Al (at.%), hereafter designated Ti-23-11, by the formation of lath-like, precipitates based on the ordered intermetallic α_2-phase, Ti_3Al [2,3]. The formation of these precipitate particles in a disordered β-phase matrix results in pronounced age hardening. For example, material in the solution-treated and peak-aged condition exhibits a yield strength about 600 MPa above that of solution-treated material. Furthermore, a significant precipitation hardening increment is retained even after overaging at temperatures as high as 675°C, or about 200°C higher than conventional β Ti alloys. At these aging temperatures it is also possible to grow the alpha-two particles to a size wherein fine, uniform slip is induced, probably due to dislocation looping of the particles [3].

The aging behavior of the Ti-23-11 alloy suggests that, after aging in the 575°C to 675°C range, the α_2-phase precipitates will likely be microstructurally stable in the 500°C to 600°C range. This should result in the retention of a high level of strength in this temperature regime. However, this is also a range of temperatures where oxidation might limit the performance of a Ti alloy. The purpose of this study is to examine the strength and deformation behavior of the Ti-23-11 alloy over a wide range of temperatures from -196°C to 650°C. The oxidation response in air at 600°C - 700°C will also be analyzed on the basis of thermogravimetric data.

Experimental Procedure

The Ti-23-11 alloy has a composition of Ti-22.8Nb-11.1 Al in at. pct. or Ti-38Nb-5Al in wt. pct. The material contained about 1580 wt. ppm oxygen, 160 wt. ppm nitrogen, and 11 wt. ppm hydrogen. Test specimens were machined from bar stock which was extruded at 1038°C using a 22:1 extrusion ratio. All heat treatments were performed by encapsulating Ta-wrapped specimens in quartz under a partial pressure of high purity argon. Prior to specimen machining, the ~25 mm diameter bar stock was solutionized at 1000°C for 1 hour and quenched in ice water. This resulted in a grain size of ≈210 μm, which was unaltered by subsequent aging treatments.

Mechanical tests were performed at elevated temperatures in high purity (99.998%) argon gas in compression on cylindrical specimens 3.17 mm in diameter and length. The compression specimens were electropolished at -40°C in a methanol, ethylene glycol, perchloric acid mixture [4]. Molydisulfide lubricant was used on the end faces to reduce friction; recent results indicate that this technique effectively produces accurate stress-strain data in compression to true strains $\geq$0.5 [5]. The tests were performed at initial engineering strain rates of $3 \times 10^{-4} s^{-1}$. "Jump" tests, wherein the strain rate was abruptly increased by $2 \times 10^{-3} s^{-1}$ and subsequently to $1 \times 10^{-2} s^{-1}$, were used to determine the strain-rate hardening exponent $m = d\ln\sigma/d\ln\dot{\epsilon}$. Stress changes on both increasing and decreasing strain rates were measured and interpreted as described in detail elsewhere [4].

Isothermal oxidation of the alloy was conducted in flowing laboratory air (≈60 cc/min) between 600° and 700°C. Temperature control was maintained to ±2°C. Sample weight was continuously monitored over the oxidation period (≤ 96 hours) by a thermogravimetric analysis device. To facilitate initial surface area measurements, oxidation samples consisted of coupons ≈3.75 mm x 3.75 mm x 10 mm.

Results and Discussion

1. Microstructure

Unlike the microstructures of recent "quasi-beta" alloys such as β-CEZ [6], the microstructure of the Ti-23-11 alloy remains predominantly beta phase even after aging. Hardening during aging at 575°C to 675°C occurs by the formation of lath-like precipitates, based on Ti₃Al, in a disordered beta matrix [2,3]. This is accomplished after a solution treatment at 1000°C, or about 40°C above the α_2 solvus temperature of $960°C \pm 15°C$. The precipitates obey a Burger's orientation relationship with the matrix and retain particle sizes on the submicron scale. For example, aging at 575°C for 6 hours, to peak hardness, results in average dimensions of the lath-like particles of length $\sim$ 170 nm, width $\sim$ 120 nm, and thickness $\sim$ 30 nm [2].

Particle coarsening due to elevated temperature exposure is relatively slow in Ti-23-11 alloy. As shown in Figure 1, the maximum dimensions of the particles grow from ~170 nm after 575°C/6 hrs to about 500 nm after 100 hrs at 675°C. The lath-like appearance of the ordered α_2 precipitates suggests coherent or semi-coherent interphase boundaries exist along the broad faces of the precipitate [7]. Since coherent or semi-coherent interfaces generally have low interface migration rates, precipitate coarsening rates are expected to be low. In addition, the coarsening rates of the α_2 precipitates may also be expected to be low due to the somewhat lower rate of diffusion in ordered precipitates. Thus, in comparison to existing β Ti alloys hardened by disordered α-phase particles, an alloy such as the Ti-23-11 alloy should exhibit superior precipitate stability. This should result in better strength retention to higher temperatures than existing β Ti alloys.

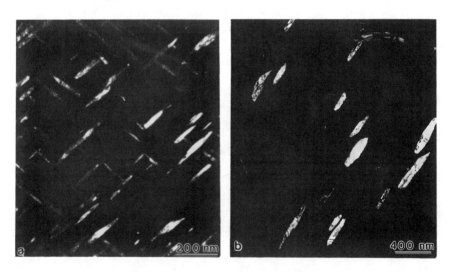

Figure 1.   CDF image of the α_2-phase precipitates formed when the Ti-23-11 alloy was aged (a) at 575°C for 6 hours and (b) at 675°C for 100 hrs.

2. Deformation Behavior

Peak hardness in the Ti-23-11 alloy can be obtained by aging solution-treated material at 575°C for about 6 hrs. However, as shown in Figure 2, a significant degree of age hardening is also possible by aging at temperatures as high as 675°C. In this case, peak hardness is achieved in two hrs, and the hardness increment, 110 VHN, is about half of that (210 VHN) obtained at 575°C.

The temperature dependence of the yield stress for the Ti-23-11 alloy from -196°C to 650°C is shown in Figure 3. It should be recalled that in all cases the alloy was solution-treated at 1000°C prior to aging. Figure 3 indicates that (a) high room temperature yield stresses are possible in this alloy; (b) like other beta Ti alloys [8], there is a strong temperature dependence

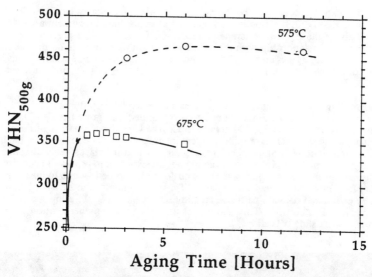

Figure 2. The age hardening response of the Ti-23-11 alloy in the solution-treated condition (1000°C/1hr, water quench).

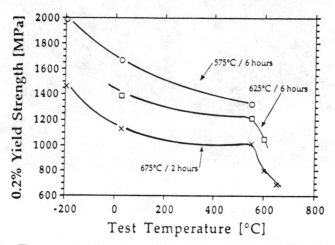

Figure 3. The temperature dependence of the yield strength for the Ti-23-11 alloy in three heat-treated conditions.

of the yield stress below room temperature; (c) most of the room temperature yield strength is retained to ~550°C such that the σ_y-T data are nearly athermal; and (d) at temperatures greater than 550°C, a rapid decrease in yield strength with increasing temperature can be observed.

The large degree of age hardening in the Ti-23-11 alloy is obtained despite the fact that there has been no alloy development performed to optimize strength. This is the only alloy of this kind that we have tested. The age hardening is obviously a result of the relatively small interparticle spacing achievable during the aging treatments, as well as the resistance to the shearing of the ordered Ti₃Al-based particles by mobile dislocations [3].

The strong temperature dependence of the yield and flow stress in bcc alloys at temperatures below room temperature is well known. Owing to a combination of the Peierls stress and interstitial solute effect [8], there is an increase in yield stress, $\Delta\sigma_y$, from 27°C to -196°C of $\Delta\sigma_y$ ~600 to 800 MPa. The larger yield stress increase occurs in the solution-treated condition and suggests the presence of thermally activated substitutional solute hardening, probably due to Al atoms. In other age-hardened β Ti alloy systems, precipitation hardening is found to be athermal and does not affect the temperature sensitivity of the yield stress (i.e., the $\Delta\sigma_y$ value) at low temperatures [9]. Finally, we note that, similar to room temperature observations [3], slip is coarse and planar at -196°C except after aging at 675°C for six hours; in this case the slip was so fine and uniformly distributed that it was unresolvable in an optical microscope.

The yield stress of the age-hardened Ti-23-11 is only weakly sensitive to temperature in the 27°-500°C range. As before, slip is relatively coarse and planar in this temperature regime except for specimens aged at 675°C. However, as depicted in Figure 4, the solution-treated material and the specimen aged at 375°C show serrated yielding (or the Portevin-LeChatlier effect) when tested at 327°C. This has been observed previously in β Ti alloys and has been attributed to dynamic strain aging [10,11], probably due to interstitial oxygen.

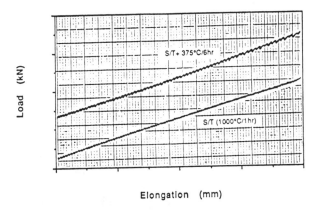

Elongation (mm)

Figure 4. Jerky flow during testing at 327°C.

An analysis of serrated yielding or jerky flow during dynamic strain aging, indicates that the diffusivity (D) of the interstitial atoms is related to strain rate at the onset of serrated yielding such that [12]:

$$D \cong 10^{-9}\, \dot{\varepsilon}. \tag{1}$$

For $\dot{\varepsilon} = 2\times10^{-4}$ s^{-1} and using values for the diffusivity of oxygen in β Ti [13], Eq. (1) predicts the onset of serrated yield to occur at 335°C. This is close to the test temperature of 327°C where the serrations are observed. We also note additional evidence for serrated yielding or jerky flow by the negative values of the strain-rate hardening exponent m = dln σ/dln ε for the solution-treated alloy (m = -0.008) and the alloy aged at 375°C (m = -0.004). The specimens aged at 575°C or 675°C did not show serrated yielding; one implication is that the aging treatment reduced the amount of oxygen in solution, decreasing the ability of oxygen to pin dislocations.

As shown in Figure 3, at temperatures greater than ≈550°C, there is a rapid decrease in yield strength with increasing temperature. Nevertheless, for the specimens aged 2 hrs at 675°C, roughly ≈70% of the room temperature yield strength is preserved to 600°C. This is not typical of existing beta-rich alloys which are comparatively much softer (400-600 MPa) in the 600°C temperature range. Thus, as suggested by the relatively good resistance to particle coarsening described previously, we believe that age hardening as a result of α_2 precipitate formation is a relatively effective means of retaining high strengths in β Ti alloys to the 500°C -600°C range. In the present case, this is achieved by overaging the alloy at 675°C. Aging at 675°C introduces relatively large α_2 precipitates which is known to induce fine, uniform slip as a result of dislocation bypass of the particles. This should be beneficial for ductility of this Ti alloy if intergranular, ductile fracture can be suppressed.

The above observations suggest that the Ti-23-11 alloy has at least short-time precipitate stability at elevated temperatures, up to and including 600°C. However, the stress-strain responses, such as Figure 5, indicate a pronounced tendency for flow softening of specimens aged at 675°C and tested at either 600°C or 650°C. Optical microscopy indicates that flow softening is accompanied by the formation of plastic flow localization wherein bands of the microstructure are heavily deformed to create pancake shaped grains with relatively equiaxed bands outside the bands; see Figure 6.

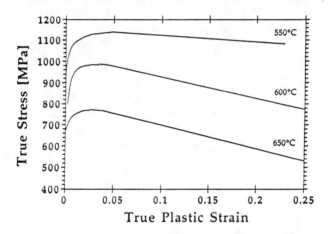

Figure 5. Stress-strain responses at 550°C exhibiting zero strain-hardening conditions, and at 600 and 650°C, respectively, exhibiting flow softening (negative strain-hardening). Specimen was solution-treated and aged at 675°C for 2 hrs.

The cause of the flow softening phenomena in Figure 5 is not well understood. The present strain rates are too slow for adiabatic heating to be the primary cause. Shear banding similar to that observed above has been observed in alpha + beta Ti alloys and was attributed to the possible generation of a softer crystallographic texture, dynamic recrystallization and/or microstructural coarsening [14,15]. In our case, the relative plastic isotropy of the beta phase minimizes any texture softening effects, and no recrystallization was observed. In addition, microhardness measurements within the shear band indicate hardnesses comparable to the adjacent matrix, suggesting no large loss of hardening due to either "damage" to the precipitate particles (or short range order hardening). Thus, although TEM was not performed, we infer from these hardness data that the precipitate microstructure appears to be relatively stable. In support of this, we also note that similar flow softening has been observed in solid solution beta-phase Ti-V alloys at 891°C at high levels of V content (25 and 30%) [16]. Thus this flow softening phenomenon in Figure 5 appears to be a characteristic of deformation dynamics behavior within the solid solution matrix and not of specific dislocation-precipitate interactions.

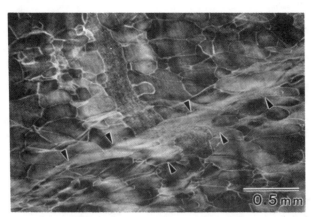

Figure 6. Optical micrograph of an as-deformed cross sectioned sample exhibiting localized flow instability within deformed pancake-shaped grains; see arrows.

The relationship between flow stress and strain rate was examined between 550°C and 650°C and strain rates of $3 \times 10^{-4} s^{-1}$ to $10^{-2} s^{-1}$. Very large (n>40) stress exponents, $n = d \ln \dot{\varepsilon}/d \ln \sigma$, were observed in all cases indicating that dislocation glide-controlled processes still dominate deformation. This is not surprising considering that 650°C is about 0.5 T_{MP}. It is consistent with the Sherby and Burke [16] observation that power law breakdown creep occurs whenever $\dot{\varepsilon}/D_L \geq 10^{13}$ m^{-2} where D_L is the self diffusion coefficient of the β Ti matrix. In the present case, assuming D_L for self diffusion in β Ti [18], $\dot{\varepsilon}/D_L \geq 1.5 \times 10^{-13}$ m^{-2} for all of the test conditions examined. Testing at temperatures higher than 650°C and at much lower strain rates, required to induce steady state creep, was preferred. However, the former was precluded by precipitate instability, while the pronounced flow softening discouraged attempts to obtain steady state creep behavior at constant stress through slow strain-rate compressive creep techniques.

3. Tensile Ductility

Owing to a very limited amount of materials, only a few tensile tests were performed. As shown in Table 1, elongation is very sensitive to heat treatment and to the flow stress.

Table I. Room Temperature Ductility Data for the Ti-23-11 Alloy.

Condition	Elongation	Fracture Mode
Solution-Treated (S/T)	22%	Ductile, microvoid fracture; transgranular
S/T + Age 575°C/6hr	~1%	Ductile, slip-band decohension evident
S/T + Age 675° C/6hr	~3%	Ductile, microvoid fracture; partly intergranular

The minimum ductility (1%) occurred in the peak hardness condition at yield stresses of ~1700 MPa. No effort was made to perform thermo mechanical treatments to optimize ductility.

4. Oxidation Behavior

In addition to creep resistance, the oxidation resistance of titanium alloys is a primary issue regarding their elevated temperature applicability. Beta titanium alloys generally possess poor oxidation resistance, due in part to the presence of the beta-phase stabilizing elements such as vanadium [19]. Recently, replacing vanadium by molybdenum (as the β-phase stabilizer), has improved the oxidation resistance of the β Ti alloy, Timetal 21S [19]. In the present case, it should be recalled that Ti-23-11 contains a significant amounts of Nb and Al, but no V.

Illustrated in Figure 7 is the weight-gain behavior of the Ti-23-11 oxidized in flowing laboratory air over the temperature range of 600°C to 700°C for times up to 96 hours.

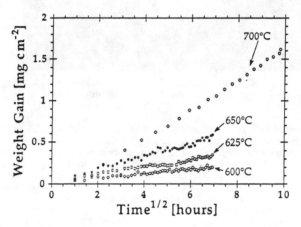

Figure 7. Weight-gain histories for Ti-23-11 oxidized in laboratory air between 600 and 700°C for up to 96 hours.

The alloy exhibits parabolic oxidation kinetics, such that weight gain during oxidation may be expressed by the relationship

$$m = \sqrt{k_p t} \tag{2}$$

where m is the weight gain per unit area of original sample, k_p is an effective parabolic rate constant, and t is the time of exposure at the oxidation condition. The continuously increasing sample weight with increasing time implies that a protective oxide scale does not form within the exposure times examined. Furthermore, the Ti-23-11 alloy did not exhibit gross discontinuities in weight gain during oxidation; i.e. no gross transitions in oxidation kinetics were recorded by the thermogravimetric analysis (TGA) equipment. This suggests single stage parabolic oxidation kinetics with no significant changes in the dominant oxidation mechanism over the temperature range and exposure times investigated.

Assuming the oxidation process to be diffusion controlled, the TGA data of Figure 7 can be analyzed in the form of an Arrhenius-type plot. Figure 8 indicates that the effective parabolic rate constant for Ti-23-11 is

$$k_p = 2.37 \times 10^{13} \exp \left(\frac{-66\,102}{RT} \right). \tag{3}$$

The activation energy for oxidation (Q = 66 kcal/mol or 277 kJ/mol) obtained from Equation 3 corresponds reasonably well with literature values (Q = 250 - 300kJ/mol) for oxidation of titanium [20]. Consistent with that behavior, this result tends to indicate that oxidation is rate controlled by oxygen diffusion through a scale.

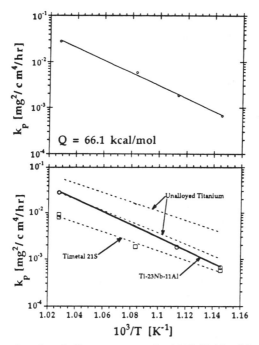

Figure 8. Arrhenius plot of parabolic rate constants for (a) Ti-23-11 oxidized in laboratory air between 600 and 700°C for up to 96 hours. Also included (b) are the parabolic rate constants for Ti-15Mo-2.7Nb-3Al-0.2Si (Timetal 21S) and unalloyed titanium after ref. 20,23,24.

Inspection of oxidized material (exposed at 600°C for 48 hours) revealed a scale of thickness on the order of ≈1 μm. This is similar to the oxide thickness (≈ 0.8 μm) which was observed in the oxidation resistant β Ti alloy, Timetal 21S, (Ti-15Mo-2.7Nb-3Al-0.2Si) which was exposed at 600°C for 72.5 hours in air [21]. The oxidized surfaces of the Ti-23-11 alloy were dark gray to black in color, with occasional patches of a white scale. X-ray diffraction identified the white patches as TiO_2. The dark gray surface scale was not identified, but it is believed to be TiO_2, based on a similar observation and identification made previously [22].

Finally, Figure 8 also shows a comparison of the parabolic rate constants for the Ti-23-11 alloy with a range of literature values for pure Ti as well as data obtained in our laboratory for the "oxidation-resistant" beta alloy Timetal 21S. The results indicate the Ti-23-11 to be somewhat better than pure Ti. When compared to Timetal 21S, the Ti-23-11 alloy shows similar oxidation behavior in air at 600°C, but it is worse at 700°C.

<u>Summary</u>

This study establishes the concept that, despite the absence of any alloy development, the β Ti alloy, Ti-23Nb-11Al, can be age-hardened by α_2-phase precipitates to a high level of strength. Furthermore, the ordered precipitates exhibit comparatively good resistance to particle coarsening phenomena, thus providing at least short-time strength retention to temperatures of ~600°C, which is much higher than conventional β Ti alloys. In this temperature range, the Ti-23-11 has similar oxidation resistance in flowing air to the recently developed Timetal 21S.

Acknowledgments

This research was supported by the Office of Naval Research.

References

1. Beta Titanium Alloys in the 1980's, R. R. Boyer and H. W. Rosenberg, eds. (Warrendale, PA: TMS, 1984).

2. L. S. Quattrocchi, D. A. Koss, and G. Scarr Scripta Metall. et Mater, 26 (1992) 267.

3. L. S. Quattrocchi and D. A. Koss, in Proc. to Seventh World Conf. on Titanium, San Diego, 1992.

4. D. Goto, M.S. Thesis, The Pennsylvania State University, 1993.

5. M. Lovato and M. G. Stout, Metall. Trans., 23A (1992) 935.

6. Y. Combres, G. Dumas, A-M. Chaze, and B. Champin in Proc. to Seventh World Conf. on Ti, San Diego, 1992.

7. R. D. Doherty, Met. Sci. 16 (1982).

8. D. A. Koss and J. Chesnutt, R. I. Jaffee and H. M. Burte, eds., Ti Science and Technology (Plenum Press, NY, 1973) p. 1097.

9. S. M. Tuominen and D. A. Koss, Metall. Trans., 6A (1975) 1737.

10. R. Zeyfand and H. Conrad, Acta Metall. 19 (1972) 985.

11. G. Hari Narayanan and T. F. Achhold, Metall. Trans. 2 (1971) 1264.

12. A. H. Cottrell, Dislocations and Plastic Flow in Crystals (Oxford, Clarendon Press, 1953).

13. P. V. Ignatov, L. F. Sokyriansky, M. S. Model and A. Yashinyaev, in Titanium Science and Technology, ed. R. Jaffee, H. M. Burte, (New York: Plenum Press, 1973), p. 2535.

14. P. Dadros and J. F. Thomas, Metall. Trans., 12A (1981) 1867.

15. S. L. Semiatin and G. D. Lahoti, Metall. Trans., 12A (1981) 1705.

16. H. Oikawa in Creep and Fracture of Engineering Materials and Structures, ed. by B. Wilshire and R. W. Evans (London, Inst. of Metals, 1991), p. 31.

17. O. D. Sherby and P. M. Burke, Prog. Mater. Sci., 13 (1967) 325.

18. H. Oikawa, K. Nishimura and M. X. Cui, Scripta Metall., 19 (1985) 825.

19. P. J. Bania, ISIJ Int., 31 (1991) 840.

20. Z. Liu and G. Welsch, Metall. Trans. A, 19A (1990) 1121.

21. T. A. Wallace, R. K. Clark and K. E. Wiedemann in <u>Seventh World Conf. on Titanium</u>.

22. J. Stringer, <u>Acta Metall.</u>, 8 (1960) 758.

23. A.M. Chaze and C. Coddet, <u>J. Less Common Met.</u>, 157 (1990) 55.

24. J. Unnam, R.N. Shenoy and R.K. Clarke, <u>Oxid. Met.</u>, 26 (1986) 231.

APPLICATIONS OF
BETA TITANIUM ALLOYS

APPLICATIONS OF BETA TITANIUM ALLOYS IN AIRFRAMES

--Keynote Lecture--

R.R. Boyer
Boeing Commercial Airplane Group
P.O. Box 3707, M/S 73-44
Seattle, WA 98124

Abstract

Beta alloys have been available since the early 1950's, but their utilization has been relatively limited on aircraft with one notable exception, the SR-71 "Blackbird". Ti-13V-11Cr-3Al (B120VCA) was used extensively on the Blackbird. The alloy is, however, difficult to work with, which had a negative impact on the cost effectiveness. Beyond this, the only significant airframe utilization of beta alloys was for springs until the 1980's. (The B-1 did have about 200 part numbers fabricated from formed Ti-15V-3Cr-3Al-3Sn sheet but beyond this the alloy was never used in significant amounts in the aerospace industry.) Ti-10V-2Fe-3Al was the next (metastable) beta alloy used, for high strength forgings, though it took about 10 years for this alloy to begin to be used in significant amounts after its introduction into production. The beta alloys have some significant advantages over the other titanium alloys with regard to the strength and toughness achievable, section sizes which can be heat treated, enhanced corrosion resistance, and ease of fabrication, such as advantages in rolling and forging. These attributes are becoming recognized, and this alloy system is being studied and used in greater amounts today. This paper will summarize production applications, along with the reasons for selection, for beta alloys used in military and commercial airframes.

Beta Titanium Alloys in the 1990's
Edited by D. Eylon, R.R. Boyer and D.A. Koss
The Minerals, Metals & Materials Society, 1993

Introduction

This is a compilation of information on the application of beta (ß) alloys on airframes from around the world. Inquiries were made in France, Germany, Austria, England, Russia, China and Japan, as well as the United States, at major airframe manufacturers and parts producers. The results indicate that ß alloys are being used to a greater extent than ever before, particularly in the United States and Russia. This is true for both military and commercial aircraft. The reasons for this are most commonly related to the higher strengths (strength refers to tensile strength unless otherwise stated throughout this text) achievable with these alloys, the excellent strength/toughness combinations which can be achieved, or high fatigue strengths. There are also applications which arose because of the enhanced processing characteristics of ß alloys, such as improved isothermal forging characteristics or formability. An attempt will be made to show as many of the various product forms as possible to provide an insight into the types of areas where beta alloys are applicable.

The primary alloys of interest include Ti-10V-2Fe-3Al (Ti-10-2-3), used for forgings, Ti-15V-3Cr-3Al-3Sn (Ti-15-3), used for sheet metal structures and castings, Ti-3Al-8V-6Cr-4Mo-4Zr (Ti-38-6-44 or ß-C), used for springs, and, in the Commonwealth of Independent States (CIS), the BT-22 alloy, which is nominally Ti-5Al-5Mo-5V-1Fe-1Cr. The BT-22 applications to be discussed involve primarily forged products. There is also a growing level of interest in a new alloy developed by Timet, ß-21S (Ti-14.7Mo-2.7Nb-3Al-0.2Si) due to its high temperature capabilities and excellent corrosion resistance, which is bill-of-material for the Boeing 777 aircraft.

Beta Titanium Alloy Applications

Historical Applications

The first major application of ß alloys was the SR-71 "Blackbird". The primary alloy used on this aircraft was Ti-13V-11Cr-3Al, an alloy which is difficult to melt, to fabricate into components, and whose properties would not be considered acceptable by today's standards. Yet the use of Ti-13V-11Cr-3Al on the SR-71 represents a real success story. The product forms used included sheet, extrusions, billet, bar and forgings. Applications included skins, frames, bulkheads, longerons, ribs, rivets and essentially the complete main and nose landing gear. Applications on the SR-71 will not be discussed as they have been reported elsewhere [1].

Following the SR-71, the next significant use of a ß alloy was the B-1B bomber built by North American Rockwell in the mid 1980's. The parts which could be readily identified today are 96 components (55 part numbers), both structural and non-structural, in the aft nacelle structure which were fabricated from Ti-15-3. Figure 1 shows the type of components involved. The components shown in this figure are either Ti-6Al-4V (Ti-6-4) or Ti-15-3; it was not attempted to denote the Ti-15-3 parts. While Ti-15-3 provided a strength increase over Ti-6-4, the principal reason for selecting Ti-15-3 was improved formability for those components which are simple sheet metal parts, i.e., cost savings drove this application [2]. Although the ingot costs of Ti-15-3 are much higher than that of Ti-6-4, the final cost of the Ti-15-3 strip product may be lower due to its greater workability. Ti-15-3 can be rolled as a coil product with fewer intermediate anneals than Ti-6-4. Ti-6-4 must be hand rolled in a pack, which is labor intensive. This advantage becomes apparent at thinner gages. Ti-15-3 strip, at gages below about 1.8 to 2 mm (0.070 to 0.080 in) can be lower cost than Ti-6-4.

At about the same time, mid-80's, engine access doors for the Fairchild-Saab FS-340 commuter aircraft were fabricated from formed Ti-15-3 built-up sheet structure. This door, which is about 5 square feet in size, was used in the solution treated condition as higher strength was not required, and thermal stability was not considered an issue for this type of application. Studies at Fairchild indicated that this structure could be fabricated at lower cost using Ti-15-3 than with Ti-6-4, again, due to the forming advantages of Ti-15-3 [3]. It should be noted that the titanium door weighed the same as the aluminum door it replaced; this was accomplished by selective chem milling to reduce the weight.

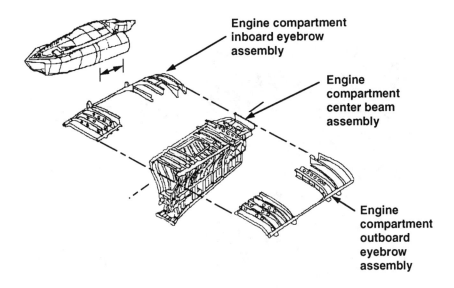

Figure 1. B-1B aft nacelle major assembly breakdown; components illustrated
represent both Ti-6Al-4V and Ti-15V-3Cr-3Al-3Sn
(Courtesy of North American Rockwell)

Current Applications

Springs McDonnell Douglas started using Ti-13-11-3 springs for their commercial aircraft in the 1970's. Titanium springs offer very large weight advantages over steel springs, up to about 70%, due to the lower shear modulus and density of titanium, and only require half the volume. In the early to mid 1980's Ti-38-6-44 springs were developed to replace the Ti-13-11-3. This new spring material offered the same properties as the latter [4], but at a much reduced cost. Lockheed also started with Ti-13V-11Cr-3Al springs on the L-1011 and ß-C was made an option when it became available. This lower cost alternative resulted in much greater utilization of titanium springs at Boeing. Boeing was slower to utilize the Ti-13-11-3 springs than other manufacturers due to high costs associated with it, and associated manufacturing problems. Typical applications for the above manufacturers include up and down locks on landing gear, door counterbalance springs, flight control springs for limits and overrides, feel and centering springs for the yoke, pedal returns (brakes) and hydraulic return springs. (This is not to infer that all aircraft manufacturers use titanium springs for each of the applications mentioned.) Between the MD-80 and MD-11, McDonnell Douglas has about 300 part numbers in Ti-13-11-3 and 150 part numbers which are Ti-38-6-44, and they are in the process of converting more of the Ti-13-11-3 springs. All of the titanium springs on the C-17 are ß-C [5]. Boeing is in the process of changing many of their stainless steel springs to ß-C; a bulk of the springs on the 777 are ß-C. (ß-C springs are used in the 1240-1450 MPa (180-210 ksi) strength range.) Dowty Aircraft Landing Gear Group and British Aerospace are beginning to utilize ß-C springs. They will be utilized on the A-330 and A-340 aircraft for the downlock springs on the wing and centerline landing gear. [5]

Ti-15-3 is being used to fabricate clock and leaf springs (at a minimum strength level of 1035 MPa (150 ksi) from strip or bar product. These types of springs and a compression spring fabricated from square ß-C wire are shown in Figure 2. These springs were used as there was not enough volume available to use a steel spring and carry the required loads. Titanium is often used where there is a volume constraint between surrounding structure.

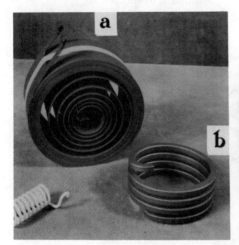
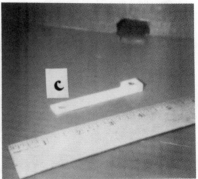

Figure 2. Titanium aircraft springs (A) A door counterbalance clock spring
fabricated from Ti-15-3 strip, (B) ß-C square wire launch bar power
unit spring for the T-45 and (C) part of a leaf spring fabricated
from Ti-15-3 RTO control stand for the Boeing 777.

Forgings and Forged products The use of Ti-10-2-3 forgings on the 757 represented the first use of a high strength (1195 MPa or 173 ksi) beta alloy forgings on commercial aircraft. These were used to save weight in replacement of steel, about 40% in replacement of 15-5PH steel in the 1240-1380 MPa (180-200 ksi) strength condition. Commercial growth of this product was fairly slow until recently, with minor forging conversions to this alloy on other Boeing aircraft and that of other aircraft manufacturers. Growth of this product received a substantial boost with the C-17, which used many Ti-10-2-3 forgings at the 1100 and 1195 MPa (160 and 173 ksi) strength levels, primarily the former. (One of the advantages of an alloy such as this is the fact that it can be heat treated over a range of strength levels to achieve the desired strength/toughness combination.) The primary uses on the C-17 were in the cargo door, nacelle and empennage areas [6].

Boeing then increased the usage of this alloy substantially on the 777 aircraft, all of it at the high strength level. In fact this may be the first commercial aircraft where Ti-6Al-4V is not the predominant titanium alloy. The most significant use of Ti-10-2-3 is in the landing gear structure. Just about everything shown in Figure 3A, except for the inner and outer cylinders, is Ti-10-2-3. The largest item in this structure is the truck beam (Figure 3B). This is fabricated by electron beam welding three pieces together. The truck beam is about 343 mm (13-1/2 in) in diameter and about 3 meters (10 ft) long. Utilization of this alloy in place of 4340M saved approximately 272.4 kg (600 lb) on the main landing gear.The steering mechanism for the nose landing gear is also Ti-10-2-3. Another major component is the wing flap tracks (Figure 4). Out of 12 flap tracks, 6 (the larger ones) have been committed to Ti-10-2-3, replacing 4330M at the 1515 MPa (220 ksi) strength level, resulting in a weight savings on the order of 41 kg (90 lb). These flap tracks are on the order of 1.3 to 1.6 m (4 to 5 ft) long, about 127 mm (5 in) deep and 63 mm (2.5 in) wide. Ti-10-2-3 is used throughout the airplane, in the wings, fuselage, doors, nacelles and cargo handling structures.

The helicopter industry is also gaining an interest in Ti-10-2-3. Bell Helicopter Textron is using the alloy for three components in the rotor system of the V-22 Osprey which they can not discuss in any detail. They will use the alloy in the 1105 MPa (160 ksi) heat treat condition. This provides superior high cycle fatigue and fracture toughness in place of Ti-6-4.

LeFiell Manufacturing Co. is fabricating some APU mount system support struts, Figure 5, for the 777. These are normally fabricated from 15-5PH. The Ti-15-3 provides about a 40% weight savings over steel, resulting in a 2 kg (4.5 lb) weight savings per airplane. Previous attempts were made to fabricate these from Ti-6Al-4V, but it could not be swaged sufficiently to enable forming the tongue ends, another example of the advantage of the improved workability of ß-alloys.

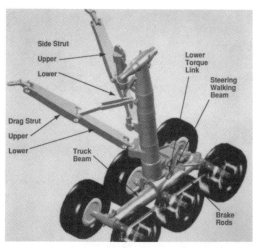

Figure 3A. Boeing 777 main landing gear, Ti-10-2-3 forging applications

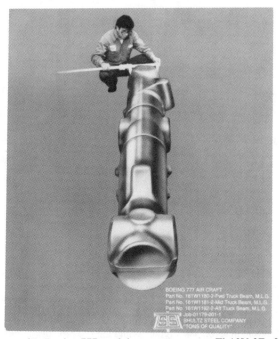

Figure 3B. Boeing 777 truck beam components, Ti-10V-2Fe-3Al

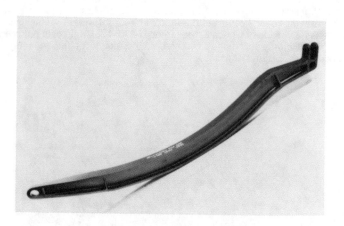

Figure 4. Boeing 777 flap track, Ti-10V-2Fe-3Al

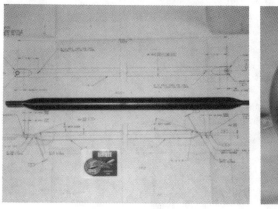

Figure 5. Ti-15V-3Cr-3Al-3Sn APU mount system support struts (Courtesy
LeFiell Manufacturing Co.) and APU fire extinguisher bottle
(Courtesy of Walter Kidde)

The CIS uses their BT-22 alloy for highly loaded forgings; applications are reported to include landing gear components, spars, beams, longerons, stringers, fasteners and springs [7]. Figure 6 illustrates the landing gear components used on the IL-86 and IL-96-300 fabricated from this alloy. They also use it for flap tracks and numerous other forged components. (Extrusions are also fabricated from this alloy.) It is used at the 1105 to 1200 MPa (160 to 175 ksi) strength strength level. They report achieving full properties in sections up to 200 mm (8 in) thick.

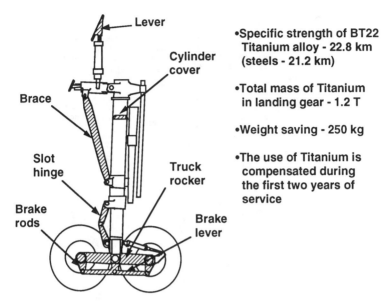

Lever

Cylinder cover

Brace

Slot hinge

Truck rocker

Brake rods

Brake lever

•Specific strength of BT22 Titanium alloy - 22.8 km (steels - 21.2 km)

•Total mass of Titanium in landing gear - 1.2 T

•Weight saving - 250 kg

•The use of Titanium is compensated during the first two years of service

Figure 6. Usage of Titanium in IL-86 and IL-96-300 landing gears

Flat product The Ti-15-3 alloy also made significant inroads on the 777. The environmental control system (ECS) pneumatic ducts were fabricated from this alloy. The ducts are 178 mm (7 in.) in diameter and the wall thickness varies from 0.5 to 1 mm (0.020 to 0.040 in), with most of it being the 0.5 mm gage. Use of Ti-15-3 permitted reducing the duct gage in half from that which would be used for commercially pure titanium. About 49 m (160 ft) of ducting will be used resulting in a weight savings of about 63.5 kg (140 lb). Ti-15-3 fire extinguisher bottles are also being specified, which are projected to provide about a 23 kg (50 lb) weight savings in place of the presently used 21-6-9 steel bottles (Figure 5). Hemispheres will be formed from Ti-15-3 sheet which will be welded together to form the bottle.

Figure 7 presents some miscellaneous clips fabricated from Ti-15-3. These applications are possible due to the high formability of the alloy. The interior window clips, used to hold the window reveal to the interior panels, are used two per window on the 747-400 (they are used in the solution treated condition). The crew microphone holders, used on the 737 and 757, are used because of the excellent spring-back of Ti-15-3. The steel holders had a problem taking a permanent set during normal use. Nut clips, for the 777, are another interesting application. These are small parts, but they will use on the order of 15 to 20,000 per airplane. Their primary use will be to attach the metallic floor structure to the composite floor beams. These are being used due to their corrosion compatibility with the composite floor structure, and there is also a weight saving over the present steel nut clips. It would not be economically viable to manufacture these from Ti-6-4. Numerous other clips and brackets are also being formed using Ti-15-3 for use in the floor structure.

The 777 nacelle will provide the first commercial usage of ß-21S [8]. This application is startling in that only about 2-3 years passed between development of the alloy and its first commercial application. Normally this will take from 10 to 20 years. The advantages of the alloy are high oxidation and corrosion resistance. It will be used in the aft cowl and the plug and nozzle of the exhaust assembly (Figure 8). The skins will be fabricated from sheet, and some of the frames will be produced from bar. It will be used in two heat treatment conditions, STA and STOA. The STA condition (for applications with temperatures of about 480C (900F) and below) will have a minimum tensile strength of 1035 MPa (150 ksi). The STOA condition will be used for higher temperatures and it will have a minimum tensile strength of 860 MPa (125 ksi); reduced strength enhances thermal stability of the alloy at high temperatures [9].

341

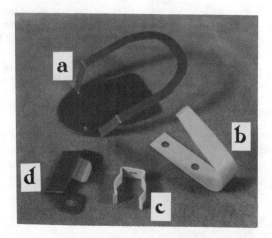

Figure 7. Miscellaneous Ti-15V-3Cr-3Al-3Sn formed sheet components. (A) crew microphone holder for Boeing 737/757, (B) access door clip (C) tool clip and (D) 747-400 interior window clip

The CIS uses their BT-22 alloy and BT-35 (Ti-15V-3Cr-1Mo-0.5Nb-3Al-3Sn-0.5Zr) in the 1100 to 1200 MPa (160 to 175 ksi) strength range to fabricate sheet metal components for their highly loaded structure. They do a lot of welding of structural components, and claim to achieve base metal properties in the welds.

Castings The use of high strength titanium castings is beginning to emerge. The 777 will represent the first use of Ti-15-3 castings at an 1140 MPa (165 ksi) tensile strength minimum. Vibration isolator mounts for the auxiliary power unit (APU) will be fabricated from Ti-15-3 castings (Figure 9). Ti-15-3 castings will also be used as cargo guides in the cargo handling system. Allied Signal-Bendix recently committed to Ti-15-3 castings for brake torque tubes (Figure 9B).

The BT-35 alloy is also used for castings in the CIS. They use it at about the 1100 MPa (160 ksi) strength level and report better ductilities than obtained for Ti-15-3.

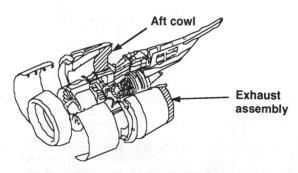

Figure 8. Beta-21S applications on the Boeing 777, Pratt and Whitney 4084 engine nacelle

Figure 9. Ti-15V-3Cr-3Al-3Sn Castings (A) Boeing 777 APU vibration isolator
mount (Courtesy of TiLine) (B) prototype brake torque tube
(Courtesy of Howmet and Allied Signal-Bendix)

Future Applications

Ti-10V-2Fe-3Al is being studied by three additional helicopter manufacturers, Westland
Helicopter Limited, Sikorsky and Eurocopter. Westland is looking seriously at the use of Ti-
10-2-3 components for their Lynx helicopter semi-rigid rotor head [10, 11]. The Lynx was
originally designed with an all-up-weight (AUW) of 3860 kg (8,500 lb) and annealed Ti-6-4
was satisfactory. Metallurgical and process optimization of the billet and forging stock
increased the performance of the Ti-6-4 sufficiently to handle an AUW of 4880 kg (10,750 lb).
They now want to increase the AUW to 5585 kg (12,300 lb) and the annealed Ti-6-4 is no
longer adequate. They studied 3 options: (1) Beta processing of Ti-6-4, (2) STOA processing
of Ti-6-4, and (3) a change from Ti-6Al-4V to one of the higher strength commercially
available alloys; they studied IMI 550 and Ti-10-2-3. Ti-10-2-3 was selected on the basis of its
enhanced high cycle fatigue life combined with its lower stiffness characteristics and superior
strength. They are using about an 1170 MPa (170 ksi) tensile strength minimum. The Ti-10-2-
3 exhibited almost twice the alternating fatigue stress capability of annealed Ti-6-4 and about a
50% improvement in comparison to STA Ti-6-4 for the endurance limit. The three components
under consideration, the disc, sleeve and mast, are illustrated in Figure 10.

Sikorsky is also studying its use for similar type components. Due to the higher fatigue
performance of Ti-10-2-3, future equipment that presently uses Ti-6-4, that is not switched to
composites, will probably utilize Ti-10-2-3.

Eurocopter is studying using Ti-10-2-3 for rear rotor head applications, again due to the high
strength and excellent fatigue properties. The fatigue advantage of this alloy is illustrated in
Figure 11 [12].

Westland is studying two other areas which could utilize ß-alloys. There is some formed sheet
metal around the engine/APU on the EH101 which is currently using ferrous alloys.

343

Conversion to either ß-21S or Ti-1100 is being studied. The key design parameter will be elevated temperature fatigue. Titanium could provide a weight savings of about 21.8 kg (48 lb).

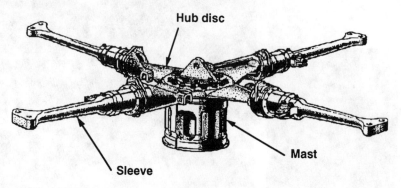

Figure 10. Lynx helicopter bolted main rotor hub illustrating Ti-10V-2Fe-3Al components under consideration are indicated by arrows.
(Courtesy Westland Helicopter Limited)

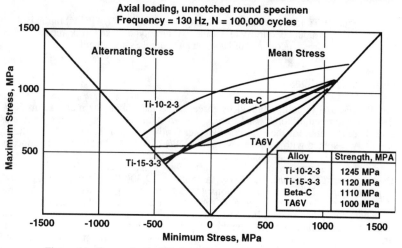

Figure 11. Comparison of Fatigue Strength of Titanium Alloys [12]

They are also studying Ti-15-3 castings for structural linkages. They recognize the advantage of the higher strength alloys and the fact that the increased strength reduces the cost per pound of weight saved. They are studying minimum tensile strengths of about 1170 MPa here also.

Boeing is also studying ß-21S castings for use in the nacelle of the 777. The aft fairing heat shields are presently fabricated from Ti-6Al-2Sn-4Zr-2Mo-Si castings. If ß-21S is castable, it could be used in this application because of its better resistance to hydraulic fluid at elevated temperatures.

The CIS reports that they are moving toward using their ß alloys at higher strengths. They are presently using them in the 1100-1200 MPa (160-175 ksi) range and they intend to increase the strength to the 1200-1300 MPa (175-190 ksi) level [7].

Timet is in the process of introducing a new low-cost alloy, Ti-6.5Mo-4.5Fe-1.5Al (Timetal LCB) initially slated for rod and wire applications. Development of this alloy was driven by the auto industry, which has a strong desire to use titanium springs, but procurement costs are too high. It is anticipated that this alloy will be less than half the price of ß-C; a very significant economic impact. Preliminary data [13] indicates that tensile strengths on the order of 1520 MPa (220 ksi) with elongations on the order of 6% can be achieved from laboratory ingots. Larger production ingots would provide a product with a more favorable working history, which could result in improved ductility. This could mean that this alloy has real promise for high strength fasteners to replace Inconel 718, and for springs.

Another high strength alloy, actually a ß-rich α/ß alloy, ß-CEZ (Ti-5Al-2Sn-4Zr-4Mo-2Cr-1Fe) could also be of interest. This alloy was really developed for engine applications for high strength-moderate temperature applications. They have reported the capability of achieving 1380 MPa (200 ksi) yield strengths with a fracture toughness of about 55 MPa-m$^{1/2}$ (50 ksi-in$^{1/2}$) [14]. This is a very attractive combination of properties which could be useful for airframe applications.

Summary

Beta alloys are beginning to play a significant role on both military and commercial aircraft. This is apparent in the last two domestic fixed wing aircraft, the McDonnell Douglas C-17 and the Boeing 777. Ti-10V-2Fe-3Al forgings have played the major role, particularly on the 777. The biggest step was utilization of this alloy for most of the major components in the main landing gear. Beta-C plays a continuing and growing role as the spring material of choice. Formed Ti-15V-3Cr-3Al-3Sn sheet structure is being utilized, primarily because of it high formability, and it is also emerging as a casting alloy where it has a significant strength advantage over the work-horse alloy, Ti-6Al-4V. ß-21S is a new alloy, which offers real weight advantages over nickel and ferrous alloys for elevated temperature applications due to its high oxidation resistance and good creep resistance. Beta-21S resistance to hydraulic fluids at elevated temperatures is another important characteristic for applications in the nacelle area, where it will be used on the Boeing 777.

The Commonwealth of Independent States has also made significant use of this alloy system. Major components of the landing gear for the IL-86 and IL-96-300, and other aircraft are fabricated from their BT-22 alloy. Other high load carrying components in their aircraft utilize BT-22 in the form of forgings, extrusions, fasteners and springs. They also use BT-35 for various product forms, including castings.

Beta alloys are also becoming more important in the helicopter industry, in the rotor hub area. Ti-10V-2Fe-3Al is an attractive alloy because of its high strength, high toughness and excellent fatigue properties.

Two alloys in the development stage, ß-CEZ and Ti-6.5Mo-4.5Fe-1.5Al, could become important because of the excellent properties they promise. The latter alloy could be particularly significant in light of its projected low cost. The economics continue to play an increasing role, making it more difficult to use titanium in applications of choice, so reduced cost could be very important. Up to this time, the use of ß-alloys has primarily been driven by their superior properties and weight savings potential. In the future, cost will become more important, so increased emphasis should be placed on lower cost alloys and/or taking advantage of improved processing capabilities of these alloys to minimize final component costs.

References

1. R.R. Boyer and H.W. Rosenberg, Beta Titanium Alloys in the 1980's (Warrendale, PA:TMS, 1984), 1-8.

2. R. Brunken, Private communication with author, Rockwell International Corporation, December 12, 1992.

3. M.E. Rosenblum, A. Shames and W.B. Treppel, Beta Titanium Alloys in the 1980's (Warrendale, PA:TMS, 1984), 307-329.

4. R.R. Boyer, R. Bajoraitis, D.W. Greenwood and E.E. Mild, ibid, 295-305.

5. C. Pepka, Private communication with author, Renton Coil Spring, January 4, 1993.

6. A. Amin, Private communication with the author. McDonnell-Douglas, February 21, 1993.

7. I.S. Polkin, Private communication with the author, VILS, January 18, 1993.

8. R.R. Boyer, "New Titanium Applications on the Boeing 777 Airplane," Journal of Metals, 44 (5) (1992), 23-25.

9. W.M. Parris and P.J. Bania, "Oxygen Effects on the Mechanical Properties of TIMETAL-21S". (Paper presented at the 7th World Conference on Titanium, San Diego, CA, June 28 to July 2, 1992.)

10. D.P. Davies, "Effect of Heat Treatment on the Mechanical Properties of Ti-10V-2Fe-3Al for Dynamically Critical Helicopter Components". (Paper presented at the 7th World Conference on Titanium, San Diego, CA, June 28 to July 2, 1992.

11. D.P. Davies, Private communication with author, Westland Helicopter Limited, November 16, 1992.

12. M.H. Campagnac, Private communication with the author, Aerospatiale Centre Commun de Recherches Louis Blériot, February 21, 1993.

13. P.J. Bania, Private communication with author, TIMET, January 14, 1992.

14. B. Prandi, J.F. Wadier, F. Schwatrz, P.E. Mosser and A. Vassel, 1990 International Confernece on Titanium Products and Applications, (Dayton, OH: TDA, 1990), 150-159.

UTILIZATION OF BETA-C™ TITANIUM COMPONENTS

IN DOWNHOLE SERVICE

D.C. Dunlap and R.W. Schutz

RMI Titanium Company
1000 Warren Avenue
Niles, Ohio 44446

ABSTRACT

The utilization of Beta-C™ titanium for various geothermal and deep oil and gas well components has expanded significantly over the past several years. This has been driven by its unique combination of elevated strength, superior resistance to high temperature sweet and sour brines, low density and modulus, and deep hardenability. This paper provides an overview of Beta-C™ alloy mechanical, physical and corrosion properties and a database developed for and relevant to downhole service. Based on these properties, several typical examples of alloy use in oil and gas drilling and production, geothermal exploitation, and wireline-conveyed evaluation are discussed.

Beta-C™ is a registered trademark of RMI Titanium Company.

Beta Titanium Alloys in the 1990's
Edited by D. Eylon, R.R. Boyer and D.A. Koss
The Minerals, Metals & Materials Society, 1993

INTRODUCTION

The Beta-C™ alloy (Ti-3Al-8V-6Cr-4Mo-4Zr) was developed by RMI Titanium Company in the late 1960's for utilization in critical aerospace applications based on its low density, low elastic modulus, and ability to be aged to very high strength levels. As a result, this alloy has been successfully designed into various aerospace springs, fasteners, fittings, and numerous missile component uses. In the early 1980's, it was also recognized that this alloy possessed superior corrosion resistance relative to conventional titanium alloys stemming from its significant molybdenum content (1). The enhanced resistance to reducing acids and hot localized chloride attack combined with its high strength-to-density characteristic motivated its consideration for the more extreme, severe downhole environments in oil, gas and geothermal exploitation.

Over the past ten years, the alloy has established a successful service record in hypersaline geothermal brine wells (2) and in deep sour gas wells (3,4). Beta-C™ titanium has recently been incorporated into ASTM product specifications as Ti Grade 19 with a UNS number of R58640, and has been approved under the NACE MR-01-75 document for sour service use since 1985. Extensive laboratory corrosion evaluation under worst-case sour gas well fluid conditions over the past three years (4,5,6) has led to further optimization of Beta-C™ alloy corrosion performance via minor alloy palladium additions. This "Pd-enhanced alloy" extends the useful stress and crevice corrosion resistance of the alloy in sour brines to at least 232°C (4,6), extending resistance beyond that of rival Ni-Cr-Mo alloys in worst-case sour gas wells.

This paper provides a cursory overview of Beta-C™ titanium alloy metallurgy and mechanical, physical, and corrosion properties which are relevant to downhole well environments. Five specific examples of alloy application in downhole service will be discussed, with emphasis on alloy selection criteria.

BETA-C™ ALLOY METALLURGY

Beta-C™, defined by the chemical composition range outlined in Table I, is a solute-rich metastable-beta titanium alloy. The chromium, vanadium, and molybdenum alloying elements serve to stabilize the beta (bcc structure) phase, such that 100% beta phase is retained upon rapid or air-cooling from temperatures above the beta transus (~750°C). The aluminum addition enhances the final aging response by stabilizing alpha phase (hcp structure) precipitates. The alloy is basically strengthened via two mechanisms: solid solution hardening derived from all alloying elements, and by aging which involves precipitation of fine alpha phase within the beta phase matrix. Due to its high beta stabilizing alloy content, the alloy has exceptional deep section age hardenability and is ageable to strengths in excess of 1380 MPa (200 ksi) depending on product form. The alloy is also available with an enhanced palladium content (typically 0.05-0.06 weight %) that serves to increase both the stress cracking and crevice corrosion resistance of the material without affecting mechanical or physical properties.

Table I   Beta-C™ Chemical Composition Range

CHEMICAL ELEMENT	PERCENT BY WEIGHT (%)
Aluminum	3.0 to 4.0
Vanadium	7.5 to 8.5
Chromium	5.5 to 6.5
Molybdenum	3.5 to 4.5
Zirconium	3.5 to 4.5
Iron	0.30 maximum
Nitrogen	0.03 maximum
Carbon	0.05 maximum
Oxygen	0.12 maximum
Hydrogen	0.02 maximum
Palladium	0.04 to 0.10
Others (each)	0.10 maximum
Others (total)	0.40 maximum

A typical Beta-C™ alloy product form utilized in downhole components is round bar or billet in the size range of 20 cm diameter or smaller. In processing material to this final size, ingot is first press or rotary forged at temperatures above 980° C to produce billet. After solution treatment, this billet is aged to achieve the final aim properties. Tightly controlling the age time and temperature allows the final strength condition to be consistently controlled over the range of possible values. It is very important to note that a final conditioning step be performed in which the surface is removed via machining or sandblast/ pickle. This will ensure that the oxide scale, alpha case, and oxygen rich surface layers incurred from hot processing and final heat treatment are completely eliminated as these hard, brittle layers may promote crack initiation and stress corrosion cracking. For downhole service above 177°C, it is vital that the Beta-C™ material be thoroughly aged to a stable metallurgical condition, thereby assuring thermal stability of strength and ductility properties with time.

Beta-C™ alloy tubular product forms are also commonly produced and utilized in downhole service. Tubulars with walls in excess of 10 -12 cm are generally beta extruded from trepanned billet to near-final product sizes. These extruded tubulars are solution treated and then aged to a strength level, followed by final surface conditioning (machining or sandblast/pickle). Thinner wall, smaller diameter pipe, such as gas well production tubing and well workover or snubbing strings, are produced most efficiently by cold pilgering after beta-extrusion to an intermediate hollow form. The final cold pilger step produces a pipe product with a highly refined microstructure after final heat treatment, possessing good ductility, uniform age, and thermal stability in these cold processed tubulars. As noted for bar/billet products, it is vital that thorough removal of oxide and oxygen-rich (alpha case) surface layers be achieved in all Beta-C™ alloy tubular products after final heat treatment.

ALLOY CORROSION BEHAVIOR

Both standard and Pd-enhanced Beta-C™ tubular and billet products have been extensively evaluated in worst-case fluid environments anticipated in geothermal brine and deep sour gas well service. The eight year service experience documented with standard Beta-C™ production casing and packers in hypersaline brines in Salton Sea geothermal wells (2) confirm the alloy's resistance to high temperature chloride brine and CO_2. Extensive laboratory stress cracking and crevice corrosion testing of Beta-C™ products (4,5,6) under simulated worst-case deep sour gas well brine fluids have also defined the alloy's useful performance window. These studies have shown that although the standard alloy's useful temperature limit may be as high as 185°C, Pd-enhanced Beta-C™ alloy can extend this resistance to temperatures as high as 218°C and 246°C for billet and production tubular products, respectively. In these tests, no negative influence of either CO_2, H_2S, or elemental sulfur on the alloy's corrosion performance was observed.

The corrosive nature of the environments typically encountered when drilling oil and gas wells is not as severe as that found in the producing environment. High pH (8-12) muds with chloride contents of 20-50 kppm are commonly used. The drilling environment also sees lower temperatures as the circulation of the drilling mud into the well tends to cool the components below the static bottomhole temperature of the well. Finally, the exposure times associated with the drilling components are much more brief. For a tool used in drilling or evaluating a well, the downhole service time will typically be only from a few hours to a few days. In contrast, production components might see continuous service for as long as 30 years.

In addition to extremely aggressive produced fluid conditions, Beta-C™ well components may also be intermittently or occasionally exposed to various fluids injected into the well for stimulation or workover purposes. Hydrofluoric, hydrochloric, acetic and formic acids could be injected to stimulate wells, whereas methanol may be used for dissolution or prevention of hydrate plugs. Although exposure to hydrofluoric acid should be avoided, Beta-C™ components are compatible with the organic acids mentioned and up to 15% HCL when properly inhibited with certain oxidizing species within appropriate temperature windows (4). A minimum water content of 5 weight % is recommended for grades of methanol injected into wells (8) so as to avoid stress cracking that is possible in lean water (i.e. anhydrous) methanol.

MECHANICAL AND PHYSICAL PROPERTIES

The synergistic combination of several physical and mechanical properties allows the Beta-C™ alloy to be readily utilized for the applications being discussed here. Table II lists several of the physical and mechanical properties that are important in these downhole applications. While other materials may possess some of the attributes listed, the presence of all of these factors in one material is unique to Beta-C™. The ability to heat treat the alloy to high strengths is very important in most applications as design engineers can optimize the tools based on the high strength material. In the solution treated condition, the yield strength of the alloy is approximately 830 MPa. While the alloy has the ability to exceed 1380 MPa yield strength after aging, typical applications utilize the material in the strength range of 932-1208 MPa. As with most materials, an increase in the strength results in a subsequent decrease in the ductility. The intermediate range that is referenced provides a good compromise of properties.

Table II Beta-C™ Alloy Physical and Mechanical Properties

Property	Value
Typical Yield Strength MPa(ksi)	932-1208 (135-175)
Typical Tensile Strength MPa(ksi)	966-1277 (140-185)
Density g/cc (lb/in^3)	4.82 (0.174)
Elastic Modulus 10^3 MPa (10^6 psi)	104 (15.0)
Typical Hardness	32 - 40 HRc

The application of the Beta-C™ alloy in the oil tool market requires that heavy sectioned components be producible from the alloy. Large diameter round billet stock (up to 502 mm) has been utilized in producing downhole components. The ability to age the material through such heavy sections has been demonstrated and consistent aging response in sections as heavy as 254 mm has been shown.

The lower density of the titanium alloys relative to the iron or nickel based materials is of definite advantage in most downhole applications. The weight savings that can be accomplished by using seamless Beta-C™ titanium tubulars can be on the order of 70,000 kg for wells with a depth of 7,620 m (4). While the weight savings that can be expected for a downhole accessory tool is small compared to that of a full tubing string, it can nevertheless be dramatically important. In the case of the wireline conveyed evaluation tools that are used to investigate newly drilled wells, the tool string must be handled and assembled by hand by the service crew at the well location. Recent changes in OSHA regulations limit the weight that an individual worker can be expected to handle. By utilizing this high strength titanium alloy, the tools can be made lighter without changing the design or function of the tool.

The low modulus of elasticity of titanium alloys can be utilized in applications where extreme flexibility is required. As the modulus of elasticity is only about half of that found in iron based alloys, Beta-C™ can successfully endure larger bending moments or greater bending stresses than other materials. It has been well documented in the past that titanium has exceptional fatigue properties compared to iron based alloys when examined in seawater environments. The exceptional fatigue resistance of Beta-C™ in chloride containing environments is also another factor that has led to the selection of the alloy for field applications.

EXAMPLES OF BETA-C™ ALLOY APPLICATIONS

Drilling Tools

The utilization of the Beta-C™ alloy for drilling tool applications can be illustrated by reviewing the use of the alloy as a flexible shaft for downhole drilling motors. When oil and gas wells are being drilled, it is necessary for power to be applied to the drill bit at the point where it is in contact with the underground formation. The transferal of power from the prime movers at the surface to the downhole bit can be accomplished in one of two ways. In traditional applications, a continuous string of pipe connects the surface power equipment to the bit. The rotation of the pipe at the surface allows torque to be transmitted downhole to the bit via the pipe. The second method utilizes a downhole motor whereby high pressure drilling mud is pumped down the drill pipe string and through the motor (Figure 1). The motor contains a rotor/ stator assembly such that the movement of the fluid through the close tolerances of the rotor/stator causes the rotor to turn. This generates sufficient torque to power the drill bit at the formation face.

The rotation of the corkscrew shaped rotor necessarily causes an eccentric rotation of the end connection of the rotor. In contrast, the rotating drill bit requires simple rotation along the centerline of the bit. It is necessary to have a connector that provides the means of transferring the torque from the eccentrically rotating rotor to the simply rotating drill bit. In past years, this has been accomplished with a universal joint assembly that is similar to what one would find on the drive shaft of an automobile. The drawback to the assembly is the limited life that the bearings are known to have. The requirement to rebuild the assembly after only a few hundred hours of service makes it desirous to find a new means by which this job can be accomplished.

Several downhole motor manufacturers are either currently using or evaluating the use of Beta-C™ as a material for their flexible shafts. The driving factors behind the selection of the alloy for this application are the low modulus of elasticity which allows for the required flexibility of the shaft, and the fatigue resistance of the material which provides for a longer life than the universal joint arrangement. Depending on the manufacturer and the motor size, the rotor can have as much as 25.4 mm of eccentricity in its rotation. The low elastic modulus minimizes the stresses that are created in the solid metallic shaft. Even with the low modulus of the material, the high amount of deflection can create fairly high stresses in the Beta-C™ shaft and necessitates the use of a high strength, aged titanium alloy for the part. The rotational life of this shaft is short in terms of calendar days but long in terms of rotational cycles. The manufacturers typically expect the part to have a life of 5-25 million cycles. Thus, the high tensile strength and the high runout fatigue strength of the alloy make it favorable for this application.

A final factor in the material selection for this part is the fracture toughness of the material. The nature of the downhole drilling operation suggests that any component used in this service will be periodically subjected to shock loads due to the uneven torque and weight applications on the bit. Thus, the material that is used in the flex shaft must have sufficient fracture toughness to withstand the uneven loading conditions that are encountered. Beta-C™ will typically exhibit a fracture toughness of 50-80 MPa√m for round bar product in the range of 51-76 mm with yield strengths of 1035-1208 MPa (9).

Geothermal Equipment

The utilization of Beta-C™ seamless production casing in geothermal wells has been previously documented (2). In these wells, the titanium tubulars are cemented into position in the well and the superheated water flows through this conduit to the electrical generation facilities at the surface. An alternative completion utilizes cement lined steel pipe as the principle flow conduit to the surface (Figure 2). However, due to the potential corrosion problems that may be seen should the cement lining fail, this pipe is not permanently installed in the well. While having the tubulars in a retrievable condition is advantageous, the disadvantage with this arrangement is that an accessory sealing mechanism must be installed on the bottom of the cement lined pipe in order to ensure that the flowing water is restricted to the ID of the pipe. The operator has also chosen to use Beta-C™ for this accessory equipment.

The equipment is designed such that an external sealing member is permanently cemented into the well at the appropriate location. This sealing member will contact the external surface of a Beta-C™ tube that is connected via a threaded coupling to the cement lined pipe. The sealing tube has three essential dimensions that must be correct. First, the outside diameter of the tube is critical in that it must be precisely machined to meet the dimensions of the permanent sealing member. If the tube is not machined to the proper OD or straightness, the assembly will fail to properly seal. Secondly, the inside diameter of the part must be carefully maintained. Any deviation from the prescribed ID will cause disturbances in the flow pattern which can help to cause premature failure of the cement lining in the connecting pipe. Finally, the tube must be of sufficient length that the thermal cycling of the steel based pipe is accounted for. As the wells have bottomhole temperatures in excess of 280°C, the well sees a dramatic temperature difference between the operating and shut in conditions.

The primary motivation for selecting Beta-C™ for this application was corrosion resistance. The exhibited capability of the alloy for dealing with the salinity of the produced fluids at the service temperatures led to the initial selection. Secondly, the high strength of the alloy allows the equipment designer to use thinner sections and less material than might be needed if other alloys were to be used. Finally, the low coefficient of thermal expansion and low modulus of elasticity of the alloy means that the flow tube can be shortened as the Beta-C™ will be more tolerant than iron or nickel based alloys of the thermal cycling that takes place in the well.

Production Tubulars

The production of hot, sour natural gas often necessitates the use of high alloy tubulars and completion equipment in order to control corrosion. As producing environments have steadily worsened, material technology has advanced solutions to the corrosion problem. Currently, nickel based alloy tubulars and tubular accessory items are being utilized in producing from the most severe environments. Titanium Beta-C™ tubulars provide an alternative that is both technically and economically competitive to the Ni-Cr-Mo alloys (Figure 3). The unique combination of mechanical and physical properties exhibited by the Beta-C™ alloy, facilitate design changes which can dramatically diminish the overall cost of drilling and completing deep sour gas wells. Critical tubular properties directly impacting tubular string design include yield strength, density, elastic modulus and coefficient of thermal expansion. With the significantly reduced density, modulus, and thermal expansion of Beta-C™ titanium compared to Ni-Cr-Mo alloys, tubing string cost can be reduced and certain ancillary equipment can be eliminated.

These improved well design concepts are readily illustrated when comparing Beta-C™ titanium and C-276 alloy string completion options for a very deep, high pressure sour gas well. In this example, a 7,620 m (25,000 ft) deep well with a bottom hole pressure of 138 MPa (20,000 psi), downhole and flowing wellhead temperatures of 224°C and 191°C, respectively, and freshwater packer fluid completion was assumed. Utilizing standard tubular design safety factors, the resultant well string design and estimated costs were prepared with the titanium tubing being a full string of 73 mm x 7 mm wall pipe. The C-276 string had to be tapered with 2,469 m of 89 mm OD x 9.5 mm wall, 945 m of 73 mm OD x 7 mm wall, and 4,206 m of 73 mm OD x 7.8 mm wall being required to meet the required design factors. For this particular case, almost $1.4 million may be saved through the selection of Beta-C™ alloy tubulars over the heavier, "telescoping" C-276 tubing string.

In addition to the reduced string cost, additional savings can be realized by downsizing or even eliminating the seal-bore assembly. The thermally induced movement of the above noted Beta-C™ string is calculated to be 4.3 m, compared to a projected movement of 5.5 m for the C-276 string. This means that the seal-bore assembly required to accommodate tubing expansion and contraction for the titanium alloy string may be shortened by 1.2 m. In fact, the thermal stress imparted into the Beta-C™ string (19,680 lbs) can be readily superimposed onto the tension load (117,750 lbs), and still maintain a tension design factor that exceeds the required minimum.

Therefore, if desired, the seal bore assembly can be totally eliminated from the string design. This eliminates a very costly component while replacing a moveable seal with a fixed one, thereby reducing the probability of leakage.

Production Accessories

It has been shown that titanium Beta-C™ tubulars provide an alternative that is both technically and economically competitive to the nickel based alloys. In addition, the use of Beta-C™ in manufacturing tubular accessory items appears to offer several advantages to the operating companies. In particular, a surface controlled subsurface safety valve (SCSSV) manufactured wholly from Beta-C™ titanium appears to offer several benefits.

Each and every well that is completed in United States Outer Continental Shelf (OCS) waters must have an SCSSV installed near the surface as a portion of the producing tubular string (Figure 4). This piece of equipment is the safety device that is used to shut in the well in the event the surface control valves are rendered inoperative, such as might happen were a ship to strike an offshore platform, for instance. As the SCSSV is an integral part of the tubing string, it must be able to withstand the pressures within the string as well as support the tensile load that is imparted by the tubing string hanging below the valve. Recent discoveries have necessitated that an SCSSV with a working pressure rating of 138 MPa (20,000 psi) be developed for wells that contain H_2S, CO_2, high chloride brines and elemental sulfur at a bottomhole temperature of 218°C. Valves with lower pressure ratings have been manufactured using nickel based alloys, but the dimensional requirements of the well design necessitated a re-examination of the valve design and the impact of material selection.

A critical point for the operator of these wells was that the SCSSV must fit inside pipe that has an internal drift of 152 mm. If the valve were larger in diameter, it would mean that the entire wellbore (7,000 m) would need to be larger which could cost an additional $6-10,000,000 per well. They also desired to have the largest possible ID in the valve as that would minimize the flow restriction in the production string. Due to the severe producing environment, only Pd-enhanced Beta-C™ and one nickel alloy were deemed suitable from a corrosion standpoint. As the nickel alloy was produced via cold working, it was not possible to obtain the material with the same high strength level in the required section thickness that was attainable with the titanium alloy. The strength differential translated to a difference in the ID of the valve of approximately 12 mm which was significant to the production characteristics of the well.

The development of this full Beta-C™ SCSSV is currently in progress. After the initial material selection was made, initial engineering work was completed. Various components within the valve have been tested in order to validate seal designs, wear coatings, etc. The delivery of the first prototype valve is anticipated in mid-to-late 1993.

Wireline Tools

After an oil or gas well is drilled, it is necessary to evaluate the underground formations in order to assess the oil and gas productive potential of the rock. Typically, this evaluation work is performed utilizing tools that are conveyed to the bottom of the well on a multi-conductor braided wire cable (Figure 6). New generations of tools have been developed that are capable of obtaining a sample of the formation fluid and bringing that fluid back to the surface where it can be analyzed in laboratories. These tools are bringing fluid samples to the surface that are under high pressure (up to 138 MPa) and that can contain H_2S, CO_2, elemental sulfur and high chloride brines. One such tool utilizes Beta-C™ as the material for manufacturing the sample transporting chamber.

The sample transport chamber is an extremely critical component as it must be handled by personnel with the fluid sample under pressure. Thus, the safety of the rig and rig crew are dependent on the reliability of the tool. Due to the critical nature of the tool, the manufacturer requires that any material utilized in this component be resistant to sulfide stress cracking and be recommended for use by the NACE MR-01-75 document, as is Beta-C™. At the time of the development of this tool, the manufacturer was able to draw from experience that had been gained from an earlier generation of the device. They were able to take a tool with a lower pressure rating and upgrade it to high pressure sour service work simply by changing materials. Only minor engineering variations were required in the thread and seal areas in order to compensate for the differing elastic modulus. The high strength capability of the alloy along with the corrosion resistance provided the optimum combination of properties for this application. Currently, well over one hundred of these tools have been manufactured and placed in service.

CONCLUSIONS

The effective utilization of titanium Beta-C™ for downhole components has been demonstrated by applications in the oil and gas drilling, oil and gas production, wireline evaluation and geothermal exploitation industrial segments. The physical and mechanical properties of the alloy, including the low density, low modulus of elasticity, high strength capability and good deep hardenability are important to the application of the material. In addition, the excellent corrosion resistance of the alloy has led to its application in many severe environments. The synergistic combination of these favorable properties makes Beta-C™ an alloy that can be utilized for many applications in nearly any type of downhole environment.

REFERENCES

1. D.E. Thomas, et. al., "Beta-C: An Emerging Titanium Alloy for the Industrial Marketplace," ASTM STP 917, ASTM, Philadelphia, PA, 1986, pp. 144-163.

2. W.W. Love, et. al., "The Use of Ti-38644 Titanium for Downhole Production Casing in Geothermal Wells," Sixth World Conference on Titanium Proceedings, Cannes, June 6-9, 1988, Societe Francaise de Metallurgie, Les Ulis Cedex, France, pp. 443-448.

3. D.E. Thomas and S.R. Seagle, "Stress Corrosion Cracking Behavior of Ti-38-6-44 In Sour Gas Environments," Titanium Science and Technology, Proceedings of the 5th International Conference on Titanium, Munich, Germany, Sept. 10-14, 1984, Deutsche Gesellschaft fur Metallkinde E.V., pp. 2533-2540.

4. D.R. Klink and R.W. Schutz, "Engineering Incentives for Utilizing Ti-3Al-8V-6Cr-4Zr-4Mo Titanium Tubulars in Highly Aggressive Deep Sour Wells," Paper No. 92063 presented at NACE Corrosion'92, Nashville, TN, Apr. 27- May 1, 1992.

5. R.W. Schutz, M. Xiao and T.A. Bednarowicz, "Stress Corrosion Behavior of Ti-3Al-8V-6Cr-4Mo-4Zr Alloy Under Deep Sour Gas Well Conditions," Paper No. 92051 presented at NACE Corrosion' 92, Nashville, TN, Apr. 27-May 1, 1992.

6. R.W. Schutz, M.Xiao and J.W. Skogsberg, "Stress Cracking and Crevice Corrosion Testing Of Pd-Enhanced Beta-C™ Titanium Pipe and Billet Products In A Worst-Case Sour Gas Well Environment," Paper presented at NACE 12th International Corrosion Congress, Houston, TX, September 1993.

7. R.W. Schutz and S.R. Seagle, "Method for Improving Aging Response and Uniformity in Beta Titanium Alloys," U.S. Patent Application #991,989, Dec. 17, 1992, RMI Titanium Company, Niles, OH.

8. R.W. Schutz and M.Xiao, "Stress Corrosion Behavior of Ti-38644 Titanium Alloy Products in Methanol Solutions," Paper No. 92148, NACE Corrosion '93, New Orleans, LA, March 8-12, 1993.

9. J.R. Wood, "Correlation of Charpy Impact, Slow Bend Precracked Charpy, and Fracture Toughness Behavior for Beta-C™ Bar Production," RMI Titanium Company Technical Memorandum #TM92-21, June 29, 1992.

Figure 1 - Drilling Motor Schematic

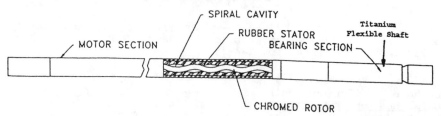

SPIRAL CAVITY

RUBBER STATOR

MOTOR SECTION

BEARING SECTION

Titanium
Flexible Shaft

CHROMED ROTOR

Figure 2 - Geothermal Completion Arrangement

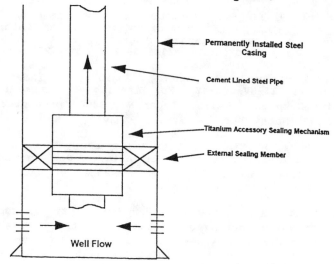

Permanently Installed Steel Casing

Cement Lined Steel Pipe

Titanium Accessory Sealing Mechanism

External Sealing Member

Well Flow

Figure 3 - Titanium Beta-C Production Tubing

Figure 4 - Safety Valve Schematic

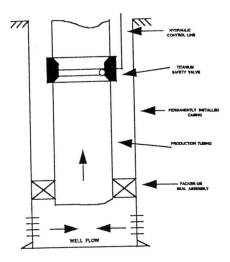

Figure 5 - Wireline Conveyed Evaluation Tool

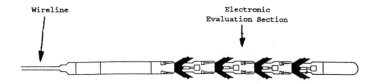

HIGH STRENGTH STRUCTURAL APPLICATIONS FOR LARGE DIAMETER

FORGINGS OF BETA-C™ (Ti-3Al-8V-6Cr-4Mo-4Zr)

P. A. Russo*, J. R. Wood*, R. B. Bhagat**, S. L. Opet***

*RMI Titanium Company
1000 Warren Avenue
Niles, OH 44446

**Applied Research Labs
Penn State University
State College, PA 16801

***Naval Air Warfare Center
Lakehurst, NJ 08733

Abstract

A program was initiated by the Navy to explore the use of Beta-C™ as a water brake cylinder for stopping the catapult on aircraft carriers. Two processing routes were explored to manufacture the water brake cylinder: 1) extrusion, and 2) trepanning of large billet and electron beam welding of an end cap. Mechanical property evaluation of material produced by both processes showed that each was capable of achieving the strength and toughness requirements. Economic considerations will dictate final process selection.

Beta Titanium Alloys in the 1990's
Edited by D. Eylon, R.R. Boyer and D.A. Koss
The Minerals, Metals & Materials Society, 1993

Introduction

The U.S. Navy has been evaluating the use of titanium for several components of the steam catapult system used aboard aircraft carriers[1]. The combination of high strength with good fracture toughness, lower weight, and excellent marine corrosion resistance make the beta titanium alloys attractive alternatives to the low alloy steels that have been used for the past 40 years. Beta-C™ has been successfully used in various applications requiring high strength and good toughness combinations. Steam driven catapults have been used since the introduction of jet aircraft following World War II. Steam power is used in the launching engine cylinders to accelerate the aircraft to its power takeoff speed. The water brake, which is a water filled tube, acts to halt the piston assembly. These cylinders have the following rough machined dimensions:

Length	2.44 m (96 in)
OD	46 cm (18 in)
ID	23 cm (9 in)
Weight	1318 kg (2900 lb)

During the initial design and alloy selection phase of the program, the Navy selected Beta-C™ as one of the alloys having potential for the achievement of the mechanical property goals of 1103 MPa (160 ksi) min. yield strength and fracture toughness of 55 MPa $\sqrt{m}$ (50 ksi $\sqrt{in}$) min. and 66 MPa $\sqrt{m}$ (60 ksi $\sqrt{in}$) aim. Beta-C™, which is classified as a metastable beta titanium alloy, is hardenable in the large section sizes, 4.5" thick wall, required for the water brake. The strength-toughness relationships for various titanium alloys were reviewed which supported the selection of Beta-C™ for the water brake application.

Once the alloy of interest was selected, consideration was given to manufacturing methods to produce the closed-end tubular product. Two methods were selected as having potential for manufacture: 1) piercing and extrusion of a closed end cylinder, and 2) trepanning of large billet and subsequent electron beam welding of an end cap. The Naval Air Warfare Center funded work through the Applied Research Laboratory at Pennsylvania State University to explore various metallurgical issues involved with the use of Beta-C™ for the water brake application. RMI was subcontracted for the extruded part of the study and for the testing phase of the electron beam weld program. The program objectives were: 1) to establish the producibility of a Beta-C™ prototype water brake by two production methods, 2) determine if the required mechanical properties could be met in large sections and in welded structures, and 3) assess structure-property relationships in large sections. This paper describes the metallurgical evaluation of a prototype water brake produced by the two methods.

Material Processing

A schematic of the potential processing routes for the water brake program is shown in Figure 1. In each case a 760 cm (30") diameter ingot of Beta-C™ was double vacuum arc melted with the chemistries shown in Table I. For the extrusion route, billet product was press forged at RMI Titanium Company and pierced and extruded at Cameron Iron, Houston, TX. After extrusion, test sections were cut from the open end of the extrusion and solution annealed and aged to explore the effect of heat treatment on strength and toughness. Final selection of heat treatment was made and was applied to the closed-end tubular. Material for the welded route was trepanned and machined to the required dimensions followed by solution annealing. The end cap was then electron beam welded by PTR-Precision Technologies, Inc., Enfield, CT, and the material returned to RMI for evaluation.

Figure 1. Proposed Processing Routes for Beta-C™ Water Brake

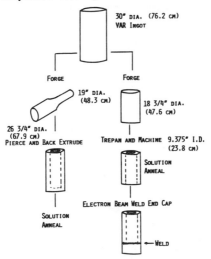

Table I. Chemical Composition of Ingots Used in Water Brake Study

Heat No.	Product	Chemical Composition, wt%								
		C	N	O	Fe	Al	V	Cr	Mo	Zr
872421	Extruded	.01	.012	.096	.10	3.4	7.6	6.0	4.3	4.3
873133	Welded	.01	.011	.087	.10	3.2	8.4	6.0	4.4	4.4

Evaluation of Extruded Water Brake

Sections of the open end of the extruded water brake were solution treated at either 816°C (1500°F) or 927°C (1700°F) for one hour and air cooled. Aging was performed at temperatures between 534°C (975°F) and 565°C (1050°F) for 24 hours. Tensiles and W=3 compact tension specimens were machined from sections as shown in Figure 2. The data generated are included in Table II and plotted in Figures 3 and 4. These data show the beneficial effect of the 927°C (1700°F) solution anneal on the strength-toughness relationship as compared to 816°C (1500°F). Much higher fracture toughness was obtained in samples that were solution treated at 927°C (1700°F) and subsequently aged than those that were solution treated at 816°C (1500°F) and aged to a similar strength. In addition, tensile elongation and reduction in area were improved by the 927°C (1700°F) solution anneal, Figure 4.

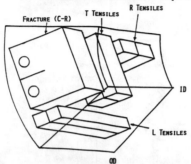

Figure 2. Sampling of Extruded Water Brake
Sections for Mechanical Properties

Table II. Mechanical Properties of Cameron Water Brake Extrusion

Solution[1,2] Anneal Temp.,°F(°C)	Aging[3] Temp.,°F(°C)	Test Dir.	YS ksi (MPa)	UTS ksi (MPa)	El %	RA %	Fracture[4] K_{Ic} (MPa) √in
1500 (816)	975 (534)	T	175.4 (1209)	183.4 (1264)	2.5	5.2	47.3 (52.0)
		L	179.3 (1236)	188.6 (1300)	4.0	7.5	
		R	177.2 (1222)	186.0 (1282)	2.5	5.1	
	1000 (538)	T	168.3 (1158)	176.7 (1218)	4.5	6.7	51.6 (56.7)
		L	172.7 (1191)	181.0 (1248)	5.0	7.9	
		R	166.9 (1151)	175.1 (1207)	4.0	7.1	
	1025 (552)	T	165.9 (1144)	173.0 (1193)	4.5	7.5	59.4 (65.3)
		L	164.8 (1136)	173.7 (1198)	7.0	9.0	
		R	162.4 (1120)	170.3 (1174)	5.0	7.9	
	1050 (565)	T	158.0 (1089)	164.6 (1135)	5.5	10.5	65.6 (72.1)
		L	155.9 (1075)	163.0 (1124)	8.5	12.0	
		R	158.1 (1090)	165.8 (1143)	5.5	8.6	
	1075 (579)	T	154.7 (1067)	160.1 (1104)	5.0	8.6	60.8 (66.8)
		L	156.3 (1078)	162.1 (1118)	5.5	9.7	
		R	149.5 (1031)	155.2 (1070)	8.0	9.4	
1700 (927)	975 (534)	T	173.9 (1199)	185.0 (1276)	7.5	9.8	65.6 (72.1)
		L	177.2 (1222)	186.4 (1285)	8.0	12.7	
		R	178.7 (1232)	187.5 (1293)	6.0	10.1	
	1000 (538)	T	167.2 (1153)	176.4 (1216)	8.0	16.2	83.7 (92.0)
		L	171.9 (1185)	180.2 (1242)	9.0	16.1	
		R	166.4 (1147)	175.5 (1210)	8.5	15.8	
	1025 (552)	T	158.3 (1091)	167.1 (1152)	9.0	20.1	91.0 (100.0)
		L	163.5 (1127)	173.0 (1193)	10.5	21.2	
		R	159.8 (1102)	168.3 (1160)	9.5	20.6	
	1050 (565)	T	154.8 (1067)	161.3 (1112)	9.5	21.2	91.1 (100.1)
		L	158.9 (1096)	166.0 (1144)	10.0	17.9	
		R	155.9 (1075)	163.0 (1124)	7.0	13.2	

1. 30 min. at temp., air cool
2. Heat No. 872421
3. Aging time 24 hours
4. W=3 C-R specimens 1.5" thick

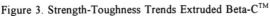

Figure 3. Strength-Toughness Trends Extruded Beta-C™

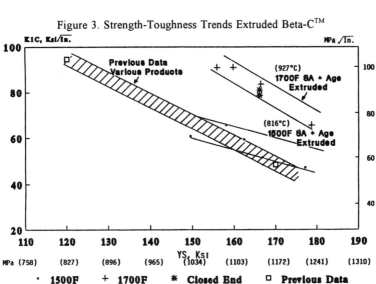

Figure 4. Strength-Ductility Relationship Beta-C™ Extruded Water Brake

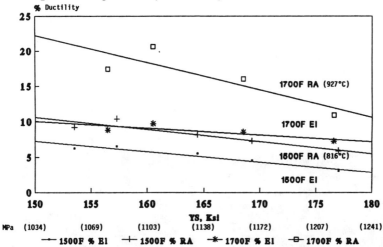

Based on this work, the extruded water brake was solution annealed at 927°C (1700°F) and aged for 24 hours at 552°C (1025°F). Properties from the closed end of the water brake are compared to the open end in Table III. The properties are shown to meet the program requirements of 1103 MPa (160 ksi) yield strength and 66 MPa $\sqrt{m}$ (60 ksi $\sqrt{in}$) aim fracture toughness. The utilization of the 927°C (1700°F) solution anneal was a key factor in allowing such high fracture toughness to be achieved. The lower ductility encountered in the closed end of the water brake, 5.8% elongation and 7.4% RA, compared to 9.7% elongation and 20.6% RA for the open end was due to much larger grain size in the closed end, 2.5 mm (0.1") diameter, as compared to the open end, 0.8 mm (0.03") diameter. These grain size differences were the result of a minimum amount of hot work in the closed end. The influence of grain size on ductility of aged Beta-C™ has been documented[2].

Table III. Mechanical Properties of Beta-C™ Extruded Water Brake

Location	YS ksi (MPa)	UTS ksi (MPa)	El %	RA %	K_Ic ksi √in (MPa √in)
Open End	160.5 (1107)	169.5 (1169)	9.7	20.6	91.0 (100.0)
Closed End	166.2 (1146)	174.7 (1204)	5.8	7.4	80.0 (87.9)
Requirements	160.0 (1103)				50 Min (55.0) 60 Aim (66.0)

1700°F(927°C)-30Min-AC
1025°F(552°C)-24Hr-AC
(18 1/8" O.D. x 9 3/8" I.D. Machined Dimensions)

Microstructures from the open end of the extrusion after solution treating and aging are shown in Figure 5. In general, these microstructures show a relatively well aged structure with some partially aged areas. These partially aged areas are encountered in beta titanium alloys and are related to thermomechanical history[3]. It was surprising that the 927°C (1700°F) solution anneal resulted in uniformly aged structure. Previous work on Beta-C™ and other beta titanium alloys in smaller section sizes has shown that high solution annealing temperatures generally result in large areas that are either unaged or partially aged[4]. This phenomenon is caused by the development of strain free recrystallized grains in which the nucleation of alpha during aging is apparently quite sluggish. It is suggested that the relatively uniform aging response in these large sections after solution treating at 927°C (1700°F) is related to both thermomechanical processing history and the relatively slow cooling rate from the solution annealing temperature.

Figure 5. Microstructures of Beta-C™ Extruded Water
Brake After Solution Treating and Aging

 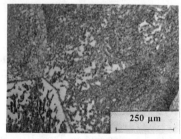

A. 1500°F(816°C)-30Min-AC   B. 1700°F(927°C)-30Min-AC
   975°F(534°C)-24Hr-AC   975°F(534°C)-24Hr-AC
                        (Open End)

Fractographs from tensiles solution annealed at both 816°C (1500°F) and 927°C (1700°F) and aged at 552°C (1025°F) are shown in Figure 6. The 816°C (1500°F) solution anneal resulted in an intergranular appearing fracture as compared to the transgranular appearance of the sample that was solution annealed at 927°C (1700°F) and aged. Similar fracture appearances were observed on fracture toughness specimens heat treated at the two solution annealing

366

temperatures studied. The improvement in both fracture toughness and tensile ductility observed by using a 927°C (1700°F) solution anneal is apparently the result of changing the fracture mode from predominately intergranular when 816°C (1500°F) was used as the solution annealing temperature to transgranular. It is felt that at least part of the differences between the fracture modes of the 816°C (1500°F) and 927°C (1700°F) solution anneal and aged specimens is due to the precipitation of silicides at beta grain boundaries[5]. It has been speculated that there is a greater tendency to precipitate silicides by using a 816°C (1500°F) solution anneal as compared to 927°C (1700°F).

Figure 6. SEM Fractographs of Beta-C™ Extruded
Water Brake Tensile Specimen

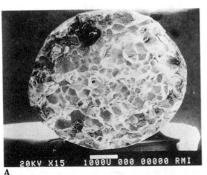
A.

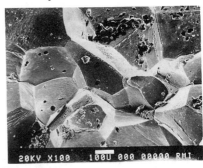
B.

1500°F(816°C)-30Min-AC + 1025°F(552°C)-24Hr

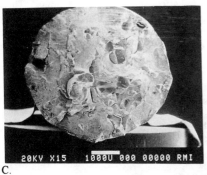
C.

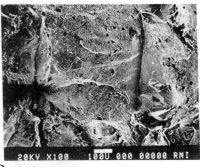
D.

1700°F(927°C)-30Min-AC + 1025°F(552°C)-24Hr
(Open End, R Direction)

Evaluation of Electron Beam Welded Water Brake

A schematic diagram showing the basic plan for evaluating the electron beam welded Beta-C™ water brake is shown in Figure 7. After electron beam welding, two routes were explored: 1) direct age, and 2) resolution anneal at 927°C (1700°F) followed by aging. Evaluation included L-R and L-C W=3 compact tension samples with the crack in the weld metal and C-L compact tension samples with the crack perpendicular to the weld. The basic layout of the test specimens is shown in Figure 8.

Figure 7. Evaluation of Beta C™
Welded Water Brake

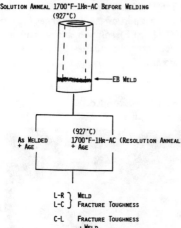

SOLUTION ANNEAL 1700°F-1HR-AC BEFORE WELDING
(927°C)

←— EB WELD

AS WELDED   (927°C)
+ AGE   1700°F-1HR-AC (RESOLUTION ANNEAL)
                   + AGE

L-R ⎫ WELD
L-C ⎭ FRACTURE TOUGHNESS

C-L   FRACTURE TOUGHNESS
       ⊥ WELD

CROSS WELD TENSILES

Figure 8. Sampling of Beta-C™
Welded Water Brake

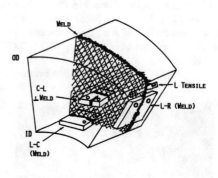

Macrostructural examination of the trepanned billet after solution treating at 927°C (1700°F) and aging showed that a macrostructural gradient existed from the outside to the inside of the piece as shown in Figure 9. This macrostructural gradient was caused by the microstructural differences shown in Figure 10. As shown, the inside locations are uniformly aged whereas the outside and middle locations are only partially aged. There are relatively large areas (white) at outside diameter and midwall locations that contain little, if any, alpha. These partially aged areas are, of course, much softer than the fully aged material. It was decided to select all test samples from inside locations in order to assure that well aged material was being tested. It was rationalized that future process development would allow a uniform microstructure to be developed across the wall.

Figure 9. Macrostructure of Solution Treated
and Aged Trepanned Beta-C™ Billet

OD

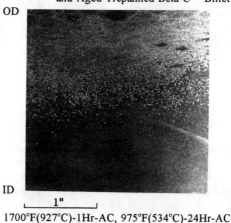

ID

|___ 1" ___|

1700°F(927°C)-1Hr-AC, 975°F(534°C)-24Hr-AC
(Base Metal, L Direction)

Figure 10. Microstructure of Heat Treated
Trepanned Beta-C™ Billet

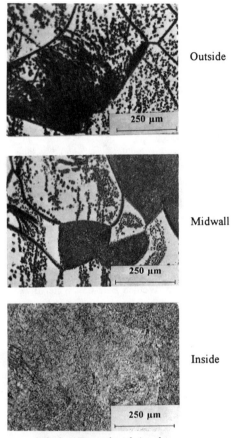

Outside

Midwall

Inside

Solution Treated and Aged
1700°F(927°C)-1Hr-AC, 975°F(534°FC)-24Hr-AC (100X)

Macrostructures of a longitudinal section after welding and after welding and aging are shown in Figures 11 and 12, respectively. The as-welded macro shows a very thin weld line which is typical for electron beam welding. After aging, a heat affected zone is apparent. This zone is thought to be related to its subtle influence on aged microstructure as opposed to major changes in microstructure that are apparent for other types of welding that have relatively large heat affected zones. Little change in the macrostructure was found when the material was resolution treated at 927°C (1700°F) after welding.

Figure 11. Photomacrograph of Longitudinal Section of
Electron Beam Welded Beta-C™ Water Brake

OD

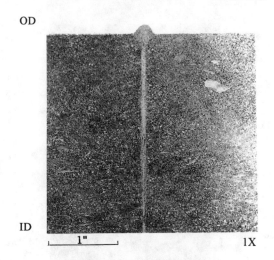

1700°F(927°C)-1Hr-AC Prior to Welding

Figure 12. Photomacrograph of Longitudinal Section of Electron
Beam Welded Beta-C™ Water Brake After Aging

OD

1700°F(927°C)-1Hr-AC, Welded
+ 975°F(534°F)-24Hr-AC
Sample B

Tensile data on the welded Beta-C™ water brake are shown in Table IV with average values
included in Table V. Average values for base metal show slightly higher yield strength for
the material aged directly after welding as compared to the resolution and aged material, 1103

MPa (160.0 ksi) vs. 1081 MPa (156.8 ksi), respectively. The resolution treated and aged base metal had better reduction in area, 19.7%, as compared to the material directly aged, 12.8%. Very little difference was noted in the cross weld tensile strength for the two heat treated conditions, although the reduction in area of the resolution treated and aged condition had a slightly higher average value than the direct aged, 19.0 vs. 13.8%, respectively. The yield strength values were slightly lower than the 1103 MPa (160.0 ksi) requirement, but the required strength can be readily achieved by a slight adjustment in aging treatment. The aging temperature in this study was adjusted to obtain strengths near the minimum required in order to maximize fracture toughness.

Table IV. Tensile and Fracture Toughness Results
on Beta-C^{TM} Welded Water Brake

Heat Treatment	.2 YS ksi (MPa)	UTS ksi (MPa)	El %	RA %	K$_{1c}$ ksi √in (MPa √in)
1700°F(927°C)-1Hr-AC + 1000° F(538°C)-24Hr-AC	156.9 (1082) (L)	167.1 (1152)	6	18.3	72.4 (79.6) (L-C) Weld
1000°F(538°C)-24Hr-AC	154.5 (1065) (L)	166.6 (1149)	8	21.9	-- Weld
1700°F(927°C)-1Hr-AC + 1000°F(538°C)-24Hr-AC	160.6 (1107) (L)	172.5 (1189)	9	14.6	58.6 (64.4) (L-R) Side Groove Weld
1700°F(927°C)-1Hr-AC + 1000°F(538°C)-24Hr-AC	--	--	--	--	79.3 (87.2) (L-R) Weld
1000°F(538°C)-24Hr-AC	157.5 (1086) (L)	170.6 (1176)	4	7.8	68.7 (75.5) (L-R) Weld
1700°F(927°C)-1Hr-AC + 1000°F(538°C)-24Hr-AC	156.9 (1082)	168.0 (1158)	4	15.4	Base Metal
1700°F(927°C)-1Hr-AC + 1000°F(538°C)-24Hr-AC	146.2 (1008)	156.4 (1078)	7	24.0	Base Metal - Sample in Partially Aged Area
1000°F(538°C)-24Hr-AC	160.7 (1108)	172.2 (1187)	5	13.1	Base Metal
1000°F(538°C)-24Hr-AC	159.3 (1098)	170.8 (1178)	4	12.4	Base Metal
1700°F(927°C)-1Hr-AC + 1000°F(538°C)-24Hr-AC					75.4 (82.9) (R-L) Notch Perpendicular to Weld
1700°F(927°C)-1Hr-AC + 1000°F(538°C)-24Hr-AC					73.3 (80.6) (C-L) Base Metal
1700°F(927°C)-1Hr-AC + 1000°F(538°C)-24Hr-AC					58.5 (64.3) (C-L) Perpendicular to Weld
1700°F(927°C)-1Hr-AC + 1000°F(538°C)-24Hr-AC	159.2 (1098)	170.4 (1175)	6	18.3	-- Weld
	158.3 (1092)	172.5 (1189)	7	11.6	-- Weld
1700°F(927°C)-1Hr-AC + 1000°F(538°C)-24Hr-AC	152.8 (1054)	163.6 (1128)	7	24.7	67.3 (74.0) (L-R) Weld
1000°F(538°C)-24Hr-AC					74.6 (82.0) (L-R) Weld
1000°F(528°C)-24Hr-AC					76.2 (83.7) (L-R) Weld

Table V. Data Summary on Beta-C™ Welded Water Brake

Sample	Heat Treatment	YS, ksi (MPa)	UTS[a], ksi (MPa)	El %	RA %	K_{Ic}, ksi √in (MPa √in)
Weld[1]	1700°F(927°C)-1Hr-AC + 1000°F(528°C)-24Hr-AC	157.4 (1067)	168.4 (1161)	7.0	19.0	75.8 (83.4)
Weld	1000°F(528°C)-24Hr-AC	156.7 (1080)	169.9 (1171)	6.3	13.8	73.2 (80.5)
Base Metal	1700°F(927°C)-1Hr-AC + 1000°F(528°C)-24Hr-AC	156.8 (1081)	162.2 (1118)	5.5	19.7	73.3 (80.6)
Base Metal	1000°F(528°C)-24Hr-AC	160.0 (1103)	171.5 (1182)	4.5	12.8	-

1. All material was solution annealed 1700°F(927°C)-1Hr-AC before welding.
2. All values are averages of available data.

Fracture toughness results for the welds are plotted in Figure 13 together with data from the extruded Beta-C™ water brake. The trend lines for the previously extruded water brake show the improvement that is gained by using a 927°C (1700°F) solution annealed compared to 816°C (1500°F) as was previously discussed. The fracture toughness data on the welds are shown to lie within the shaded region between the 927°C (1700°F) and 816°C (1500°F) solution annealed and aged tests from the extruded water brake. The base metal data for the welded tests lie within the grouping of the weld data indicating little difference between the weld and base metal with respect to their strength-toughness relationship. There did not appear to be significant differences between the L-C and L-R test directions in the welds, although the amount of test data was limited. The side grooved sample and the C-L test perpendicular to the weld had the lowest fracture toughness within the welded test group, 64.4 MPa √m (58.6 ksi √in) and 64.3 MPa √m (58.5 ksi √in), respectively. Note that the C-L test perpendicular to the weld was from the outside location in the partially aged region. Neither of these test data was included in Figure 16. The fracture toughness data for the directly aged welds and the welds resolution annealed at 927°C (1700°F) and aged appeared to lie within the same data population indicating no significant difference between the two treatments. It is suggested that the slight adjustment in strength of the welded material required to achieve the goal of 1103 MPa (160 ksi) yield strength would still result in fracture toughness well in excess of the 66 MPa √m (60 ksi √in) based on trend lines following the same slope as those established for the extruded product. Weld fracture paths for all samples appeared to follow a straight line path from the specimen precrack plane. The fracture path did not deviate into the weld heat affected zone.

Figure 13. Strength-Toughness Trends Beta-C™ Water Brake

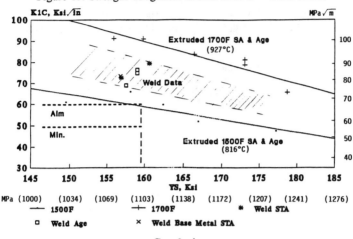

Conclusions

1. Beta-C™ is capable of achieving the mechanical properties required for the Navy water brake cylinder, i.e., 1103 MPa (160 ksi) min. yield strength and 66 MPa √m (60 ksi √in) aim K_{Ic}.

2. Both extruded and electron beam welded processes have potential for manufacturing the water brake based on the development of the required yield strength and fracture toughness.

3. Process development will be required to achieve uniform billet properties, if the trepanned and EB weld route is selected as the process of choice for water brake production.

Acknowledgment

The authors would like to acknowledge Bill Headland of RMI for the process engineering work required for billet manufacturing and monitoring of the processing of the extruded water brake.

References

1. Stephen L. Opet, Jr. and David F. Peters, "Beta Titanium in the United States Surface Navy Improved Water Brake", presented at the Seventh World Conference on Titanium, June 28-July 2, 1992, San Diego, CA.

2. J. G. Ferrero, P. A. Russo, and J. R. Wood, "Microstructure/Mechanical Property Relationships in Bar Products of Beta-C™ (Ti-3Al-8V-6Cr-4Mo-4Zr)", presented at this conference.

3. F. H. Froes, et al, "The Processing Window for Grain Size Control in Metastable Beta Titanium Alloys", Beta Titanium Alloys in the 80's AIME, Warrendale, PA, 1984, pp. 161-184.

4. G. A. Bella, J. G. Ferrero, P. A. Russo, and J. R. Wood, "Effects of Processing on Microstructure and Properties of Ti-3Al-8V-6Cr-4Mo-4Zr (Beta-C™)", in Microstructure/Property Relationships in Titanium Aluminides and Alloys, TMS, Warrendale, PA, 1990, pp. 493-510.

5. S. Ankem, et al, "Silicide Formation in Ti-3Al-8V-6Cr-4Mo-4Zr", Met. Trans. A, vol.18, no.12, December 1987, pp. 2015-2025.

INVESTMENT CASTING OF BETA TITANIUM ALLOYS

FOR AEROSPACE APPLICATIONS

D.A. Wheeler *, R.G. Vogt **, and M.S. Cianci *

Howmet Corporation
Ti-Cast Operations, Whitehall, MI   *
Engineering Group, Hampton, VA   **

Abstract

The process of investment casting offers the ability to produce complex titanium components with minimal finish machining, thereby reducing their overall manufacturing cost. While aerospace applications for cast titanium have focused primarily on alpha+beta alloys, recent interest in higher strength beta alloys has prompted an examination of their suitability for investment casting. In this paper, the processing characteristics and mechanical properties of Ti-15V-3Cr-3Al-3Sn, Ti-3Al-8V-6Cr-4Mo-4Zr, and Ti-15Mo-3Nb-3Al-0.2Si (wt.%) will be discussed. It will be shown that all three alloy compositions are readily processed using only slight modifications from current Ti-6Al-4V (wt.%) production operations. In addition, the mechanical properties of the cast product form can be manipulated through heat treatment and compare quite favorably with typical properties obtained in wrought beta titanium products. Finally, several demonstration castings are reviewed which illustrate the shape-making capabilities of the investment casting approach for beta titanium alloys.

Beta Titanium Alloys in the 1990's
Edited by D. Eylon, R.R. Boyer and D.A. Koss
The Minerals, Metals & Materials Society, 1993

Introduction

While one of the principle characteristics that make beta titanium alloys attractive for structural applications is its workability, benefits can still be gained for many components through the use of an investment casting approach. In particular, the ability to produce net or near-net shapes for complex geometries with minimal final machining can yield significant cost and cycle time reductions during the manufacture of titanium hardware (1). Recent advances in titanium investment casting can also provide reduced manufacturing costs for multiple-piece assemblies by producing them as single-piece castings (2). As interest in beta titanium alloys grows for selected applications, so will the potential for use of beta alloy castings. In an effort to gain experience with this class of alloys, the processing characteristics of Ti-15-3 (Ti-15V-3Cr-3Al-3Sn wt.%), BetaC (Ti-3Al-8V-6Cr-4Mo-4Zr wt.%), and Beta21S (Ti-15Mo-3Nb-3Al-0.2Si wt.%) were evaluated and compared to standard processing sequences used for alpha+beta alloy Ti-6-4 (Ti-6Al-4V wt.%). The mechanical properties of these alloys in the investment cast product form were also characterized. The results of these evaluations, including several demonstration castings, are reviewed below.

Investment Cast Processing

While the basic process of investment casting is used for many alloy systems, the highly reactive nature of molten titanium creates several unique aspects to its processing. These aspects can also vary with alloy type. As such, the processing of beta titanium alloys was evaluated with respect to the more commonly used Ti-6-4.

Fluidity / Fill

Cold-hearth or skull melting practices are commonly used for the production of titanium castings due to the reactivity of molten titanium. Unfortunately, these melting methods provide minimal melt superheat, resulting in a decreased ability to fill thin sections. Because superheat is not a viable process variable, the fluidity of new alloy compositions must be characterized. This was done for Ti-15-3 and BetaC using a standard fluidity plate mold. The results shown in Figure 1 for nominal casting conditions indicate that beta alloys have a slightly decreased ability to fill sections less than 3 mm when compared to Ti-6-4. However, it should be noted that these measurements do not suggest that beta alloys cannot be used for thin-walled components. Rather, the results suggest that gating techniques must be slightly more robust for beta alloys to ensure the complete fill of such components.

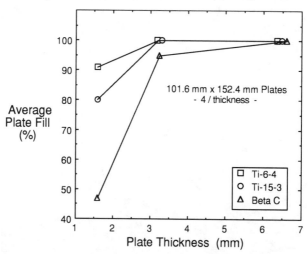

Figure 1: Relative Fluidity of Several Beta Titanium Alloys vs. Ti-6Al-4V (wt%)

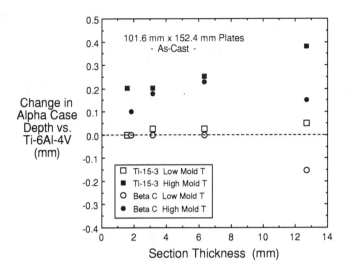

Figure 2: Relative Alpha Case Penetration of Several Beta Titanium Alloys vs. Ti-6Al-4V for Low and High Mold Preheat Levels

Alpha Case Formation

The ceramic mold systems used for titanium investment casting are specifically designed to minimize reaction with molten titanium, yet a dimensional allowance must still be made to the starting component patterns to account for the formation of an oxygen-stabilized alpha case on the surface of the casting. The formation of this case is dependent on many processing features, including alloy composition and section size. Alpha case depths were measured for Ti-15-3 and BetaC under nominal casting conditions and compared to Ti-6-4 as a function of cast section size. The results, shown in Figure 2, reveal little difference in penetration for a "low" mold preheat temperature, but slight increases were observed when the mold temperature was increased. This deeper case is significant since such higher temperatures may be used to facilitate the fill of beta castings with thin sections. While not detrimental to the final casting, this result suggests that additional stock must be added to starting wax patterns to account for the additional alpha case.

Chemical Milling

The surface alpha case formed during casting is a hard, brittle layer which must be removed from the component prior to its use. This removal is usually done chemically through immersion of the component within an acid bath for a predetermined length of time depending on case depth. The active ingredient commonly used in these chemical milling baths is hydrofluoric acid (HF). Evaluation of Ti-15-3 and BetaC within a "standard" bath for Ti-6-4 resulted in significant uptakes of hydrogen and severe intergranular attack (IGA) of the beta alloy castings. Baths containing inhibitors or otherwise diluted have been successfully used to chemically mill these alloys without such problems, but the metal removal rates are quite slow. As such, additional work is required to optimize the removal rate and removal uniformity of beta alloy chemical mill baths without causing hydrogen uptake or IGA. The highly corrosion-resistant nature of the Beta21S alloy brings its own special problems when trying to chemically mill the surface of castings. Attempts to chemically mill castings from this alloy in current baths have been generally unsuccessful, prompting the need for process development in this area.

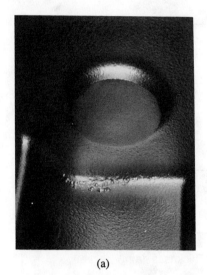 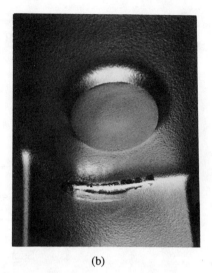

(a) (b)

Figure 3: Typical Weld Repair of Fillet Porosity in Ti-15V-3Cr-3Al-3Sn (wt%) Casting
(a) As-cast fillet porsity
(b) After weld repair

Weld Repair

The excellent weldability of titanium alloys allows for the repair of minor casting defects, such as flowlines or fillet porosity. Tungsten inert gas (TIG) welding with appropriate filler metal is commonly used for this operation. Successful weld repair has been done on all three beta alloys discussed in this paper using similar-alloy filler metal. An example of the weld repair of fillet porosity in a Ti-15-3 casting is shown in Figure 3. Minimal deviations from the standard Ti-6-4 welding conditions were required for these beta alloys, and the x-ray and surface penetrant inspection of the weld regions were quite good. Initial work performed by Boeing has shown that weld repair results in slight but acceptable reductions in mechanical properties of cast beta alloys (3). Further work is needed to fully characterize the heat treatment response and mechanical properties of welded-repaired beta alloy castings to assure their viability for structural applications.

Table I - Typical 70°F Tensile Properties of Cast Beta Titanium Alloys
after Direct Aging Treatments

Alloy	Aging Treatment	Yield Strength (MPa)	Ultimate Strength (MPa)	Plastic Elongation (%)
Ti-15-3	538°C / 8 hrs	1112	1226	6.6
Beta C	538°C / 24 hrs	1103	1200	6.0
Beta 21S	649°C / 8 hrs	1035	1108	6.1

Table II - Effect of Direct Aging Temperature on Properties of Cast Ti-15Mo-3Nb-3Al-0.2Si

Aging Temperature (°C)	Aging Time (hrs)	Yield Strength (MPa)	Ultimate Strength (MPa)	Plastic Elongation (%)
593	8	1113	1195	3.3
621	8	1072	1159	4.4
649	8	1035	1108	6.1
677	8	969	1017	10.5

Mechanical Properties

Typical room temperature tensile properties of these cast beta alloys are summarized in Table I. As seen, direct aging treatments can result in an attractive combination of strength and ductility for the cast product form. It is encouraging to note that these properties are quite comparable to the properties specified for wrought beta alloy applications (4). The properties of these alloys can also be readily manipulated through direct aging heat treatment to provide a range of strength and ductility levels, as shown in Table II for Beta21S. This characteristic of beta alloys allows their properties to be specifically tailored for the required application through relatively simple aging treatments. The high strength levels attained in the beta alloys promote quite good fatigue properties, as shown in Figure 4.   The relatively high fatigue limits measured for these alloys are very encouraging considering that the larger microstructural feature size of typical cast products is not expected to provide optimum resistance to fatigue loadings. Overall, all of the above data suggests that investment casting can provide beta alloy components without any debit in mechanical properties when compared to wrought product.

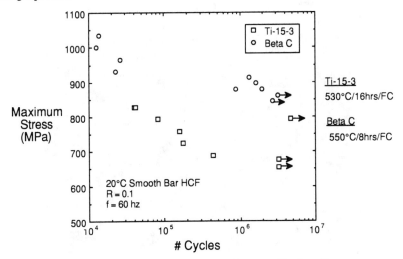

Figure 4: Smooth Bar HCF Properties of Cast Beta Titanium Alloys

379

Component Applications

To take advantage of their excellent strength-to-weight ratio, beta alloys are being evaluated for numerous structural castings in aerospace systems. An example of such an application is the Ti-15-3 brackets shown in Figure 5. By using a casting approach for this type of component, the need for final machining is minimized and the overall production costs can be reduced. The use of weld repair to correct minor casting imperfections is also illustrated in this example. Investment casting can also be used to manufacture larger structural parts from beta alloys, as shown by the BetaC engine stub frame in Figure 6. The manufacture of this geometry as a single-piece casting can lead to overall cost reductions when compared to a multiple-piece wrought assembly. The relatively thin support struts between the inner and outer hubs also give an indication as to the overall castability of these alloys. Finally, investment casting can be used to manufacture even larger components where other methods may become difficult, such as the Beta21S heat shield shown in Figure 7. The successful production of this piece illustrates the scale-up potential of investment cast beta alloys for aerospace applications.

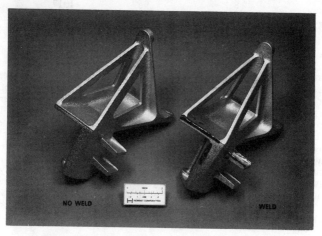

Figure 5: Cast Ti-15V-3Cr-3Al-3Sn (wt%) Airframe Brackets

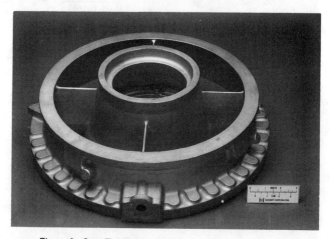

Figure 6: Cast Ti-3Al-8V-6Cr-4Mo-4Zr (wt%) Stub Frame

Figure 7: Cast Ti-15Mo-3Nb-3Al-0.2Si (wt%) Airframe Heat Shield

Conclusions

Investment casting has been demonstrated as a viable approach for the manufacture of complex components from beta titanium alloys. The manufacture of such shapes requires only minor modifications to the existing methods and technology used for more common alloys like Ti-6Al-4V (wt.%). In particular, slightly decreased fill capabilities and slightly increased alpha case penetration depths need to be acknowledged when processing beta titanium alloy components. In addition, alternative chemical milling baths need development to more effectively remove the alpha case from these alloys. The mechanical properties of cast, HIP'ed, and heat treated beta alloys have been shown to compare quite favorably with the wrought product form. Because of these properties, a number of applications have been identified for cast beta alloys and activities are currently underway to manufacture selected components by investment casting for the aerospace industry.

References

(1) M.J. Donachie, Jr., Titanium - A Technical Guide, ASM International, 1988
(2) M.S. Cianci, R.G. Vogt, & G.N. Colvin, "Recent Advances in Titanium Investment Casting, Proceedings of 7th World Conference on Titanium, San Diego, 1992, in press
(3) R.R. Boyer, "High Strength Titanium Castings", Boeing Commercial Airplane Group, Internal Report SR8383, 1991
(4) R.R. Boyer and H.W. Rosenberg, eds., Beta Titanium Alloys in the 1980's, TMS-AIME Publications, Warrendale, PA, 1984

LASER WELDING OF TI-6AL-4V TO BETA-21S

P. S. Liu,* K. H. Hou,* W. A. Baeslack III* and J. Hurley**

*Department of Welding Engineering
The Ohio State University
Columbus, OH 43210 USA

**Edison Welding Institute
Columbus, OH 43212 USA

Abstract

CO_2 laser welds were produced autogenously between Ti-6Al-4V and Beta-21S sheet. Three different nominal fusion zone chemical compositions were obtained by varying the laser beam location relative to the joint centerline and thereby melting different quantities of each base metal. Fusion zone microstructures exhibited relatively fine beta grains comprised of retained-beta phase and martensite, with the proportion of martensite increasing with an increase in the quantity of Ti-6Al-4V nominally in the fusion zone. The location of these phases within the fusion zone was influenced by macrosegregation which originated from incomplete mixing of the melted base metals and the occurrence of transverse-solute banding during solidification. Postweld aging heat treatment at 482°C/20 hr and 538°C/8 hr resulted in extremely fine alpha precipitation within the retained-beta regions and tempering of the martensite to an extremely fine alpha + beta structure. These fusion zone microstructures exhibited high hardness and strength superior to that of the Ti-6Al-4V and Beta-21S base metals (ie., 100% joint efficiency was achieved), but very low ductility. An increase in the aging temperature to 593°C/4 hr promoted fusion zone transformation to a coarser intragranular and grain-boundary alpha + beta structure which exhibited strengths superior to those of the base metals and acceptable ductility. An increase in the proportion of Ti-6Al-4V within the weld fusion zone increased average hardness but had no effect on ductility. For comparable postweld aging conditions, the laser welds exhibited ductilities superior to those of coarse-grained gas tungsten-arc welds. Fracture analysis indicated a transition from transgranular fracture in the as-welded and low-temperature aged conditions to mixed transgranular/intergranular fracture following aging at higher temperature; this transition was attributed to an increase in the thickness and continuity of alpha phase at beta grain boundaries.

Beta Titanium Alloys in the 1990's
Edited by D. Eylon, R.R. Boyer and D.A. Koss
The Minerals, Metals & Materials Society, 1993

Introduction

Beta-21S is a relatively new, metastable-beta titanium alloy (nominal composition Ti-15wt%Mo-2.7wt%Nb-3wt%Al-0.2wt%Si) which exhibits superior oxidation resistance and mechanical properties at elevated temperatures, and excellent corrosion and hydrogen resistance (1). The effective integration of Beta-21S into structural components may require its dissimilar welding to alpha-beta and near-alpha alloys, such as Ti-6Al-4V and Ti-6Al-2Sn-4Zr-2Mo-0.1Si, respectively. To date, relatively little work has been performed regarding the dissimilar-alloy welding of metastable-beta titanium alloys. Studies of gas tungsten-arc (GTA) welds produced autogenously between Ti-6Al-4V and Ti-15V-3Al-3Cr-3Sn sheet (equal mixing from each base metal) determined that the weld microstructure and mechanical properties are dependent on both the weld cooling rate and postweld heat treatment (2). In addition, it was determined that macrosegregation within the fusion zone resulting from transverse-solute banding during solidification affected the tendency for transformation to orthorhombic martensite in the fusion zone on cooling, and correspondingly on the postweld heat treatment response (2). This type of macrosegregation is attributed to changes in the solid-liquid interface velocity during weld solidification, typically due to instantaneous changes in the arc power. It was found that postweld aging resulted in high fusion zone strengths but ductilities well below those of either the Ti-6Al-4V or Ti-15V-3Al-3Cr-3Sn base metals. This property degradation was attributed to the coarse beta grain size in the weld fusion zone and the nature of the transformed-beta microstructure.

Based on this previous work, two approaches were considered for improving the ductility of dissimilar-alloy welds between Ti-6Al-4V and Beta-21S: 1) the application of a low heat input welding process to minimize the beta grain size and 2) modification of the fusion zone chemical composition to allow greater microstructural optimization through postweld aging. Correspondingly, two objectives were set forth for this study: 1) to investigate the potential of utilizing laser beam welding to generate a fine-grained weld microstructure with a controlled fusion zone chemical composition and 2) to determine the influence of fusion zone chemical composition and postweld heat treatment on the weld structure, properties and fracture behavior.

Experimental Procedure

The Ti-6Al-4V and Beta-21S sheets evaluated in this study were provided in the mill-annealed and solution heat-treated conditions, respectively. The 1.5 mm thick sheets were machined into coupons 150 mm x 50 mm with the coupon length oriented perpendicular to the sheet rolling direction. Prior to welding, the coupons were degreased in acetone.

Full-penetration butt welds were produced using a GE Fanuc C 3000 CO_2 laser equipped with a 190 mm plano-convex ZnSe lens. A laser power of 3000 watts and a beam traversing rate of 42.3 mm/s were utilized. In order to vary the fusion zone chemical composition, three beam locations relative to the joint centerline were utilized: 1) directly at the centerline; 2) offset 0.18 mm to the Ti-6Al-4V side and 3) offset 0.18 mm to the Beta-21S side. Based on these beam locations and assuming uniform mixing of both base alloys, it was anticipated that fusion zones exhibiting Ti-6Al-4V/Beta 21S volumetric proportions of 50/50, 64/36 and 36/64 would be achieved. Laser welding was performed in a helium-purged collapsible chamber in order to prevent atmospheric contamination. Automatic GTA welds were produced autogenously with equal base metal melting for comparative purposes (70 amperes, 8.5 volts, 4.2 mm/s).

Following welding, coupons were sectioned and heat treated at 482°C, 538°C and 593°C for 2, 4, 8 and 16 hours. Based on metallographic analysis, diamond-pyramid hardness (DPH) testing and heat treatment information developed previously for Beta-21S (1, 3), three postweld heat treatments were selected for more detailed characterization: 1) 482°C/20 hr; 2) 538°C/8 hr; and 593°C/4 hr.

Microstructural analysis of the as-welded and postweld heat-treated welds included light and analytical-electron microscopy, and electron-probe microanalysis (EPMA). Mechanical testing included DPH hardness testing, three-point bend testing of longitudinal-weld oriented specimens and tensile testing of transverse-weld oriented specimens (12.7 mm gage length, 1.5 mm x 3.2 mm gage cross section). Hardness testing involved both the generation of traverses across the weld zone (200 or 500 gram load), and the determination of an "average" fusion zone hardness by measuring hardness at four consistent locations within the fusion zone (1000 gram load). Fractographic examination of the bend specimens was performed using scanning-electron microscopy (SEM).

Results and Discussion

Microstructure and Compositional Analysis

CO_2 laser welds produced in the present study were characterized by an "hourglass" shape typical of laser welds in thin titanium sheet (Fig. 1A). An increase in energy input into the weld zone (via a reduction in travel rate) relative to welds produced previously in the same thickness Beta-21S sheet (3) provided an increased weld width and greater flexibility in composition control while precluding the generation of lack of fusion defects. Examination of the as-welded fusion zones indicated narrow heat-affected zones (HAZ's) and the epitaxial nucleation and growth of columnar beta grains in the fusion zone directly from beta grains in the near-HAZ. Near the center of the weld thickness a distinct transition from cellular growth near the fusion boundary to cellular-dendritic growth at the weld center was observed.

Figure 1B shows chemical compositions at locations across the as-welded near-HAZ's and fusion zone for the weld produced with the laser beam located at the joint centerline. Compositional gradients for each respective alloying element were observed across the fusion zone which steepened particularly near the Ti-6Al-4V fusion boundary. Fluctuations in the gradients were attributed to macrosegregation effects. By averaging the chemical compositions to obtain a nominal fusion zone chemical composition, it was determined that contributions from each base metal were not equal (ie., the measured composition differed from the calculated composition of Ti-4.5Al-2.0V-7.5Mo-1.3Nb-0.1Si, with a 55-60% contribution from the Beta-21S and a 40-45% contribution from the Ti-6Al-4V. Although this imbalance may be related to limitations in specimen/laser beam alignment, it is of interest to note that a similar effect has recently been observed in dissimilar GTA welds produced using relatively low energy inputs between Ti-6Al-4V and the metastable-beta alloy Ti-15V-3Al-3Cr-3Sn (4). The greater extent of melting of the metastable-beta alloy, despite location of the arc at the joint centerline, may be attributed differences in the thermal properties of the alloys.

Regions of both retained-beta phase and martensite were observed within the weld fusion zone, with the proportion of martensite increasing with a greater quantity of Ti-6Al-4V in the fusion zone, and nearer to the Ti-6Al-4V fusion boundary. In Figure 1A, dark-etching regions within the fusion zone are comprised primarily of martensite, whereas white-etching regions are entirely retained-beta phase. The observed microstructural variations across the fusion zone were associated with compositional variations originating from incomplete, nonuniform mixing of the melted base metals and transverse-solute banding during solidification. Figure 2A shows the Ti-6Al-4V side of the weld fusion zone at increased magnification, and reveals solute bands of retained-beta phase in the predominantly martensitic matrix. Figure 2B shows a TEM bright-field micrograph of the martensitic region. Auto-partitioning of the beta grains during martensite transformation results in a wide range of martensite plate sizes. Based on the fusion zone chemical compositions, in the context of previous phase transformation studies on Ti-Mo (5,6) and Ti-V (7) binary alloys, the relatively coarse martensite plate size and extensive twinning in the martensite plates, an orthorhombic versus hexagonal crystal structure would be anticipated.

A comparison of the weld microstructure with the chemical compositions measured across the fusion zone showed very good correlation with results of the aforementioned dissimilar-alloy welding study (2) and with an investigation into the GTA welding of the near-beta alloy Ti-10-Al-2Fe-3Al (8). These studies determined that an alloy chemistry of approximately 3-4.5 wt% Al and 11-12 wt% beta stabilizer (eg., V, Fe) defined the boundary between retained-beta and orthorhombic martensite formation on rapid weld cooling. An increase in the beta stability from this nominal chemistry (as in transverse-solute bands) promoted retention of the beta-phase, while a decrease promoted beta decomposition to martensite. A comparison of Figures 1A and 1B indicate a similar relationship across the dissimilar-alloy laser weld.

Welds produced with the laser beam offset to the Beta-21S sheet showed an entirely retained-beta microstructure, except for bands of martensite adjacent to the Ti-6Al-4V fusion boundary. This observation was consistent with the strong beta-stability of the composite fusion zone (calculated fusion zone chemistry of Ti-4.0Al-1.4V-9.9Mo-1.6Nb-0.13Si). Conversely, the weld produced with the laser beam offset to the Ti-6Al-4V sheet exhibited a nearly entirely martensitic fusion zone, with retained-beta phase present only as solute bands adjacent to the Beta-21S fusion boundary. Again, this observation was consistent with the relatively weak beta stability of this composite fusion zone (calculated fusion zone chemistry of Ti-5.0Al-2.6V-5.1Mo-0.8Nb-0.07Si).

Although not apparent in Figures 1 and 2, it should be noted that microsegregation on a dendritic scale was observed based on the etching response. The occasional presence of martensite at the boundaries of retained-beta dendrites (which were depleted in Mo and Nb) was observed.

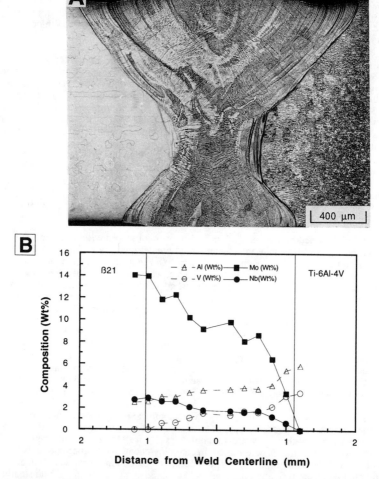

Figure 1 Laser weld between Ti-6Al-4V and Beta-21S (equal base metal melting): (A) light macrograph; (B) EPMA traverse across weld zone (2 micron beam size) 0.5 mm from weld top surface.

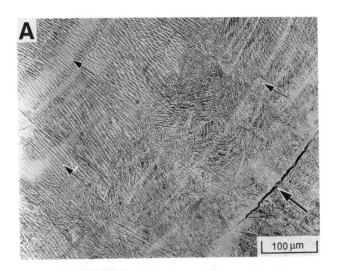

Figure 2　Laser weld between Ti-6Al-4V and Beta-21S (equal base metal melting): (A) light micrograph of fusion zone near Ti-6Al-4V fusion boundary (large arrow), small arrows indicate retained-beta phase in transverse-solute bands; (B) TEM bright-field micrograph of martensite.

The effect of laser beam location on the nominal fusion zone chemical composition and macrosegregation effects on the as-welded and postweld aged microstructures is revealed more distinctly in Figure 3A for a weld produced with the beam located directly at the joint centerline and postweld aged at 482°C/20 hr. In this photomicrograph, regions which were originally retained-beta phase appear dark, while regions which were martensite as-welded appear white. Evidence of incomplete mixing and transverse-solute bands in promoting macrosegregation and correspondingly microstructural variations in the fusion zone is apparent. Figures 3B and 3C show light micrographs of the specimen shown in Figure 3A at the fusion zone center and at the Ti-6Al-4V fusion boundary. Examination of the white-etching regions clearly revealed the original martensite structure, while the featureless, dark-etching regions were consistent with an extremely fine alpha structure in a beta matrix. Note the difference in martensite morphology for the fusion zone and Ti-6Al-4V near-HAZ.

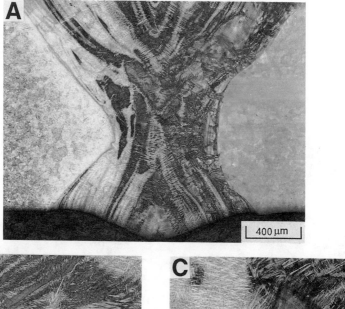

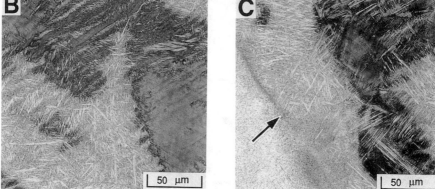

Figure 3   Laser weld between Ti-6Al-4V and Beta-21S (equal base metal melting) aged at
482°C/20 hr: (A) macrograph; (B) fusion zone center; (C) fusion zone at Ti-6Al-4V
fusion boundary (arrow).

Figure 4A shows a light macrograph of the laser weld produced with the laser beam directly at
the joint centerline and postweld aged at 593°C/4 hr. The Ti-6Al-4V side of the fusion zone was
comprised of primarily martensite and the Beta-21S side of aged beta phase. Figures 4B-E show
light and TEM bright-field micrographs of the fusion zone regions which were comprised of
martensite (B,C) and retained-beta phase (D,E) in the as-welded condition. The morphology of
the original martensite plates are still readily apparent. As indicated above, the relatively coarse
size and extensive twinning exhibited by these platelets, in conjunction with the high alloying
content in this region, suggest their origin as orthorhombic martensite. TEM analysis of the
structure indicates the presence of beta phase at prior plate and twin boundaries. Analysis of the
Beta-21S side of the weld shows fine plates of alpha phase in a beta matrix. These platelets were
appreciably finer than those observed in the Beta-21S near-HAZ or base metal, which is
consistent with the appreciably greater hardness in this region. Increased evidence of nearly
continuous grain boundary alpha was observed in the TEM micrographs.

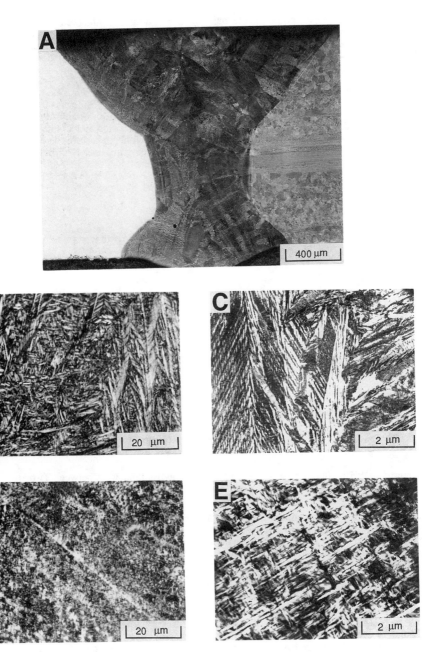

Figure 4 Laser weld between Ti-6Al-4V and Beta-21S (equal base metal melting) aged at 593°C/4 hr: (A) macrograph; (B,C) tempered martensite; (D,E) aged beta phase.

As indicated above, variations in the base metal dilution levels markedly influenced the proportions of phases (retained beta versus martensite) present in the as-welded microstructure. However, these compositional changes did not visibly influence the microstructural characteristics of these structures (ie., plate coarseness, extent of grain boundary alpha formation).

Mechanical Properties

Figure 5 shows a hardness traverse across the fusion zone of a dissimilar-alloy laser weld produced with equal melting of both base metals in the as-welded and 593°C/4 hr postweld heat treated conditions. In the as-welded condition, the retained-beta phase in the Beta-21S base metal and HAZ exhibited the lowest hardness, while the alpha-prime near-HAZ in the Ti-6Al-4V exhibited the highest hardness. Following postweld heat treatment, the Ti-6Al-4V base metal exhibited the lowest hardness, with the fusion zone on the Ti-6Al-4V side exhibiting the highest hardness. Similar trends were observed for other base metal dilutions.

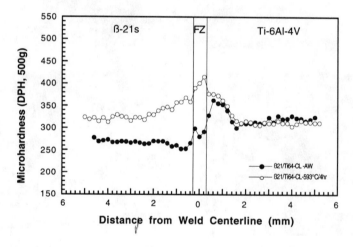

Figure 5   DPH hardness traverse (500 gram load) across laser weld between Ti-6Al-4V and Beta-21S (equal base metal melting).

Figure 6 shows microhardness traverses (200 gram load) across the fusion zone of welds produced with the laser beam at the joint centerline in the as-welded and heat-treated conditions. In the as-welded condition, negligible hardness differences were observed with compositional variations across the fusion zone, or between the martensite and retained-beta microstructures. The low hardness (275-295 DPH) was consistent with an orthorhombic versus hexagonal structure. In contrast, the aged specimens showed a distinct variations in hardness across the fusion zone. For the weld aged at 482°C/20 hr, hardness increased to a maximum at the centerline. This was attributed to the presence of fine plates of tempered martensite, versus coarser tempered martensite plates near the Ti-6Al-4V fusion boundary and aged beta phase near the Beta-21S fusion boundary. The sharp decrease in hardness near the Beta-21S fusion boundary was associated with the relatively lower hardness of aged beta versus the tempered martensite. The weld aged at 593°C/4 hr showed a gradual increase in hardness toward the Ti-6Al-4V fusion boundary. Since the hardnesses of the aged beta phase and tempered martensite were not observed to differ significantly for this heat treatment condition, the increase is attributed to a compositional effect on the fine alpha/beta microstructures present in these regions.

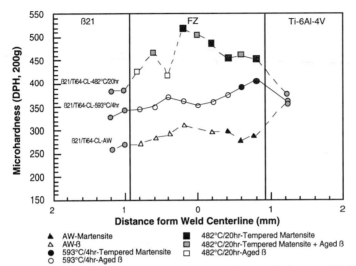

Figure 6   DPH hardness traverses (200 gram load) across laser welds between Ti-6Al-4V and Beta-21S (equal base metal melting) as-welded and aged at 482°C/20 hr and 593°C/4 hr. Open symbols represent regions which were retained-beta phase as-welded, solid symbols represent regions which were martensite as-welded.

Figure 7 shows a plot of average fusion zone hardness versus aging time for the three base metal dilution combinations and for aging temperatures of 482°C and 593°C. The weld fusion zone hardnesses were appreciably greater than that of the Beta-21S near-HAZ and base metal, which was consistent with the finer transformed-beta microstructures observed in these regions. An increase in the proportion of Ti-6Al-4V nominally present in the fusion zone resulted in an increase in hardness for the welds aged at 593°C. Interestingly, this trend was consistent with the microhardness traverse shown above in Figures 5 and 6, which showed an increase in hardness at the Ti-6Al-4V side of the weld.

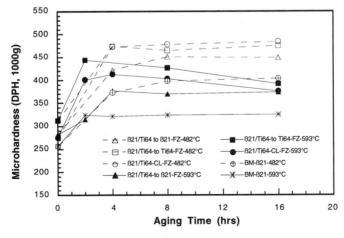

Figure 7   Average DPH hardness (1000 gram load) of Beta-21S base metal and fusion zone of laser welds between Ti-6Al-4V and Beta-21S postweld aged at 482°C and 593°C.

Consistent with the hardness results, aged transverse-weld oriented tensile specimens failed exclusively in the Ti-6Al-4V base metal. Table I summarizes the results of three-point bend testing for the aged dissimilar-alloy laser and GTA welds. Postweld aging at 482°C/20 hr resulted in low fusion zone and base metal ductilities (elastic + plastic strain at crack initiation) of <1.5% and 3.3%, respectively. Postweld heat treatment at 538°C/8 hr improved base metal ductility to 8.8%, but only slightly increased fusion zone ductility to 1.7%. Heat treatment at 593°C/4 hr promoted a significant improvement in bend ductility to 3.8%.

Table I also shows that the bend ductility of GTA welds also increased with postweld aging temperature, but remained well below that of the laser welds.

Table I. Three-point bend ductilities of laser and GTA welds between Ti-6Al-4V and Beta-21S.

Weld Type/Heat Treatment	Minimum Fracture Strain (%)	Fracture Location
Base Metal/482°C,20 hr	3.3	—
LW-FZ-CL/482°C,20hr	<1.5	FZ
Base Metal/538°C,8 hr	8.8	—
LW-FZ-CL/538°C,8 hr	1.7	FZ
GTAW-FZ-CL/538°C,8 hr	<1.3	FZ
Base Metal/593°C, 4 hr	>10.7	—
LW-FZ-CL/593°C, 4 hr	3.8	FZ
LW-FZ-to ß21/593°C, 4 hr	3.8	FZ
LW-FZ-to Ti-6-4-/593°C, 4 hr	3.8	FZ
GTAW-FZ-CL/593°C, 4 hr	2.3	FZ

Fracture Analysis

Figures 8A-D show fracture surfaces of longitudinal-weld oriented bend specimens laser welded with equal base metal mixing and postweld heat treated at 482°C/20 hr and 593°C/4 hr. As indicated, fracture of specimens heat treated at low temperatures occurred transgranularly in the fusion and near-HAZ regions. Despite the relatively low bend ductility of the weld fusion zone, examination of the fracture surface at increased magnification (Fig. 8B), indicated a microscopically ductile appearance on the facet surfaces. As shown in Figures 8C and 8D, heat treatment at 593°C/4 hr promoted a transition from transgranular to mixed transgranular/ intergranular fracture. A similar transition in weld zone fracture morphology with an increase in postweld heat treatment temperature was observed for laser welds in Beta 21S (3), and was attributed to increased alpha phase at beta grain boundaries, which is consistent with microstructural observations. The increase in ductility associated with this heat treatment, despite the presence of increased grain boundary alpha phase and intergranular fracture, was associated with increased intragranular deformation prior to fracture due to a coarser, softer microstructure. Note that effects of compositional and microstructural variations within the fusion zone on the fracture surfaces were not apparent.

Figures 8E and 8F compare the fracture surfaces for the near-HAZ on the Beta-21S side of laser and GTA welds postweld heat treated at 593°C/4 hr. The appreciably greater grain size observed for the GTA weld certainly contributed to its poorer bend ductility.

Conclusions

1. CO_2 laser welding effectively produces fine-grained welds between Ti-6Al-4V and Beta-21S sheet. Control of the laser beam position relative to the joint can be utilized to vary the nominal fusion zone chemical composition. However, macrosegregation in the fusion zone due to incomplete mixing of the base metals and transverse solute banding during solidification result in local variations in the tendency for beta decomposition on weld cooling to martensite, and correspondingly, on the postweld heat-treated microstructures.

2. As-welded fusion zone microstructures exhibit a combination of retained-beta phase and martensite, with the proportion of martensite increasing with an increase in the proportion of Ti-6Al-4V nominally in the fusion zone and locally due to macrosegregation effects. Postweld aging results in alpha precipitation within the retained beta phase and decomposition of the martensite to a fine alpha/beta structure.

3. Although the as-welded fusion zone hardness is low, postweld aging promotes significant increases in hardness and strength. A maximum hardness in the weld fusion zone promotes tensile failures exclusively in the unaffected base metal. Weld zone ductilities are below those of the base metal and although not markedly influenced by chemical composition, they do increase with an increase in aging temperature. Laser welds consistently exhibit greater ductility than GTA welds.

4. A transition in fracture mode from transgranular in the as-welded and low-temperature postweld aged conditions to mixed transgranular/intergranular for high temperature postweld aged conditions is attributed to an increase in the thickness and continuity of grain boundary alpha phase.

Acknowledgement

The authors are indebted to Dr. Paul Bania of TIMET Co., Henderson, Nevada for providing the Beta-21S sheet material and to Mr. Troy Paskell of EWI for producing the GTA welds. Appreciation is also expressed to Mr. Min Kuo of OSU for assisting in SEM analysis.

References

1. P. J. Bania and W. M. Paris, "Beta-21S: A High-Temperature Metastable-Beta Titanium Alloy," Titanium 90-Products and Applications, (Dayton, OH: Titanium Development Association, 1990), 784.

2. W. A. Baeslack III, "Effect of Solute Banding on Solid-State Transformations in Titanium Alloy Weldments," Journal of Materials Science Letters, 1 (1982) 229.

3. P. S. Liu, K. H. Hou, W. A. Baeslack III and J. Hurley, "Laser Welding of an Oxidation Resistant, Metastable-Beta Titanium Alloy - Beta-21S," Proceedings of 1992 International Titanium Conference, San Diego (in press).

4. I. Harris, Unpublished Research, Edison Welding Institute (1993).

5. R. Davis, H. M. Flower, D. R. F. West, "Martensitic Transformations in Ti-Mo Alloys," Journal of Materials Science, 14 (1979) 712.

6. R. Davis, H. M. Flower, D. R. F. West, "The Decomposition of Ti-Mo Alloy Martensites by Nucleation and Growth and Spinodal Mechanisms," Acta Metallurgica, 27 (1979) 1041.

7. L. A. Bagiariatskii, G. I. Nossova and T. V. Tagunova, Sov. Phys. Doklay Eng. Trans., 3 (1959) 1014.

8. S. Boston and W. A. Baeslack III, "Heat Treatment Effects on the Microstructure and Properties of GTA Welds in Ti-10V-2Fe-3Al," Technical Memorandum, Air Force Materials Laboratory, WPAFB, OH (1980).

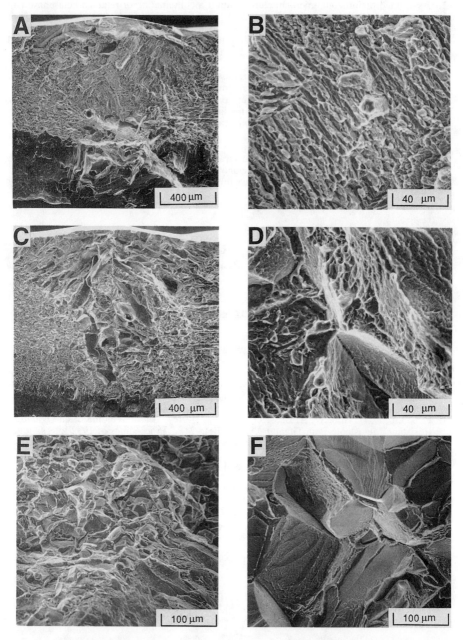

Figure 8   SEM fractographs of longitudinal-weld oriented three-point bend specimens: (A,B) laser weld aged at 482°C/20 hr; (C,D) laser weld aged at 593°C/4hr; (E) laser weld and (F) GTA weld Beta-21S near-HAZ aged at 593°C/4 hr. (B) and (D) show center of fusion zone.

DATA SHEETS OF
BETA TITANIUM ALLOYS

TIMETAL®21S PROPERTY DATA

J. C. Fanning

TIMET Henderson Technical Lab
PO Box 2128
Henderson, NV 89009

ABSTRACT

TIMETAL®21S, which has the nominal weight percent composition 15Mo,3Al, 3Nb, and 0.2Si, is a metastable beta titanium alloy that offers a unique combination of high strength, good elevated temperature properties, and extraordinary environmental degradation resistance. It was developed by TIMET in 1988 as a foil matrix material for titanium metal matrix composites for the NASP, but currently monolithic applications are of much more significance.

Among the alloy's unique properties are a high resistance to attack by commercial aircraft hydraulic fluids (commonly referred to as Skydrol™, which will be the term used in this paper) at all temperatures, which has led to its use in nacelle components on the Boeing 777 and other commercial aircraft.

This paper provides an overview of the physical and mechanical properties of *TIMETAL*®21S.

Beta Titanium Alloys in the 1990's
Edited by D. Eylon, R.R. Boyer and D.A. Koss
The Minerals, Metals & Materials Society, 1993

PHYSICAL PROPERTIES

Chemical Composition

Typical chemical composition ranges are given in Table I. Versions of the alloy are also available with palladium additions (when extremely high stress corrosion resistance is required) and without aluminum (for orthopedic implant applications).

Table I

Element	Weight, %	
	Min.	Max.
Molybdenum	14	16
Niobium (Columbium)	2.4	3.2
Aluminum	2.5	3.5
Silicon	0.15	0.25
Iron	-	0.40
Oxygen	0.11	0.17
Carbon	-	0.05
Nitrogen	-	0.05
Hydrogen	-	0.015
Titanium	Remainder	

Table II.   Typical Tensile Properties of *TIMETAL*®21S.

Direction	Ultimate Tensile Strength, ksi (MPa)	0.2% Yield Strength, ksi (MPa)	Elongation, %
Aged 1000°F [538°C] for 8 hrs.			
L	193 (1331)	181 (1248)	5.0
L	189 (1303)	179 (1234)	6.5
T	193 (1331)	179 (1234)	4.5
T	201 (1386)	187 (1289)	4.0
Aged 1100°F [593°C] for 8 hrs.			
L	164 (1131)	155 (1069)	8.5
L	168 (1158)	159 (1096)	10.0
T	172 (1186)	161 (1110)	10.0
T	172 (1186)	162 (1117)	10.0

Physical Properties

Table III summarizes the physical properties of **TIMETAL®21S** in the solution treated plus aged condition.

Table III. Physical Properties of **TIMETAL®21S** After Solution Treating Plus Aging at 1000°F[538°C] for 8 Hours.

Property	T (°F)	T (°C)	Value	Value (SI)
Density	72	22	.178 lb./cu.in	4.93 g/cu.cm
Beta Transus	1485	807		
Thermal Conductivity	91	33	4.4 Btu/(hr.ft.°F/ft.)	7.6 w/m°k
	498	259	6.8	11.8
	1024	551	9.8	16.9
	1520	827	12.0	20.8
Specific Heat	75	24	.117 Btu/lb °F	.117 cal/g °C
	500	260	.128	.128
	1000	538	.142	.142
	1500	816	.155	.155
Elec. Resistivity	75	24	$5.31 (10)^{-5} \Omega in$	$135.0 (10^{-4})$ onm.m
	500	260	5.60	142.3
	1000	538	5.81	147.8
	1500	816	5.84	148.4
Coef. of Thermal Expansion	100	38	$3.93 \cdot 10^{-6}/°F$	$7.07 \cdot 10^{-6}/°k$
	200	93	4.41	7.9
	400	204	4.75	8.6
	600	316	4.95	8.9
	800	427	5.11	9.2
	1000	538	5.28	9.5
	1500	816	5.73	10.3
Modulus of Elasticity				
Solution Treated	75	24	10.5 to 12×10^6 psi	72 to 85×10^4 MPa
Aged 1000°F (538°C)/8 hrs	75	24	15 to 16×10^6 psi	103 to 110×10^4 MPa
Overaged	75	24	14 to 15×10^6 psi	96 to 103×10^4 MPa

MECHANICAL PROPERTIES

Tensile Properties

Typical and minimum mechanical properties are given in Tables II and IV. The alloy is capable of ultimate tensile strengths as high as 210 Ksi [1470 MPa] with ductility still above 2%. For applications where service temperatures are less than 800°F (427°C), **TIMETAL®21S** is usually aged at 1100°F (593°C) or 1150°F (621°C) to produce ultimate strength levels of 160-180 Ksi (1120-1260 MPa). For elevated temperature applications (service temperatures up to about 1200°F [650°C]), a duplex age of 1275°F (691°C) followed by 1200°F (650°C) is used to provide maximum long-term thermal stability. Unaged material should not be used above about 400°F (204°C) because the potential exists for embrittlement by the precipitation of omega phase or very fine alpha. Care should be used during aging to avoid heating or cooling too slowly, because this can result in very high strength with concomitant low ductility.

The effect of elevated test temperature on tensile mechanical properties is shown in Figures 1 and 2.

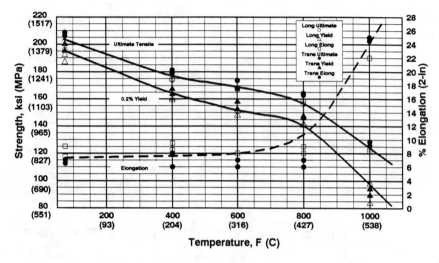

Figure 1. Effect of test temperature on tensile properties of *TIMETAL* ®21S strip aged at 1000 F (538 C) for 8 hours.

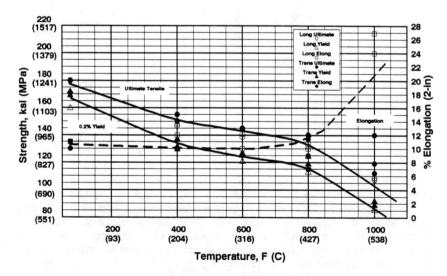

Figure 2. Effect of test temperature on tensile properties of *TIMETAL* ®21S strip aged at 1100 F (593 C) for 8 hours.

Table IV. Guaranteed Tensile Properties for Strip and Sheet to 0.1875-in. (4.8-mm) Thick (1550°F Solution Heat Treatment, Air Cool)

Aging Temp.°F(°C) Aging Time	Ultimate Tensile Strength, ksi (MPa)	0.2% Yield Strength, ksi (MPa)	Elongation, %
—	115–140 (793–965)	110–135 (759–931)	12 min.
1000 (538), 8 hrs.	170 (1172) min.	160 (1103) min.	4 min.
1100 (593), 8 hrs.	150 (1034) min.	140 (965) min.	6 min.
1275 (691), 8 hrs.	125 (862) min.	115 (793) min.	10 min.
+1200 (649), 8 hrs.			

Fracture Toughness and Crack Growth Resistance

The crack growth resistance of *TIMETAL*®21S is not significantly affected by strength level, orientation or test temperature. Typical crack growth properties are given in Figure 11. Typical fracture toughness values for 0.050-in. (1.3 mm) strip are as follows:

Table V

Condition	Ultimate Strength Ksi (MPa)	0.2% Yield Strength, Ksi (MPa)	% Elong.	Kc, Ksi√in, (MPa√m)
Unaged	128 (896)	125 (875)	15	98 (108)
Aged 1000F(538C), 8 Hrs, Air Cool	191 (1337)	178 (1246)	6	67 (74)
Aged 1100F(593C), 8 Hrs, Air Cool	164 (1148)	151 (1057)	8	92 (101)

Metal Matrix Composites

Typical Properties are given in Table VI.

Creep Resistance

TIMETAL®21S has excellent creep resistance for a metastable beta alloy. Although not as resistant to creep as near alpha alloys such as *TIMETAL*®6242 and *TIMETAL*®1100, it is better than other metastable beta alloys and*TIMETAL*®6-4. Figure 6 gives the Larson Miller curve for 0.2% creep strain.

FORMING

TIMETAL®21S is formed in the solution heat treated (as supplied) condition, then aged to the desired strength level. Many of the forming characteristics of *TIMETAL*®21S are similar to those of *TIMETAL*®15-3.

Typical 90° minimum bend radii are 1.5-2.0 times the thickness.

Cold reductions greater than 80% are possible in most compressive operations, including rolling, spinning, and swaging.

Because of relatively low work hardenability, maximum tensile deformations are achieved when strains are uniform, such as in hydroforming and bulge-forming.

The alloy is strain rate sensitive; therefore forming should be performed as slowly as practical.

Springback is relatively severe, but can be compensated for by over-forming or by forming at higher temperatures.

Elevated forming temperatures (400-1400°F [204-760°C]) increase deformation capability and reduce springback.

Intermediate anneals can be used between forming operations to restore workability. However, it is essential to choose a combination of cold work and solution heat treatment that produces a high degree of recrystallization with minimal grain growth, such as suggested in Table VII. Surface contamination (alpha case) must always be removed prior to further forming operations.

Excess flanges on hemispheres, cups, and other hydroformed or drawn parts should be left on until after aging to minimize distortion due to springback.

Machining should only be performed after aging to avoid a brittle surface that can result from the enhanced aging response of the machining-induced severely cold worked layer.

Hot forming and hot sizing are best done at the aging temperature (they are then counted as part of the total aging cycle). Exposure of solution heat treated material to temperatures of 500-800°F (260-427°C) should be kept to less than one hour to avoid the possibility of embrittlement. If forming temperatures exceed the beta transus (about 1485°F [807°C]), time at temperature should be minimized to avoid excessive grain growth.

Surface contamination (alpha case), when present, must <u>always</u> be removed prior to forming.

Chemical milling and pickling solutions must be selected carefully to avoid excessive hydrogen absorption. A 5% HF-35% HNO_3 solution is recommended (see Figure 8).

ENVIRONMENTAL DEGRADATION RESISTANCE

General Corrosion

TIMETAL®21S has outstanding corrosion resistance; it even surpasses the corrosion resistance of commercially pure titanium in many instances. Like all titanium, it is immune to corrosion in water and seawater. The corrosion resistance of *TIMETAL*®21-S in hydrochloric acid is compared to other titanium alloys in Figure 7.

Oxidation

TIMETAL®21S is designed for good oxidation resistance; it unquestionably has the best oxidation resistance of any metastable beta alloy. The resistance of *TIMETAL*®21S is compared to other titanium alloys in Figures 9 and 10.

Skydrol™

Traditionally, titanium aircraft structures have been prohibited when the possibility existed for Skydrol™or Skydrol™-type hydraulic fluids (used on all commercial aircraft) to accumulate on the titanium structures above 275°F (135°C). This limitation was needed because above about 275°F(135°C) these hydraulic fluids decompose to form an organophosphoric acid which will chemically attack all titanium except for *TIMETAL*®21S. The general corrosion and pitting are usually accompanied by excessive hydrogen absorption. Thus, the traditional use of titanium in the nacelle areas of aircraft has been severely restricted. However, *TIMETAL*®21S shows exceptionally good resistance to this type of attack. *TIMETAL*®21S not only resists both the initial surface attack and hydrogen absorption, but, because it is a metastable beta alloy, would not be expected to suffer any detrimental effects at levels of up to a thousand parts per million if conditions were somehow created that caused hydrogen absorption. Alpha-beta and near alpha alloys, such as *TIMETAL*®6-4 and *TIMETAL*®6-2-4-2, can experience hydrogen embrittlement at hydrogen levels as low as 250 ppm. Table VIII compares *TIMETAL*®21S to *TIMETAL*®6-4 with regards to resistance to attack by Skydrol™.

Considerable weight savings on the Boeing 777 nacelle, has been made possible by the unique Skydrol™ resistance of *TIMETAL*®21S (7).

Thermal Stability

Table IX shows the long term thermal stability of *TIMETAL*®21S in an air environment. Specimens were tested both with exposed surface intact and after pickling off the exposed surface in order to differentiate between bulk stability effects and bulk stability effects plus surface contamination effects. Significant changes in residual tensile properties after prolonged thermal exposure are usually the result of grain boundary alpha formation. Since the grain boundary alpha is relatively soft, strain localization at the grain boundary alpha during deformation results in macroscopically low ductility. Overaging minimizes the strength difference between the grain boundary alpha and aged matrix and thus more ductility is retained following thermal exposure.

Table VI. SiC/*TIMETAL*®21S Metal Matrix Composite Typical Tensile Mechanical Properties with 4-Ply Unidirectional Laminate and SCS-6 Fibers (1)

Temp. °F (°C)	Density lb./in.^{3} (g/cm^3)	Strength ksi (MPa)	Modulus Msi (GPa)
70 (21)	0.153 (4.24)	250 (1724)	27 (186)
1000 (649)	—	165 (1138)	—
1500 (816)	—	75 (517)	—

Table VII. Recrystallization Annealing (Re-solution Heat Treatment)

Cold Work, %	Annealing Temperature, °F(°C)	Annealing Time, Minutes
20 - 40	1550	30
40 - 60	1525	30
60+	1500	30

Table VIII. Skydrol™ Resistance, Specimens Partially Submerged for 48 Hrs; Skydrol™ Replenished Twice (2)

Test Temp. °F (°C)	*TIMETAL*® Alloy	Hydrogen Absorbed (ppm)	Comments
350 (177)	6–4	0	0.0004 in (0.01 mm) of Thinning
350 (177)	21S	12	No attack
450 (232)	6–4	>1000	Thinned to knife-edge
450 (232)	21S	97	Scattered pits ~ 0.0002 in (0.005 mm) deep
550 (288)	6–4	>1000	0.028 in (0.71 mm) of Thinning
550 (288)	21S	138	Scattered pits ~ 0.0014 in (0.036 mm) deep; 0.0018 in (0.046 mm) of Thinning
650 (343)	6–4	>1000	0.021 in (0.53 mm) of Thinning
650 (343)	21S	83	Scattered pits ~ 0.002 in (0.05 mm) deep; 0.0018 in (0.046 mm) of Thinning

Table IX
Thermal Stability Data for Two gages of TIMETAL 21S Sheet
Solution Treated then Aged at 1275F (691C), 8 hrs, plus 1200F (649C), 8 hrs,
Prior to Exposure in Freely Circulating Air at Indicated Temperature
Residual Tensile Properties Determined at Room Temperature

Exposure Temp. F	C	Sheet Gage in	mm	Exposure Time hrs.	Surface Pickled? (1)	Ultimate Strength ksi	MPa	0.2% Yield Strength ksi	MPa	Elong %
none	none	0.052	1.32	none		133.0	917	123.0	848	19
		0.023	0.58	none		139.6	963	128.4	885	12
950	510	0.052	1.32	500	no	145.4	1003	132.3	912	15.0
					yes	145.0	1000	133.3	919	13.0
				1000	no	147.3	1016	136.5	941	13.0
					yes	146.5	1010	134.5	927	14.0
				3000	no	145.3	1002	136.1	938	13.5
				5000	no	143.9	992	135.8	936	11.0
		0.023	0.58	500	no	141.6	976	132.9	916	7.0
					yes	145.2	1001	133.7	922	7.0
				1000	no	144.2	994	137.1	945	5.5
					yes	151.0	1041	139.0	958	12.0
				3000	no	142.7	984	135.0	931	9.0
				5000	no	147.6	1018	130.8	902	6.5
1050	566	0.052	1.32	500	no	139.2	960	130.3	898	11.0
					yes	136.5	941	126.5	872	13.5
				1000	no	137.4	947	129.6	894	10.0
					yes	137.6	949	127.2	877	17.0
				3000	no	135.2	932	130.3	898	6.0
				5000	no	135.2	932	128.8	888	5.5
		0.023	0.58	500	no	144.6	997	136.7	943	4.0
					yes	141.2	974	132.5	914	8.0
				1000	no	142.9	985	136.4	940	5.5
					yes	145.5	1003	133.3	919	13.5
				3000	no	138.3	954	133.1	918	2.5
				5000	no	144.3	995	131.5	907	3.0
1140	616	0.052	1.32	500	no	132.8	916	127.5	879	5.0
					yes	137.6	949	126.9	875	12.5
				1000	no	132.1	911	126.8	874	2.5
					yes	136.5	941	121.5	838	15.0
				3000	no	bby (2)				
				5000	no	bby				
		0.023	0.58	500	no	139.2	960	134.2	925	0.0
					yes	142.6	983	132.4	913	9.0
				1000	no	141.0	972	136.9	944	2.5
					yes	139.3	960	125.1	863	13.5
				3000	no	bby				
				5000	no	bby				

(1) Descaled and pickled to remove 0.004-in (0.1-mm) from gage area.
(2) bby = broke before yield

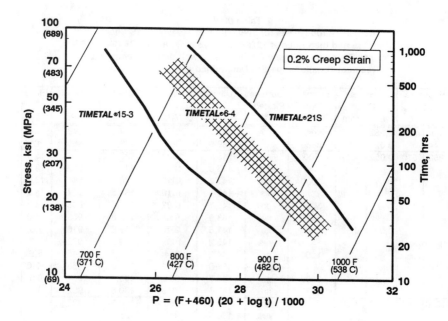

Figure 6. Larson-Miller 0.2% Creep Comparison.

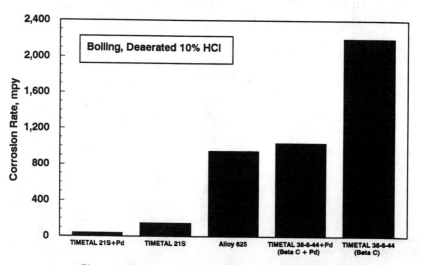

Figure 7. Relative Corrosion Rates in Hydrochloric Acid (6).

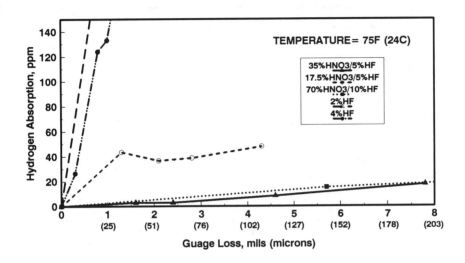

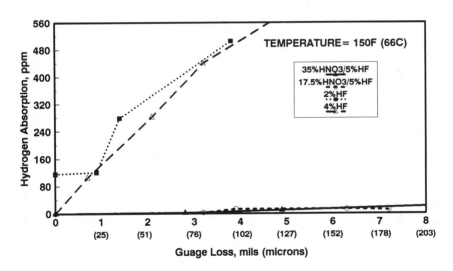

Figure 8. Effect of chemical milling/pickling solution on hydrogen absorption
vs. guage removal at 24 C (TOP) and 66 C (BOTTOM) [4].

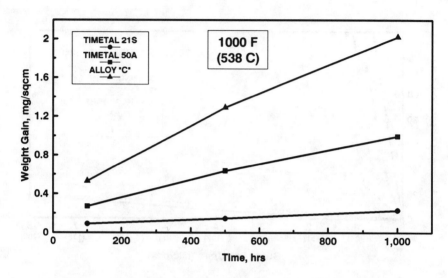

Figure 9. Weight gain vs. time when exposed at 538 C in circulating air
for indicated time (3).

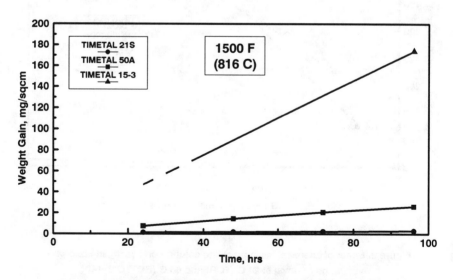

Figure 10. Weight gain vs. time when exposed at 816 C in circulating air
for indicated time (5).

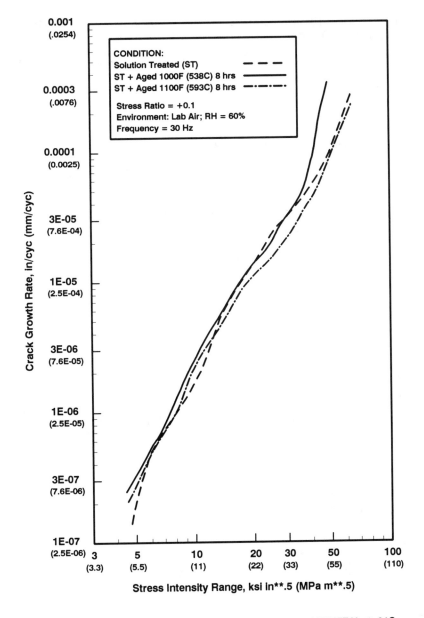

Figure 11. Fatigue crack growth resistance of *TIMETAL* ⊚ 21S
in three heat treat conditions.

REFERENCES

1. J. Sorensen, "Titanium Matrix Composites - NASP Materials and Structures Augmentation Program", AIAA-90-5207, 1990.

2. R. Boyer, Private Communication, Boeing Commercial Airplane Company, 1993.

3. W. M. Parris, "Comparison of *TIMETAL*®21S and Alloy C Properties", TIMET Internal Report, 1990.

4. J. S. Grauman, "Hydrogen Absorption Behavior of Ti Alloys in Pickling Solutions", TIMET Technical Report No. 13, 1991.

5. P. J. Bania and W. M. Parris, "*TIMETAL*®21S: A High Temperature Metastable Beta Titanium Alloy", TIMET, 1990.

6. J. S. Grauman, "Corrosion Behavior of *TIMETAL*®21S for Non-Aerospace Applications", presented at 7th World Conference - San Diego, CA, 1992.

7. R. R. Boyer, "New Titanium Applications on the Boeing 777 Airplane", JOM, May, 1992.

TIMETAL®15-3 PROPERTY DATA

J. C. Fanning*

TIMET Henderson Technical Lab
PO Box 2128
Henderson, NV 89009

ABSTRACT

TIMETAL®15-3 (Ti-15V-3Cr-3Al-3Sn) is a strip producible, cold formable metastable beta titanium alloy designed to reduce the weight of aerospace and other structures while minimizing processing and fabrication costs. It is heat treatable to strengths from 130 ksi (896 MPa) to 190 ksi (1310 MPa) tensile ultimate. Introduced and first applied in 1982, *TIMETAL*®15-3 is the product of a number of U.S. Air Force and Navy programs to develop an economical metastable beta alloy (1-8). Formability, weldability, air hardenability, and environmental degradation resistance were all important; however, cold rollability and cold formability were the main development criteria.

TIMETAL®15-3 is used on numerous aircraft programs, including the B1 and B2 strategic bombers, the C-17 military transport and the Boeing 777 commercial transport. Although originally developed for use in the aerospace industry, non-aerospace applications, such as the Easton "TIPHOON" softball bat, are rapidly developing.

The alloy is currently (January 1993) available in strip, sheet, tube, plate, bar, wire and castings. MIL HNBK 5 allowables and an AMS specification have been published for strip and sheet.

*This is an updated version of a paper originally published by H. W. Rosenberg in 1983.

Beta Titanium Alloys in the 1990's
Edited by D. Eylon, R.R. Boyer and D.A. Koss
The Minerals, Metals & Materials Society, 1993

COMPOSITION

The composition of *TIMETAL*®15-3 is given in wt. % below:

	V	Cr	Al	Sn	O	N	C	H	Fe	Ea	Other Tot
Max	16	3.5	3.5	3.5	0.13	0.03	0.03	0.015	0.30	0.10	0.30
Min	14	2.5	2.5	2.5	-	-	-	-	-	-	-

Titanium constitutes the remainder.

PHYSICAL PROPERTIES

Most physical properties of structural metals and alloys, including *TIMETAL*®15-3, depend on the direction in which they are taken. This feature arises primarily because single crystals of beta titanium are anisotropic and *TIMETAL*®15-3 tends to be slightly textured crystallographically after cold rolling. In general, therefore, any physical property that is not a simple scaler quantity will exhibit at least some anisotropy. Those physical properties that depend on direction are so indicated in what follows.

Density

The density of *TIMETAL*®15-3 is 0.170 lbs./in.^{3} (4.71 g/cm^3).

Thermal Expansion

Thermal expansion is a directional physical property. Yet, *TIMETAL*®15-3 is fairly isotropic in this regard. Typical data follow:

Temperature °F (°C)	L/Lo-% L	L/Lo-% T	10^{-6} in/in/°F (10^{-6} m/m/°C) L	10^{-6} in/in/°F (10^{-6} m/m/°C) T
73 (23)	0	0	-	-
212 (100)	.0647	.0653	4.67 (8.41)	4.71 (8.48)
392 (200)	.1582	.1537	4.97 (8.95)	4.82 (8.68)
572 (300)	.2540	.2514	5.09 (9.16)	5.04 (9.07)
752 (400)	.3558	.3559	5.24 (9.43	5.24 (9.43)
932 (500)	.4622	.4614	5.38 (9.68)	5.37 (9.67)
1112 (600)	.5701	.5763	5.49 (9.88)	5.55 (9.99)
1292 (700)	.7054	.7085	5.79 (10.42)	5.82 (10.48)
1472 (800)	.8046	.8080	6.07 (10.93)	6.09 (10.96)

Specific Heat Capacity

The specific heat of *TIMETAL*®15-3 is a scaler quantity having the following typical values:

Temperature-°F(°C)	BTU/Lb F	J/Kg C
77 (25)	.121	.508
392 (200)	.137	574
752 (400)	.155	649
1112 (600)	.173	724
1400 (800)	.187	784

Thermal Conductivity

Thermal conductivity is a directional quantity for which typical values follow:

Temperature-°F(°C)	BTU IN/H FT² F	W/m K
75 (25)	56.0	8.08
500 (260)	83.1	11.99
1000 (538)	115.4	16.64
1400 (760)	137.2	19.79

Electrical Resistivity

This directional property has the following typical values:

Temperature-°F(°C)	OHM-Meters x 10^{-8}
77 (25)	147.6
500 (260)	155.0
1000 (538)	160.2

Transformation Temperature

The precise transformation temperature depends only on alloy composition at ambient pressure. This temperature is also called the beta transus in titanium metallurgy. For *TIMETAL*®15-3, it typically lies in the range 1385-1415°F (750-770°C).

Heat Treatment

Heat treatment for *TIMETAL*®15-3 is straight forward. Typical ranges are:

Solution Heat Treat (Solution Anneal)	1450F-1550F, 4-30 Min, AC
Age	900-1100F (8-16 Hrs)

Use of the longest aging times is recommended when it is desireable to minimize property scatter. In many cases, use of the shortest times will be quite satisfactory. See the following section for the effects of age temperature on strength.

MECHANICAL PROPERTIES

Tensile Properties

Solution Annealed Condition

Table I presents the typical tensile properties for *TIMETAL®*15-3. Analyses of the data for sources of variance show that the main factors contributing to scatter are test direction and lot-to-lot differences, the latter being larger. Test direction differences typically are small but can amount to about 4 ksi (30 MPa); the lot-to-lot difference can be up to about twice as much. These values are small, however, when compared with the scatter encountered in unalloyed titanium strip or alpha beta titanium alloy sheet.

Table 1. Typical Tensile Properties of Solution Treated *TIMETAL®*15-3 300+ Tests

	Average	STD DEV
UTS, ksi(MPa)	117.6 (811)	2.0 (14)
YS, ksi (MPa)	115.3(795)	2.0 (14)
EL, %	16	2.7

Aged Condition

*TIMETAL®*15-3 can be aged to a tensile strength of at least 190 ksi (1310 MPa) while maintaining adequate ductility. Between 950F and 1000F (510-540°C), the fully aged strength of *TIMETAL®*15-3 is a linear function of aging temperature. Table II shows the aged strength capability of *TIMETAL®*15-3. By selecting the proper aging temperature, any desired ultimate tensile strength up to about 190 ksi (1310 MPa) can be realized with ductility still in excess of 4%. Table III gives typical compression, shear and bearing properties.

Possible aging temperatures range from 850F(455C) to 1100F(593C). As would be expected, higher aging temperatures produce lower strengths with concomitant ductility increases. The most common aging treatments are 1000F(538C) for 8 hrs, 900F(482C) for 16 hrs, or 925F(496C) for 8 hrs. These conditions have AMS specification coverage.

Compressive Yield Strength

Typical STA (1000F[538°C]-8 hr. age) compression data follow:

		0.2% Compressive YS	
Temperature°F(°C)	Direction	ksi	MPa
-60 (-51)	L	181.7 (4.9)*	1253 (34)
	T	187.4 (6.2)	1292 (43)
75 (24)	L	158.4 (10.0)	1092 (69)
	T	169.0 (11.9)	1165 (82)
400 (204)	L	139.4 (6.6)	961 (46)
	T	139.5 (4.6)	962 (32)

*Values in parens. are standard deviations for two or three samples.

Table II. Effect of Aging Temperature on Tensile Properties of *TIMETAL*®15-3 Typical (Average) Values from Ten Lots of 0.032-0.094 in. (0.81-2.39)

Age Treatment	900F/8 Hrs (482C)		900F/16 Hrs (482C)		925F/8 Hrs (496C)		950F/8 Hrs (510C)		1000F/8 Hrs (538C)	
Test Direction	L	T	L	T	L	T	L	T	L	T
UTS, ksi (MPa)	193 (1331)	198 (1365)	194 (1338)	200 (1379)	188 (1296)	190 (1310)	179 (1234)	184 (1269)	160 (1103)	165 (1138)
YS, ksi (MPa)	180 (1241)	181 (1248)	181 (1248)	184 (1269)	174 (1200)	176 (1214)	164 (1131)	169 (1165)	149 (1027)	148 (1020)
El, %	8.4	7.5	8.7	8.2	9.1	8.4	10.3	10.0	12.8	12.0

Table III. Typical (Average) Compression, Shear, and Bearing Strength, Ksi (MPa); and Typical Moduli, Msi (GPa)

Age:	Solution Treated(19)		900°F/16 Hrs (482°C/16 Hrs)		925°F/8 Hrs (496°C/8 Hrs)(20)		1000°F/8 Hrs (538°C/8 Hrs)	
	L	T	L	T	L	T	L	T
Compression Yield Strength	119.6 (825)	124.4 (858)	192.8 (1329)	195.8 (1350)	184.1 (1269)	189.6 (1307)	152.6 (1052)	156.3 (1078)
Shear Ultimate Strength	90.2 (622)	89.2 (615)	-	124.8 (860)	113.7 (784)	115.9 (799)	105.8 (729)	106.4 (734)
Bearing Ultimate Strength e/D=1.5	198.1 (1366)	198.8 (1370)	284.6 (1962)	285.8 (1971)	300.1 (2069)	300.6 (2073)	251.0 (1731)	253.3 (1746)
e/D=2.0	258.9 (1785)	257.0 (1772)	323.6 (2231)	-	329.9 (2275)	316.3 (2181)	316.2 (2180)	321.1 (2214)
Bearing Yield Strength e/D=1.5	160.8 (1109)	161.2 (1111)	269.5 (1858)	267.4 (1843)	284.1 (1959)	283.6 (1955)	223.8 (1543)	223.5 (1541)
e/D=2.0	179.7 (1239)	182.1 (1256)	296.5 (2044)	294.0 (2027)	292.3 (2015)	294.9 (2033)	253.1 (1745)	256.3 (1767)
Tensile Modulus	11.4	12.4	15.7	16.0	15.3	15.4	15.2	15.7
Compressive Modulus	12.5	13.4	16.0	16.3	14.8	15.0	15.3	16.0

Cold Deformation Effects on Aging

Cold deformation has two primary effects: a) increases aging rate to full strength and b) increases both the aged and unaged strength levels. The aged or unaged strength increases are a linear function of the amount of cold deformation over the range from 20% to 60%. Furthermore, for the aged condition, cold deformation up to at least 40% does not significantly affect the relationship between aged strength and ductility. These features are shown in Figure 1. The effect of cold work on annealed properties is shown in Table IV. What is more, hydrogen up to 417 ppm has little influence on cold worked and aged tensile properties. The data available are shown in Figure 2.

Table IV. Effects of Cold Rolling on Unaged Tensile Properties

% Cold Work	Regressed Values		
	UTS-ksi(MPa)	YS-ksi(MPa)	El-%
0	114.4(789)	110.7(763)	16.0
20	129.5(893)	123.4(851)	13.1
40	144.5(996)	136.1(938)	10.2
60	159.5(1100)	148.8(1025)	7.2

Test Temperature Effects

Figure 3 illustrates the effect of test temperature on the STA tensile, shear, compression, and bearing strengths.

Fracture Toughness

Typical (average of 3 values) plane stress fracture toughness values for *TIMETAL*®15-3 strip aged at 1000°F for 8 hrs. appear below (8):

Gauge, in. (mm)	Test Temp., F(C)	Dir.	0.2% TYS, ksi (MPa)	Kc, ksi in.$^{.5}$ (MPa m$^{.5}$)
.050 (1.25)	75 (24)	L	154.7 (1067)	91.3 (100.4)
"	75 (24)	T	158.7 (1094)	91.3 (100.4)
"	-60 (-51)	L	170.3 (1174)	74.0 (81.4)
"	-60 (-51)	T	177.0 (1220)	69.3 (76.2)
.070 (1.75)	75 (24)	L	152.0(1048)	102.7 (113.0)
"	75 (24)	T	154.3 (1064)	97.0 (106.7)
"	-60 (-51)	L	171.7 (1184)	83.3 (91.6)
"	-60 (-51)	T	172.0 (1186)	78.0 (85.8)

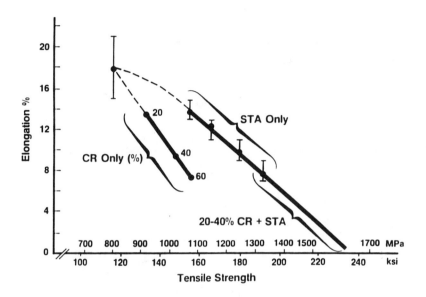

Figure 1. Properties of coldworked unaged *TIMETAL* ®15-3 (7).

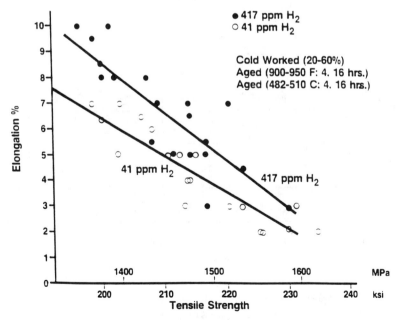

Figure 2. Effect of hydrogen on coldworked plus aged *TIMETAL* ®15-3 (7).

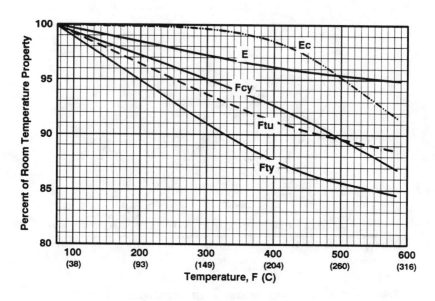

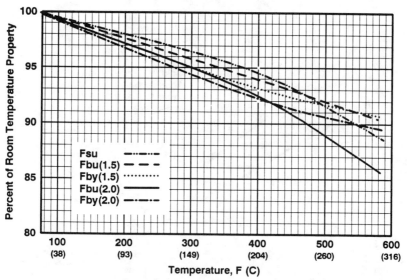

Figure 3. Retention of Mechanical Properties at Elevated
Temperature for STA Strip.

Low Temperature Properties

At lower test temperatures, *TIMETAL*®15-3 exhibits increases in tensile and yield strengths with concomitant decreases in ductility and toughness. However, abrupt ductile-to-brittle transition behavior commonly found in other bcc materials, such as ferritic steels, has not been observed in *TIMETAL*®15-3.

Typical (average of 3 values) tensile data are given in Table V.

Table V. Effect of Test Temperature on Tensile Mechanical Properties

Condition	Test Temp. °F (°C)	0.2% YS, ksi (MPa)	UTS, ksi (MPa)	% El	Notched UTS
Solution Annealed	-320(-196)	148.2(1022)	248.5(1713)	7.4	264.9(1826)
"	-110(-79)	144.0(1993)	157.7(1087)	12.0	180.4(1244)
"	75(24)	114.4(789)	114.7(791)	14.4	126.7(874)
"	250(121)	92.4(637)	101.3(699)	21.7	111.2(767)
"	600(316)	78.4(541)	91.7(632)	19.8	101.8(702)
Aged 1000°F/8 hr. (538°C/8 hr.)	-320(-196)	195.5(1348)	269.9(1861)	6.5	284.1(1959)
"	-110(-79)	185.7(1280)	195.8(1350)	9.5	213.9(1475)
"	75(24)	155.2(1070)	168.5(1162)	10.2	182.0(1255)
"	250(121)	140.2(967)	157.2(1084)	9.6	170.2(1173)
"	600(316)	122.8(847)	144.5(996)	7.4	158.7(1094)
Aged 900°F/16 hr. (482°C/16 hr.)	-320(-196)	249.8(1722)	286.6(1976)	2.9	261.3(1802)
"	-110(-79)	200.6(1383)	214.8(1481)	0.9	243.3(1678)
"	75(24)	188.0(1296)	202.1(1393)	6.3	214.3(1478)
"	250(121)	167.2(1153)	183.3(1264)	6.5	201.1(1387)
"	600(316)	131.6(907)	170.9(1178)	5.6	185.2(1277)

Compressive Tangent Modulus

Figures 12 and 13 illustrate the compressive tangent modulus behavior for *TIMETAL*®15-3 at temperatures from room temperature to 800°F(427°C).

ENVIRONMENTAL DEGRADATION RESISTANCE

General Corrosion

TIMETAL®15-3 is virtually immune to general corrosion in the environments a structural material might encounter, including atmospheric, sea water, steam (to 600°F[316°C]), aviation fuel, bodily wastes, sump tank water, and salt spray.

Compatibility with Propellants

TIMETAL®15-3 is fully compatible with common rocket propellants. No surface attack, cracking, pitting, or deterioration of tensile properties occurred in 12-month compatibility tests, the results of which are given in Table VI. Also, no stress corrosion cracking or other attack was detected in specimens loaded to 90% of yield strength during exposure(11).

Compatibility with Fire Suppressants

TIMETAL®15-3 is fully compatible with fire suppressants such as Halon 1301. Fire extinguisher bottles made from *TIMETAL*®15-3 show no attack whatsoever when filled with Halon 1301 and then held at 160°F(71°C) and an internal pressure of 1150 psig(7.9 Mpa) for an extended period (12).

Alpha Case Formation

Like other titanium alloys, *TIMETAL*®15-3 will absorb oxygen (and, to a lesser extent, other interstitial elements, such as nitrogen) when exposed to air at temperatures above 1000°F (538°C). The oxygen diffuses into the surface to form a brittle alpha-stabilized layer known as "alpha case". Even though an alpha case layer is rarely more than a few mils deep, it can substantially reduce the formability and the fatigue life of any titanium alloy, and therefore should always be removed by pickling or machining.

Figure 4 shows the rate of alpha case formation on *TIMETAL*®15-3, which is slightly faster than for Ti-6Al-4V(13).

Burn Resistance

TIMETAL®15-3 has good burn resistance(14). As seen in Figure 5, which shows percent damage by self-sustained combustion (induced by a high energy ignition source) versus air pressure and temperature, *TIMETAL*®15-3 is less combustible than Ti-6Al-4V (at least for this particular set of test conditions).

Note that high air pressure, high air velocity, and high air temperature combined with a very high energy ignition source are necessary for self-sustained combustion of bulk titanium. That is, bulk titanium will not burn in static air even if a high energy ignition source is present; nor will it burn in high pressure/high velocity/high temperature air unless a very high energy ignition source is available.

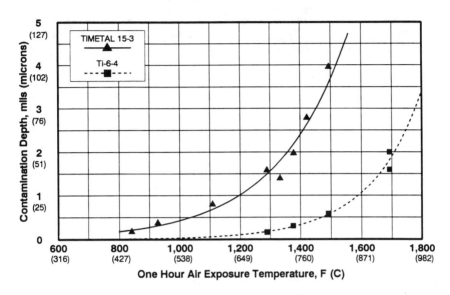

Figure 4. Rate of alpha case formation (13).

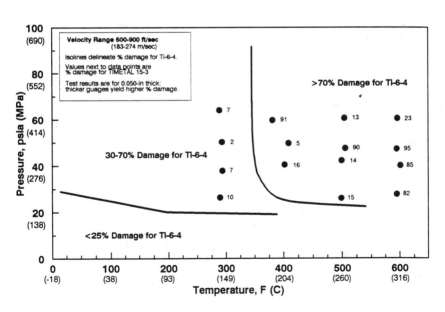

Figure 5. Mass Loss (% Damage) for self sustained combustion (14).

Table VI. Residual Tensile Properties of *TIMETAL*®15-3 After a 12-Month
Exposure to Rocket Propellant(11)

Exposure	Condition	UTS, ksi(MPa)	YS, ksi(MPa)	El, %
N₂O₄ (Mon-1) 160F(71C)	ST 1450F(788C) AC EB welded - aged 950F(510C) 8 Hrs.	185(1276)	163(1124)	5
	ST 1450F(788C) AC	114(786)	113(779)	17
	ST 1450F(788C) AC aged 950F(510C) 8 Hrs.	179(1234)	170(1172)	12
Hydrazine 140F(60C)	ST 1450F(788C) AC EB welded - aged 950F(510C) 8 Hrs.	185(1276)	173(1193)	6
	ST 1450F(788C) AC	118(807)	117(807)	18
	ST 1450F(788C) AC aged 950F(510C) 8 Hrs.	178(1227)	165(1138)	11
Monomethyl Hydrazine 140F(60C)	ST 1450F(788C) EB welded - aged 950F(510C) 8 Hrs.	186(1283)	168(1158)	6
	ST 1450F AC	117(807)	115(793)	19
	ST 1450F(788C) AC aged 950F(510C) 8 Hrs.	178(1227)	164(1131)	11

THERMAL STABILITY AND CREEP

Creep Properties

TIMETAL®15-3, which was not developed for elevated temperature (>550°F [288°C]) use,
has slightly lower creep resistance than *TIMETAL*®6-4. Larson-Miller Curves are given
in Figure 6.

Thermal Stability

The ability of a material to undergo extended elevated temperature exposure is important
in many applications. Tables VII and VIII illustrate that the stability of *TIMETAL*®15-3
depends on the prior heat treatment given. The more fully aged, the less exposure changes
mechanical properties.

Table VII. Metallurgical and Dimensional Stability [600F(316C) - 150 Exposure]

Process	Gage-in(mm)	Direc-tion	Ftu ksi(MPa)	Fty ksi(MPa)	El %	Dimen-sional Change %
Lab.Sim.Strip	.050(1.27)	L	173(1193)	159(1096)	11	+ .05
	.050(1.27	T	174(1208)	158(1089)	12	+ .06
	.100(2.54)	L	172(1186)	156(1076)	12	- .17
	.100(2.54)	T	192(1324)	176(1214)	5	-
Sheet Sim.Strip	.050(1.27)	L	170(1172)	154(1062)	12	- .47
	.050(1.27)	T	172(1186)	155(1069)	14	- .25
	.100(2.54)	L	176(1214)	161(1110)	8	- .39
	.100(2.54)	T	184(1269)	168(1158)	8	- .15
Hot & Cold Roll	.050(1.27)	L	177(1220)	162(1117)	11	- .13*
	.050(1.27)	T	180(1241)	166(1145)	11	- .18*
	.100(2.54)	L	174(1200)	159(1096)	15	- .13
	.100(2.54)	T	171(1179)	157(1083)	15	- .22

*Aged 950°F-16 hours, all others aged 1000°F-8 hours.

Table VIII. Metallurgical and Dimensional Stability (600°F - 1000 Hour Exposure)*

Age Temp F(°C)	Age Time Hrs.	As Aged			As Exposed			Dimen-sional Change %
		FTU ksi(MPa)	Fty ksi(MPa)	El %	FTU ksi(MPa)	Fty ksi(MPa)	El %	
950(510)	2	129(889)	119(821)	12	185(1276)	166(1145)	2	- .02
	8	171(1179)	158(1089)	10	186(1282)	169(1165)	9	- .02
	16	180(1241)	168(1158)	10	188(1296)	175(1207)	8	+ .03
	24	178(1227)	168(1158)	8	186(1282)	172(1186)	10	+ .03
1050(566)	2	128(883)	117(807)	17	174(1103)	160(1103)	6	- .03
	8	145(1000)	132(910)	12	158(1089)	144(993)	13	+ .05
	16	146(1007)	132(910)	14	155(1069)	141(972)	17	+ .02
	24	141(972)	128(882)	13	153(1055)	139(958)	16	0
1150(621)	2	117(807)	110(758)	17	161(1110)	146(1007)	6	0
	8	129(889)	117(807)	15	145(1000)	133(917)	15	+ .09
	16	125(862)	113(779)	19	140(965)	127(876)	18	0
	24	125(862)	114(786)	21	138(952)	125(862)	18	+ .01

*As aged properties determined from duplicate longitudinal properties.
1400F-4 min. solution treatment used in all cases.

Fatigue

TIMETAL®15-3 exhibits excellent fatigue properties. See the section on fatigue data published by Fanning elsewhere in this book. Some typical results are given below:

Table IX. Smooth and Notched Typical Results for *TIMETAL*®15-3 Strip Aged
8 Hours at 1000°F(538°C)

	Runout Stress - ksi(MPa)*	
Temp-°F	Smooth	Notched $K_t=3$
-60	105(724)	30(207)
75	95-110(655-758)	30-35(207-241)
400	95-100(655-690)	32-35(221-241)
*Runout > 10^7 cycles, R=0.1, maximum stress shown.		

Fatigue Crack Propagation

TIMETAL®15-3 exhibits slightly faster crack growth characteristics than mill annealed *TIMETAL*®6Al-4V. However, *TIMETAL*®15-3 is not sensitive to environment; salt water has virtually no influence on growth characteristics. This feature is shown in Figure 8. These data were generated using R=0.1 on single edge notch specimens. The tests in air were run at 20 Hz, those in salt water at 5 Hz. A frequency effect is thus nested with the environmental one. In all, 24 specimens were tested. Evaluating the data at a ΔK = 20ksi $\sqrt{in}$ (22MPa $\sqrt{m}$) shows the crack growth rate increases slightly as sheet gage increases. The combined salt water plus frequency effect is just at the "detection limit" statistically. The data are:

		da/dN (Δ = 20 ksi $\sqrt{in}$; 22 MPa $\sqrt{m}$)	
		$in x 10^{-6}$	$mm x 10^{-6}$
Environmental	Air	9.12	232
Effect	Salt	9.80	249
Gage Effect	.050"	8.58	218
	.100"	10.33	262

Test error for these data was estimated to be 0.49 x 10^{-6} in/cycle (12 x 10^{-6} mm/cycle).

Cold Rolling

TIMETAL®15-3 is readily cold rolled on a Sendzimir mill. *TIMETAL*®15-3 strip, in gages below about 0.100" (2.54mm), competes favorably cost wise with *TIMETAL*®6-4 off hot roll mills. This economic benefit arises directly from the alloy's cold rollability which reduces costs while improving dimensional tolerance. Although *TIMETAL*®15-3 requires higher rolling forces than do the unalloyed grades, it is easier to roll than most other beta alloys.

The forming properties of *TIMETAL*®15-3 are rather insensitive to minor variations in reduction, annealing time and temperature. Aging kinetics are the main feature affected by such variations.

Foil Capability

TIMETAL®15-3 can be readily rolled to foil. No intermediate anneals are required upon rolling from .050"(1.27mm) directly to .003"(.076mm), a 94% cold reduction. Typical tensile properties are (15):

% Cold Roll	Age	UTS, Ksi (MPa)	El, %
0	None	112	19
20	None	150	7
60	None	195	3
0	480°C/24 Hrs.	190	Not Avail.
20	480°C/24 Hrs.	205	Not Avail.
60	480°C/24 Hrs.	235	Not Avail.

The minimum gage capable of being produced is about .0001 in (.0025mm). Because *TIMETAL*®15-3 is cold rollable and available as strip (coil), the cost of *TIMETAL*®15-3 foil is typically less than one-fifth the cost of *TIMETAL*®6-4 foil.

Castings

TIMETAL®15-3 has similar casting attributes to *TIMETAL*®6-4, but has marginally better fluidity, slightly less shrinkage, and forms a thinner skull(16).

Typical HIP cycle is 1750°F (954°C) at 15 Ksi (103 MPa) for two hours. As HIP-ed tensile properties of 110 Ksi (758 MPa) ultimate strength and 20-25% elongation enable room temperature straightening of castings(16).

Slow cooling rates from the mold, HIP, or solution heat treatment can result in heavy grain boundary alpha, which results in low tensile ductility because of strain localization (17).

Typical tensile mechanical properties for *TIMETAL*®15-3 castings are given in Table X. Minimum tensile properties after aging at 950°F(510°C) for 12 hours, are:

UTS	165 Ksi (1138 (MPa)
TYS	155 Ksi (1069 MPa)
% El.	4 (in 4D)

Forged Products

When processed properly, *TIMETAL*®15-3 bar shows similar properties to *TIMETAL*®15-3 strip, but with larger grain size (ASTM No. 3 for bar vs. ASTM No. 6 for strip). Typical tensile mechanical properties are given in Table XI.

Rolled rings, die forgings, and isothermal forgings can also be made from *TIMETAL*®15-3.

Table X. Room Temperatue Tensile Properties, Separatly Cast Coupons(16)

Condition		TYS		UTS			
		Ksi	MPa	Ksi	MPa	% EL	%RA
As-Cast, HIPed		100	688	108	745	22	29
		102	702	110	760	25	57
Aged 900°F (Air)	2 Hrs	129	889	139	960	7	16
	4 Hrs	166	1145	182	1255	8	12
	8 Hrs	179	1231	192	1327	6	6
		173	1191	184	1269	4	4
	12 Hrs	174	1200	188	1296	6	11
	16 Hrs	183	1259	196	1349	4	4
		186	1285	199	1371	4	3
Aged 950°F (Air)	2 Hrs	141	972	160	1105	11	13
		143	986	159	1096	6	10
	4 Hrs	152	1045	168	1156	11	18
	8 Hrs	155	1069	170	1174	10	19
		156	1074	171	1179	10	19
Aged 950°F (Vacuum)	12 Hrs	167	1148	181	1249	5	10
		167	1149	183	1259	7	9
	12 Hrs	158	1088	173	1191	10	19
	16 Hrs	180	1241	190	1307	3	4
		168	56	181	1249	6	12
Aged 1000°F (Vacuum)	2 Hrs	135	932	148	1020	11	18
		139	958	152	1046	9	9
	4 Hrs	137	943	148	1023	13	23
	8 Hrs	146	1008	158	1087	4	6
		143	987	154	1058	5	4
	12 Hrs	159	1096	174	1198	7	9
	16 Hrs	148	1018	160	1103	11	9
		149	1028	152	1051	2	

Table XI. Tensile Mechanical Properties of *TIMETAL*®15-3 Bar

Thick., in (cm)	Orient.	Solution Treated				Solution Treated plus Aged 1000F(538C) 8 Hrs			
		UTS, Ksi(MPa)	TYS, Ksi(MPa)	% El	% RA	UTS, Ksi(MPa)	TYS, Ksi(MPa)	% El	% RA
2.25(5.7)	L	114(786)	112(772)	22	66	160(1103)	147(1014)	13	30
	T	113(779)	111(765)	24	65	172(1186)	159(1096)	13	23
3(7.6)	L	112(772)	109(752)	22	56	156(1076)	143(986))	14	26
	T	112(772)	108(745)	21	57	162(1117)	151(1041)	10	19
4.4(11.2)	L	110(758)	108(745)	16	40	166(1145)	154	8	13
	T	109(752)	108(745)	15	39	163(1124)	153	7	19

426

Table XII. Typical Formability of Annealed *TIMETAL*®15-3

Bend radius	2.0 x thickness
Cup Height	0.30 in (7.62mm)
Draw	46%*
Flange Stretch Shrink	20% 1%**
Joggle L/d d/t	2 5
Spring Back	15° at 90°
*Limit of dies, not of material. **Special procedures can improve this significantly.	

FABRICATION

TIMETAL®15-3 approaches the formability of the commercially pure grades. It can be hydroformed, shear spun, flow formed, stretch formed, and chem milled. Press brake, joggling, drilling and dimpling are also operations readily performed on *TIMETAL*®15-3. Table XII gives some typical results. Compared with the unalloyed grades, *TIMETAL*®15-3 requires higher forming forces and greater compensation for springback to achieve the same degree of forming. The alloy can also be hot formed. The forming notes provided in "*TIMETAL*®21S Property Data" by Fanning elsewhere in this book are applicable to *TIMETAL*®15-3.

Care should be exercised during chem milling because *TIMETAL*®15-3 readily picks up hydrogen from certain titanium chem milling baths. Bath oxidizing power must be kept high. A solution of 5% HF-35% HNO_3 is recommended. The principal effect of hydrogen below about 1500 ppm is to suppress aging response.

Springback data are provided in Figure 9 and Table XIII.

Strain Hardening

TIMETAL®15-3 in the solution annealed condition does not follow the usual strain hardening laws. The plastic portion of the stress-strain curve is relatively flat. Typical tensile and compressive stress-strain curves are shown for solution treated strip in Figures 10 and 11.

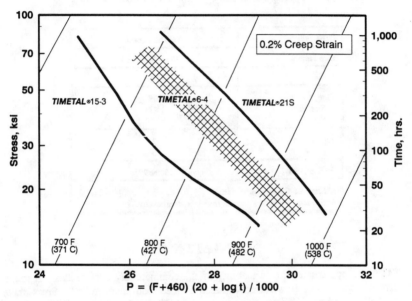

Figure 6. Larson-Miller 0.2% Creep Comparison.

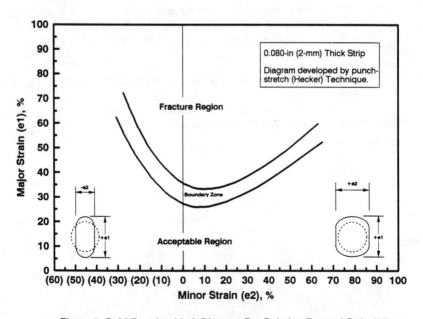

Figure 7. Cold Forming Limit Diagram For Solution Treated Strip (10).

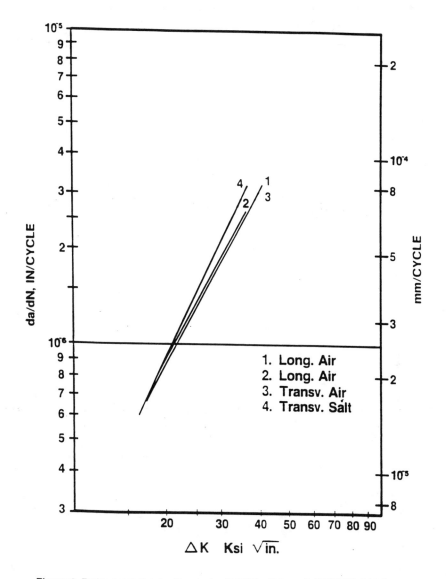

Figure 8. Fatigue crack growth rate for 0.050-in (1.3-mm) *TIMETAL* ®15-3 aged 950 F (510 C) for 16 hours.

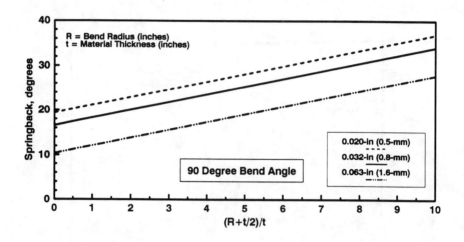

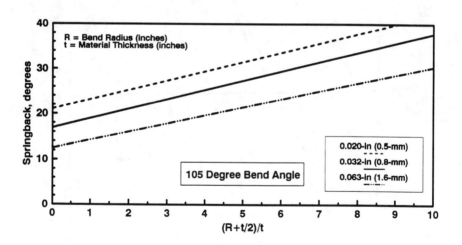

Figure 9. Springback of solution treated *TIMETAL* ®15-3 after brake forming at room temperature to indicated bend angle.

Table XIII. Springback, as Formed and After Aging [1000°F (540°C), 8 Hrs.](9)

		Bend Radius 1.5T								Bend Radius 3.0T					
		Bend Angle 90°				Bend Angle 105°				Bend Angle 90°			Bend Angle 105°		
		0.020" (0.0508)	0.050" (0.127)	0.070" (0.178)	0.125" (0.3175)	0.020" (0.0508)	0.050" (0.127)	0.070" (0.178)	0.125" (0.3175)	0.020" (0.0508)	0.050" (0.127)	0.070" (0.178)	0.020" (0.0508)	0.050" (0.127)	0.070" (0.178)
As Formed															
	L	25	21	24.5	21.5	19.5	12	15	10.5	25.5	29	28	18	15.5	15
	T	24.5	21	23	20.5	19.5	12	15	10	25.5	27.5	27	19	15	14
After Age															
	L	28.5	23	26	23	22.5	13	16	11	28.5	31	29.5	20	17.5	16.5
	T	28	22.5	24.5	21.5	23	13	17	10.5	39.5	29.5	29	20	17	16

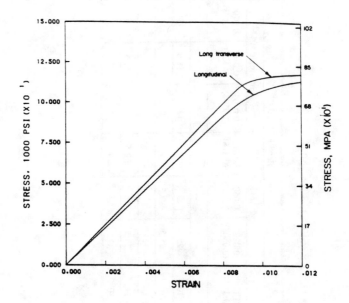

Figure 10. Tensile stress-strain curve for *TIMETAL* ®15-3 ST strip (19).

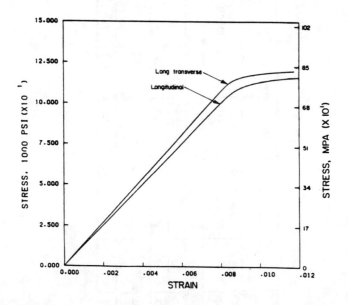

Figure 11.Compressive stress-strain curve for *TIMETAL* ®15-3 ST strip (19).

WELDING

Welding Methods

TIMETAL®15-3 can be welded by Gas Tungsten Arc (GTA), Electron Beam (EB), and all other welding processes generally applicable to titanium. *TIMETAL*®15-3 weldments have tensile strength, notched tensile strength, fracture toughness, and crack growth resistance equivalent to, or superior to, *TIMETAL*®6-4 weldments (18).

As with the welding of any titanium, it is critical to have clean surfaces and good inert gas shielding in order to produce sound welds. When filler material is required, *TIMETAL*®15-3 wire is recommended.

Heat Treatment of Weldments

Weldments can be directly aged after welding, or can be re-solution heat treated before aging (this has been shown to improve the high cycle fatigue life).

Greater uniformity of properties can be obtained in weldments by pre-aging weldments prior to final aging. For example, pre-aging at 800°F(427°C) for 4 hours creates numerous nucleation sites at which the alpha grows during the 1000°F(538°C) final age cycle. If the pre-aging cycle is not performed, alpha might precipitate and grow at the grain boundaries only, rather than throughout the matrix.

Mechanical Properties of Weldments

Mechanical properties of weldments are similar to those of the parent metal, but usually the strength is slightly higher and ductility is slightly lower.

Typical properties are given in Tables XIV and XV.

Table XIV. Mechanical Properties of Weldments in 0.100 In. TIMETAL®15-3 Sheet(18)

Condition	Rolling and Weld Direction	Ultimate Tensile Strength		Yield Strength		Elongation (%) 2 in. or 50 mm	Bend (XT)	
		ksi	MPa	ksi	MPa		Pass	Fail
SA + EB	L	118.0	813.6	101.0	696.4	13	0.6	·
	L	117.7	811.5	102.8	708.8	14	0.6	·
	L	117.9	812.9	104.8	722.6	13	0.6	·
	T	113.8	784.7	113.7	784.0	15	0.6	·
	T	114.6	790.2	113.7	784.0	15		
	T	114.6	790.2	113.7	784.0	15		
SA + EB + 900F, 8 hrs	L	190.6	1314.2	171.9	1185.3	6.5	9	8.7
	L	191.1	1317.6	174.2	1201.1	6	9	8.7
	L	188.8	1301.8	171.0	1179.0	6	10	9
	T	194.7	1342.5	179.5	1237.7	6	12.5	11.2
	T	194.6	1341.8	181.6	1252.1	6.5		
SA + EB + 950F, 8 hrs	L	183.2	1263.2	165.2	1139.1	7.5	7.5	7.5
	L	181.8	1253.5	163.9	1130.1	5.5	8.7	7.5
	L	181.2	1249.4	162.4	1119.7	6.5	8.7	7.5
	T	177.4	1223.2	165.7	1142.5	8.5	9	8.7
	T	175.5	1210.1	163.4	1126.6	9		
	T	175.3	1208.7	167.1	1152.2	8		
SA + GTA	L	117.7	811.5	108.9	750.9	12	0.8	0.6
	L	118.7	818.4	106.8	736.4	13	0.6	·
	L	119.0	820.5	107.6	741.9	10	0.7	0.6
	T	116.8	805.3	111.6	769.5	12.5	0.8	0.6
	T	117.6	810.9	111.5	768.8	12		
	T	117.3	808.8	112.3	774.3	10.5		
SA + GTA + 900F, 8 hrs	L	197.1	1359.0	181.7	1252.8	2.5	15	11.3
	L	193.8	1336.3	177.0	1220.4	4.0	10	8.7
	L	194.2	1339.0	179.8	1239.7	3.5	11.3	10
	T	194.1	1338.3	176.5	1217.0	5.0	12.5	12.2
	T	194.2	1339.0	166.7	1149.4	6.5		
	T	193.3	1332.8	176.8	1219.0	5.5	7	
SA + GTA + 950F, 8 hrs	L	183.9	1268.0	170.0	1172.2	4	7.5	6.9
	L	182.3	1257.0	170.6	1176.3	4.5	9	8.7
	L	176.6	1217.7	162.4	1119.7	3.5	8.7	7.5
	T	170.1	1172.8	154.6	1066.0	7.5	8.7	7.5
	T	173.1	1193.5	·		6.5		
	T	171.4	1181.8	159.1	1097.0	8		
SA + GTA (filler)	L	108.0	744.7	102.8	708.8	15	0.8	0.6
	L	111.8	770.9	110.3	760.5	15	0.6	·
	L	107.1	738.5	103.4	712.9	13	0.8	0.6
	T	111.1	766.0	101.6	700.5	15	0.6	·
	T	110.8	764.0	106.8	736.4	15		
	T	110.7	763.3	102.4	706.0	17		
SA + GTA (filler) + 900F, 8 hrs	L	192.7	1328.7	181.4	1250.8	1.5	8.7	7.5
	L	182.9	1261.1	174.3	1201.8	1.0	7.5	7.0
	L	192.5	1327.3	178.1	1228.0	3.0	9.4	8.7
	T	177.5	1223.9	163.5	1127.3	1.5	·	15
	T	187.9	1295.6	172.8	1191.5	3.5		
	T	186.6	1286.6	167.5	1154.9	3.0		
SA + GTA (filler) + 950F, 8 hrs	L	182.7	1259.7	·		4.0	8.7	7.5
	L	178.8	1232.8	169.3	1167.3	3.0	7.0	5.6
	L	183.5	1265.2	168.5	1161.8	·	8.7	7.5
	T	162.8	1122.5	153.5	1058.4	3.0	9.4	8.7
	T	173.2	1194.2	159.0	1096.3	8.5		
	T	170.3	1174.2	150.9	1040.5	7.5		

Abbreviations:
SA   Solution Anneal
EB   Electron Beam Weld
GTA   Gas Tungsten Arc Weld

Table XV. Mechanical Properties of Weldments in 0.300-IN *TIMETAL*®15-3 Plate (18)
(Toughness tests performed along welds)

Condition	Rolling and Weld Direction	Ultimate Tensile Strength		Yield Strength		Elongation (%) 2 in. or 50 mm	KIc	
		Ksi	MPa	Ksi	MPa		Ksi(in)$^{1/2}$	MPa(m)$^{1/2}$
SA+EB	L	117.2	808.1	102.3	705.4	18	75.9*	83.4*
	L	116.1	800.5	101.0	696.4	17		
	L	116.6	804.0	102.0	703.3	16.5	81.5*	89.6*
	T	119.5	821.9	110.3	760.5	14		
	T	119.5	824.0	110.7	763.3	12		
	T	119.4	823.3	110.9	764.7	15		
SA+EB +900F, 8 Hrs	L	190.9	1316.3	178.9	1123.5	3	45.4	49.9
	L	187.8	1294.9	171.7	1183.9	2.5		
	L	187.7	1294.2	177.5	1223.9	2.5	39.8	43.7
	T	192.9	1330.0	177.0	1220.4	2.5		
	T	186.5	1285.9	173.8	1198.4	2.5		
	T	180.2	1242.5	171.4	1181.8	3.0		
SA+EB +950F, 8 Hrs	L	178.4	1230.1	159.9	1102.5	4.5	46.0	50.5
	L	180.1	1241.8	158.1	1090.1	5.0		
	L	176.6	1217.7	154.7	1066.7	6.5	45.1	49.6
	T	182.9	1261.1	167.0	1151.5	3.5		
	T	179.4	1237.0	164.6	1134.9	3.5		
	T	180.3	1243.2	162.3	1119.1	4.5		

*K$_Q$

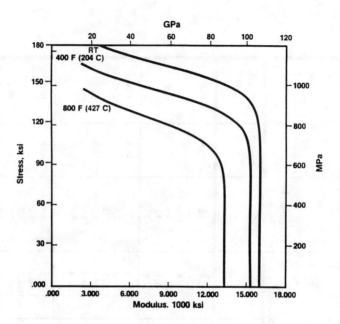

Figure 12. Compressive tangent modulus curves for *TIMETAL* ®15-3 STA strip
[Transverse Orientation] (6)

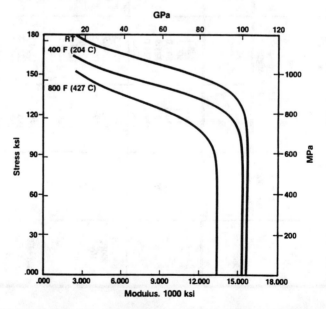

Figure 13. Compressive tangent modulus curves for *TIMETAL* ®15-3 STA strip
[Longitudinal Orientation] (6)

REFERENCES

1. Wood, R. A., Williams, D.N., Boyd, J. D., Rothman, R. L., and Bartlett, E. S., Battelle Memorial Institute, Technical Report AFML Tr-70-257, USAF Contract No. F33615-69-C-1890, December, 1970.

2. Stemme, H. W., Lockheed-Georgia, Co., Technical Report AFML-TR-73-296, USAF Contract No. F33615-72-C-1696, January, 1974.

3. Wardlaw, T. L., Rosenberg, H. W., and Parris, W. M. TIMET, Technical Report AFML-TR-73-296, USAF Contract No. F33615-72-C-1696, January, 1974.

4. Day, D. L., Rockwell International, Columbus Aircraft Division, Report NR75H-162 under USAF Contract No. F33615-74-C-5063, December, 1975.

5. Lenning, G. A., TIMET, Technical Report AFML-TR-76, USAF Contract No. F33615-74-C-5063, 1976.

6. Deel, Omar, "Collected Engineering Data Sheets", Battelle, AFML-TR-78-179, USAF Contract F33615-77-C-5009, pp. 378-385, 435-456, December, 1978.

7. Parris, W. M. and Hall, J. A., "Cold Rolled and Aged Ti-15V-3Cr-3Al-3Sn Strip for Springs", Navy Contract 66604-9169/2290, Final Report, January, 1980.

8. Lenning, G. A., Hall, J. A., Boham, G. A., Porterfield, C. and Wongiwat, K., "Cold Formable Titanium Sheet", Interim Technical Report 1R-110-8-VII, 1 June 1980-31 August 1980, Air Force Contract F33615-78-C-5116, Section 1, 72 pp., Section 2, 42 pp., September, 1980.

9. Lenning, G. A., Hall, J. A., Rosenblum, M. E., and Trepel, W. B., "Cold Formable Titanium Sheet", Final Report, AFWAL-TR-82-4174 Air Force Contract F33615-78-C-5116, December, 1982, 451 pp.

10. Kaneko, R. S., and Woods, C. A., "Low Temperature Forming of Beta Titanium Alloys", NASA Contract Report 3706, Contract NASI-15568, September, 1983, 196 pp.

11. Leach, A.E., and Szpakowski, R. J., "Low Cost Titanium Propellant Tankage", Bell Aerospace Textron, AIAA-82-1228, June, 1982.

12. Spitzner, A., Walter Kidde Aerospace, Private Communication, February 3, 1993.

13. Davies, D. P., "Assessment of Ti-15-3 Sheet for Main Rotor Blade Erosion Shield Applications", Westland Helicopters, June, 1990.

14. Elrod, C. W., "Self-Sustained Combustion Tests on Ti-15-3", WPAFB Memorandum, October, 1982.

15. Hicks, A. G., and Rosenberg, H. W., "Ti-15-3 Foil Properties and Applications", Beta Ti Alloys in 1980's, ed. R. Boyer, TMS, 1984.

16. McKenzie, R.M., "A Study of the Cast Beta Titanium Alloy Ti-15V-3Sn-3Cr-3Al, TiTech International, 1988.

17. Boyer, R. R., et.al., "Microstructure/Properties Relationships in Ti-15V-3Cr-3Al-3Sn High Strength Castings", Boeing Commercial Airplane Group, 1991.

18. Leach, A. E., and McDonough, J. D., "Weldability of Formable Sheet Titanium Alloy - Ti-15V-3Cr-3Al-3Sn", AFML-TR-78-199, Contract F33615-75-C-5232, Bell Aerospace Textron, Buffalo, New York, 1978.

19. Ruff, P. E., "Effect of Manufacturing Processes on Structural Allowables - Phase 1", Battelle Columbus Division, AFWAL-TR-85-4128, 1985.

20. Lockheed Report, Data submitted in support of AMS 4914, Rev. A, J. Pengra, November 5, 1985.

FATIGUE DATA FOR *TIMETAL*®15-3

J. C. Fanning

TIMET Henderson Technical Lab
PO Box 2128
Henderson, NV 89009

ABSTRACT

This report consists of a compilation of all known, meaningful, releasable fatigue data for *TIMETAL*®15-3 strip and castings. Data for strip are included for different heat treatments (925°F [496°C] age, 930°F [500°C] age, 1000°F [538°C] age, and unaged), different thicknesses (0.032-in [0.8-mm] to 0.113-in [2.87-mm]), different test temperatures (-60°F to 550°F), and different prior thermal exposures (550°F [288°C] to 1340°F [727°C]). For castings, data are included to show the effects of different HIP cycles, aging treatments (900°F[482°C] to 1000°F[538°C]), surface condition, and weld repair. Of course, not every combination of variables has been tested, but the data nevertheless should prove extremely useful to users of the alloy.

DATA PRESENTATION

Fatigue data for strip are plotted in Figures 1-23. Figures 24-35 give fatigue data for castings. All stresses are net unless otherwise indicated.

DISCUSSION

The figures, as presented, speak for themselves, so no detailed discussion is necessary. However, the following trends are noteworthy:

1. Fatigue strength tends to be directly related to tensile and yield strengths, and inversely related to ductility, but the correlations are weak.

2. Fatigue strengths at elevated temperatures are comparable to (or better than) fatigue strengths at room temperature.

3. Fatigue strength of solution treated (unaged) or as-formed material is very low. Fatigue strength of hot formed material can be fully restored by aging.

4   The degradation of fatigue life is directly proportional to the stress intensity factor (that is, an SIF of 3 with roughly decrease the fatigue life by a factor of 3).

5. Weld-repair of castings results in a slight fatigue debit, but the overall fatigue life is still outstanding.

LIST OF FIGURES

(1)Exposed after aging.
(2)Formed before aging

ABBREVIATIONS:

F	=	Degrees Farenheit
C	=	Degrees Celsius
Kt	=	Stress Intensity
Ctr Hole	=	Center Hole Fatigue Specimen
DEN	=	Dougle Edge Notch Fatigue Specimen

REFERENCES

1. Lockheed Advanced Development Company, Report Dated December 1, 1986, Data submitted in support of AMS 4914 Rev. A, J. Pengra.

2. G. A. Lenning and J. A. Hall, Cold Formable Titanium Sheet, Contract F33615-78-C-5116, Interim Report, 1980.

3. Douglas Aircraft Company, Long Beach, CA, 1986.

4. D. P. Davies, Assessment of Ti-15V-3Cr-3Al-3Sn Sheet for Main Rotor Blade Erosion Shield Applications, Westland Helicopters, Yeovil UK, January 6, 1990.

5. P. E. Ruff, Effect of Manufacturing Processes on Structural Allowables - Phase 1, AFWAL-TR-85-4128, Columbus, OH, January, 1986.

6. R. M. McKenzie, A Study of the Cast Beta Titanium Alloy Ti-15V-3Sn-3Cr-3Al, TiTech International, December, 1988.

7. R. R. Boyer, "High Strength Titanium Castings", Boeing Commercial Airplane Group, SR 7790, Seattle, WA, 1990.

8. R. R. Boyer, "High Strength Titanium Castings", Boeing Commercial Airplane Group, SR 8383, Seattle, WA, 1991.

9. R. M. McKenzie, "A Study of the Cast Beta Titanium Alloy Ti-15V-3Sn-3Cr-3Al, TiTech International, 1988, (As reported in Ref. 7).

10. J. Dippel, T. Serfozo and R. Kaneko, "A Comparison of Beta C, Ti-15-3, and Ti-6Al-4V Investment Castings, Joint Lockheed-TiTech Report. (As reported in Ref. 7).

11. D. Eylon, L. S. Steel and D. J. Evans, "Study of Aging Response and Fatigue Crack Initiation Mechanisms in High Strength Titanium Alloy Castings", Semi-Annual Status Report, 1989 (As reported in Ref. 7).

Figure 1

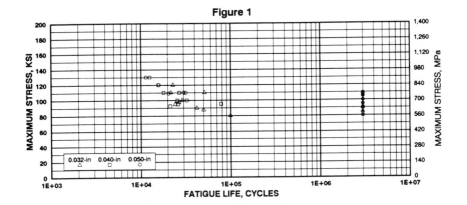

Product Form: Strip
Heat Treat: Aged 8 hrs, 925 F
Properties: UTS 187.6 ksi
          TYS 174.2 ksi
          Elong 9.1%
          Temp RT

Specimen: Kt=1.0
Stress Ratio: +0.1
Loading: Axial
Frequency: 20 Hz
Solid Marker = No Failure

Test Temperature: RT
Prior Exposure: None
Test Environment: Air
File: FAT1
Source: Lockheed (1)

Figure 2

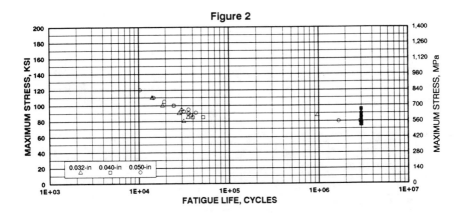

Product Form: Strip
Heat Treat: Aged 8 hrs, 925 F
Properties: UTS 185.2 ksi
          TYS 170.6 ksi
          Elong 9.1%
          Temp RT (after exposure)

Specimen: Kt=1.0
Stress Ratio: +0.1
Loading: Axial
Frequency: 20 Hz
Solid Marker = No Failure

Test Temperature: RT
Prior Exposure: 550 F,
                    1500 hrs, Air
Test Environment: Air
Source: Lockheed (1)

Figure 3

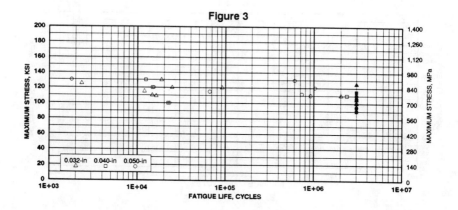

Product Form: Strip
Heat Treat: Aged 8 hrs, 925 F
Properties: UTS 168.4 ksi
        TYS 147.0 ksi
        Elong 5.7%
        Temp 550 F

Specimen: Kt=1.0
Stress Ratio: +0.1
Loading: Axial
Frequency: 20 Hz
Solid Marker = No Failure

Test Temperature: 550 F
Prior Exposure: None
Test Environment: Air

Source: Lockheed (1)

Figure 4

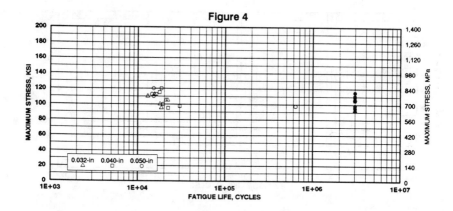

Product Form: Strip
Heat Treat: Aged 8 hrs, 925 F
Properties: UTS 167.3 ksi
        TYS 144.5 ksi
        Elong 5.9%
        Temp 550 F (after exposure)

Specimen: Kt=1.0
Stress Ratio: +0.1
Loading: Axial
Frequency: 20 Hz
Solid Marker = No Failure

Test Temperature: 550 F
Prior Exposure: 550 F,
        1500 hrs, Air
Test Environment: Air
Source: Lockheed (1)

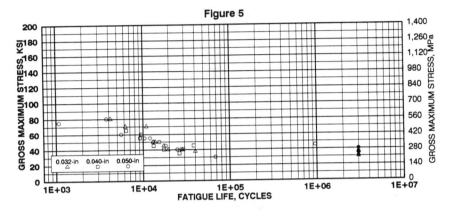

Figure 5

Product Form: Strip
Heat Treat: Aged 8 hrs, 925 F
Properties: UTS 187.6 ksi
         TYS 174.2 ksi
         Elong 9.1%
         Temp RT

Specimen: Kt=2.7 Ctr Hole
Stress Ratio: +0.1
Loading: Axial
Frequency: 20 Hz
Solid Marker = No Failure

Test Temperature: RT
Prior Exposure: None
Test Environment: Air

Source: Lockheed (1)

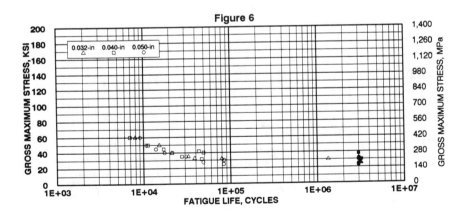

Figure 6

Product Form: Strip
Heat Treat: Aged 8 hrs, 925 F
Properties: UTS 185.2 ksi
         TYS 170.6 ksi
         Elong 9.1%
         Temp RT (after exposure)

Specimen: Kt=2.7 Ctr Hole
Stress Ratio: +0.1
Loading: Axial
Frequency: 20 Hz
Solid Marker = No Failure

Test Temperature: RT
Prior Exposure: 550 F,
         1500 hrs, Air
Test Environment: Air
Source: Lockheed (1)

Figure 7

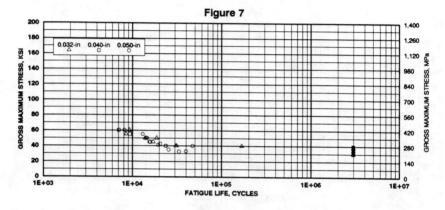

Product Form: Strip
Heat Treat: Aged 8 hrs, 925 F
Properties: UTS 168.4 ksi
TYS 147.0 ksi
Elong 5.7%
Temp 550 F

Specimen: Kt=2.7 Rnd Hole
Stress Ratio: +0.1
Loading: Axial
Frequency: 20 Hz
Solid Marker = No Failure

Test Temperature: 550 F
Prior Exposure: None
Test Environment: Air

Source: Lockheed (1)

Figure 8

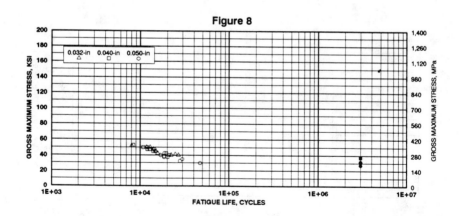

Product Form: Strip
Heat Treat: Aged 8 hrs, 925 F
Properties: UTS 167.3 ksi
TYS 144.5 ksi
Elong 5.9%
Temp 550 F (after exposure)

Specimen: Kt=2.7 Ctr Hole
Stress Ratio: +0.1
Loading: Axial
Frequency: 20 Hz
Solid Marker = No Failure

Test Temperature: 550 F
Prior Exposure: 550 F,
1500 hrs, Air
Test Environment: Air
Source: Lockheed (1)

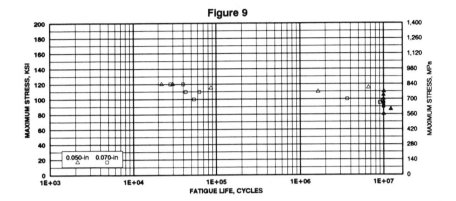

Figure 9

Product Form: Strip
Heat Treat: Aged 8 hrs, 1000 F
Properties: UTS 167.3 ksi
             TYS 154.9 ksi
             Elong 10.5%
             Temp RT

Specimen: Kt=1.0
Stress Ratio: +0.1
Loading: Axial
Frequency: 30 Hz
Solid Marker = No Failure
Orientation: Transverse

Test Temperature: RT
Prior Exposure: None
Test Environment: Air

Source:
    AFWAL-TR-82-4174 (2)

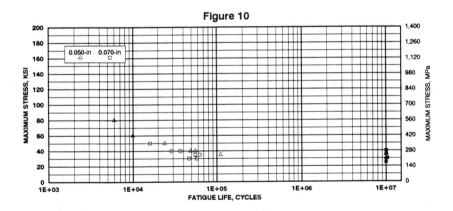

Figure 10

Product Form: Strip
Heat Treat: Aged 8 hrs, 1000 F
Properties: UTS 167.3 ksi
            TYS 154.9 ksi
            Elong 10.5%
            Temp RT

Specimen: Kt=3.0 DEN
Stress Ratio: +0.1
Loading: Axial
Frequency: 30 Hz
Solid Marker = No Failure
Orientation: Transverse

Test Temperature: RT
Prior Exposure: None
Test Environment: Air

Source:
    AFWAL-TR-82-4174 (2)

Figure 11

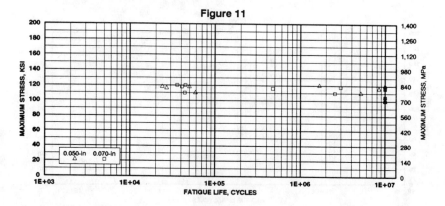

Product Form: Strip
Heat Treat: Aged 8 hrs, 1000 F
Properties: UTS 184.4 ksi
TYS 172.8 ksi
Elong 10.5%
Temp -60 F

Specimen: Kt=1.0
Stress Ratio: +0.1
Loading: Axial
Frequency: 30 Hz
Solid Marker = No Failure
Orientation: Transverse

Test Temperature: -60 F
Prior Exposure: None
Test Environment: Air

Source:
AFWAL-TR-82-4174 (2)

Figure 12

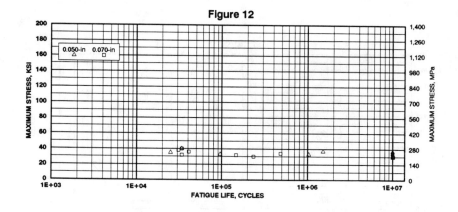

Product Form: Strip
Heat Treat: Aged 8 hrs, 1000 F
Properties: UTS 184.4 ksi
TYS 172.8 ksi
Elong 10.5%
Temp -60 F

Specimen: Kt=3.0 DEN
Stress Ratio: +0.1
Loading: Axial
Frequency: 30 Hz
Solid Marker = No Failure
Orientation: Transverse

Test Temperature: -60 F
Prior Exposure: None
Test Environment: Air

Source:
AFWAL-TR-82-4174 (2)

Figure 13

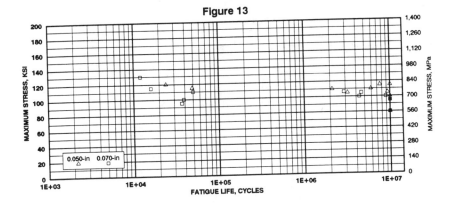

Product Form: Strip
Heat Treat: Aged 8 hrs, 1000 F
Properties: UTS 149.3 ksi
           TYS 125.3 ksi
           Elong 8.6%
           Temp 400 F

Specimen: Kt=1.0
Stress Ratio: +0.1
Loading: Axial
Frequency: 30 Hz
Solid Marker = No Failure
Orientation: Transverse

Test Temperature: 400 F
Prior Exposure: None
Test Environment: Air

Source:
AFWAL-TR-82-4174 (2)

Figure 14

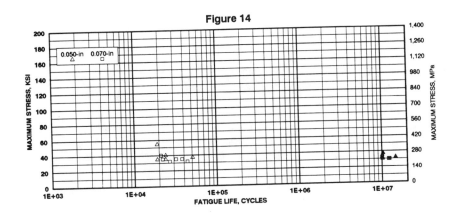

Product Form: Strip
Heat Treat: Aged 8 hrs, 1000 F
Properties: UTS 149.3 ksi
           TYS 125.3 ksi
           Elong 8.6%
           Temp 400 F

Specimen: Kt=3.0 DEN
Stress Ratio: +0.1
Loading: Axial
Frequency: 30 Hz
Solid Marker = No Failure
Orientation: Transverse

Test Temperature: 400 F
Prior Exposure: None
Test Environment: Air

Source:
AFWAL-TR-82-4174 (2)

Figure 15

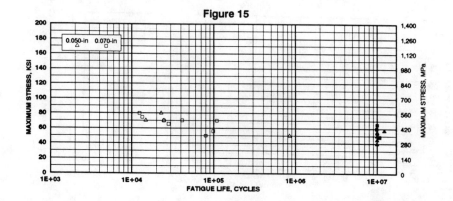

Product Form: Strip
Heat Treat: Aged 8 hrs, 1000 F
Properties: UTS 165 ksi
           TYS 155 ksi
           Elong 10%
           Temp RT

Specimen: Kt=2.53 DEN
Stress Ratio: +0.1
Loading: Axial
Frequency: 30 Hz
Solid Marker = No Failure

Test Temperature: RT
Prior Exposure: None
Test Environment: Air

Source: Douglas (3)

Figure 16

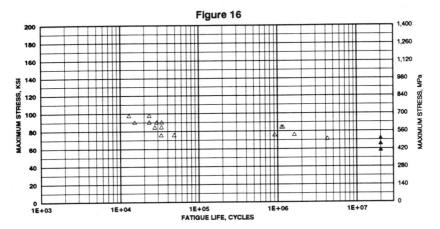

Product Form: Strip
Heat Treat: As-formed (Not Aged)
Properties: UTS 131 ksi
              TYS 117 ksi
              Elong 13%
              Temp RT

Specimen: Kt=1.0
Stress Ratio: +0.1
Loading: Axial
Frequency: Not Reported
Solid Marker = No Failure

Test Temperature: RT
Prior Exposure:
  Formed at 1060 F
Test Environment: Air
Source: Westland (4)

Figure 17

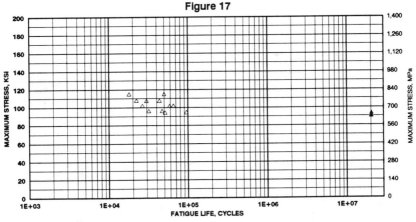

Product Form: Strip
Heat Treat: Aged 8hrs, 930F, After Forming
Properties: UTS 176 ksi
              TYS 160 ksi
              Elong 10%
              Temp RT

Specimen: Kt=1.0
Stress Ratio: +0.1
Loading: Axial
Frequency: Not Reported
Solid Marker = No Failure

Test Temperature: RT
Prior Exposure:
  Formed at 1060 F
Test Environment: Air
Source: Westland (4)

Figure 18

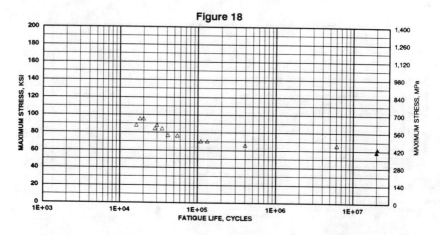

Product Form: Strip
Heat Treat: As-formed (Not Aged)
Properties: UTS 128 ksi
TYS 116 ksi
Elong 13.5%
Temp RT

Specimen: Kt=1.0
Stress Ratio: +0.1
Loading: Axial
Frequency: Not Reported
Solid Marker = No Failure

Test Temperature: RT
Prior Exposure:
Formed at 1250 F
Test Environment: Air
Source: Westland (4)

Figure 19

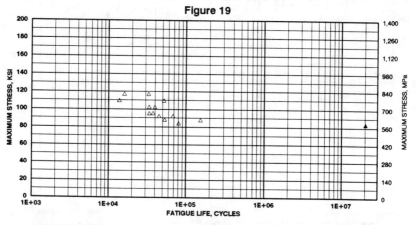

Product Form: Strip
Heat Treat: Aged 8hrs, 930F, After Forming
Properties: UTS 176 ksi
TYS 160 ksi
Elong 8.5%
Temp RT

Specimen: Kt=1.0
Stress Ratio: +0.1
Loading: Axial
Frequency: Not Reported
Solid Marker = No Failure

Test Temperature: RT
Prior Exposure:
Formed at 1250 F
Test Environment: Air
Source: Westland (4)

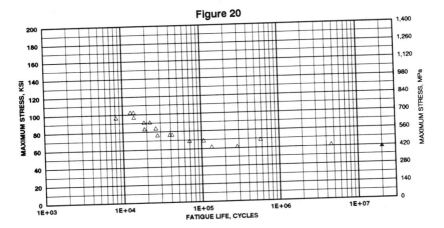

Figure 20

Product Form: Strip
Heat Treat: As-formed (Not Aged)
Properties: UTS 120 ksi
           TYS 115 ksi
           Elong 10%
           Temp RT

Specimen: Kt=1.0
Stress Ratio: +0.1
Loading: Axial
Frequency: Not Reported
Solid Marker = No Failure

Test Temperature: RT
Prior Exposure:
    Formed at 1340 F
Test Environment: Air
Source: Westland (4)

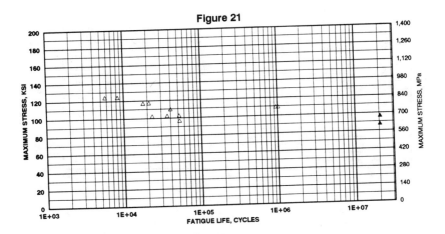

Figure 21

Product Form: Strip
Heat Treat: Aged 8hrs, 930F, After Forming
Properties: Not Reported

Specimen: Kt=1.0
Stress Ratio: +0.1
Loading: Axial
Frequency: Not Reported
Solid Marker = No Failure

Test Temperature: RT
Prior Exposure:
    Formed at 1340 F
Test Environment: Air
Source: Westland (4)

Figure 22

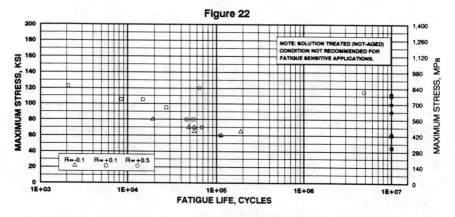

Product Form: Strip, 0.113-in
Heat Treat: Solution Treated (Not Aged)
Properties: UTS 119.7 ksi
TYS 118 ksi
Elong 13.3%
Temp RT

Specimen: Kt=1.0
Stress Ratio: As Indicated
Loading: Axial
Frequency: 20 Hz
Solid Marker = No Failure
Orientation: Transverse

Test Temperature: RT
Prior Exposure: None
Test Environment: Air

Source:
AFWAL-TR-85-4128 (5)

Figure 23

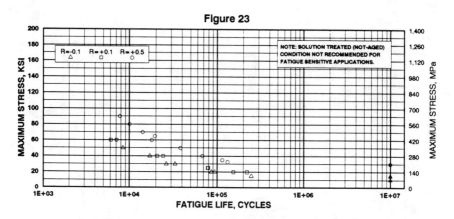

Product Form: Strip, 0.113-in
Heat Treat: Solution Treated (Not Aged)
Properties: UTS 119.7 ksi
TYS 118 ksi
Elong 13.3%
Temp RT

Specimen: Kt=3.0
Stress Ratio: As Indicated
Loading: Axial
Frequency: 20 Hz
Solid Marker = No Failure
Orientation: Transverse

Test Temperature: RT
Prior Exposure: None
Test Environment: Air

Source:
AFWAL-TR-85-4128 (5)

Figure 24

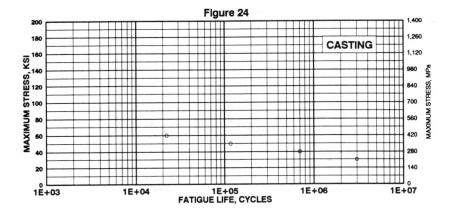

Product Form: Investment Casting
HIP: 1700 F, 2 hrs, 15 ksi
Heat Treat: Aged 8 hrs, 900 F
Properties: UTS 188 ksi
          TYS 176 ksi
          Elong 5 %
          Temp RT

Specimen: Kt=2.7
Stress Ratio: +0.1
Loading: Axial
Frequency: 25 Hz

Test Temperature: RT
Prior Exposure: None
Test Environment: Air
Source: TiTech (6)

Figure 25

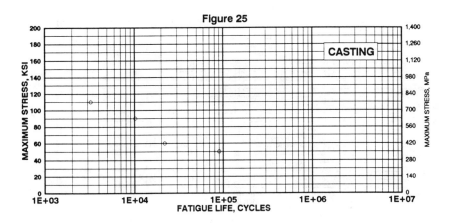

Product Form: Investment Casting
HIP: 1700 F, 2 hrs, 15 ksi
Heat Treat: Aged 12 hrs, 1000 F
Properties: UTS 174 ksi
          TYS 159 ksi
          Elong 7 %
          Temp RT

Specimen: Kt=2.7
Stress Ratio: +0.1
Loading: Axial
Frequency: 25 Hz

Test Temperature: RT
Prior Exposure: None
Test Environment: Air
Source: TiTech (6)

Figure 26

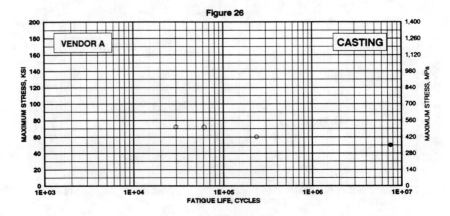

Product Form: 0.5-in thick Cast Plate
HIP: 1750 F, 2 hrs, 15 ksi
Heat Treat: Aged 12 hrs, 975 F
Properties: UTS 180 ksi
            TYS 164 ksi
            Elong 7 %
            Temp RT

Specimen: Kt=2.4
Stress Ratio: +0.1
Loading: Axial
Frequency: 30 Hz
Solid Marker = No Failure

Test Temperature: RT
Prior Exposure: None
Test Environment: Air
Source: Boeing (7)

Figure 27

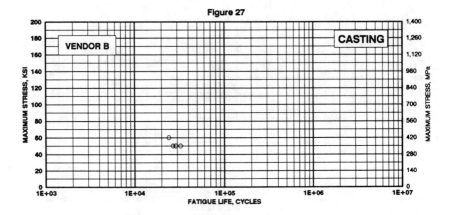

Product Form: 0.15-in thick Cast Plate
HIP: 1650 F, 2 hrs, 15 ksi
Heat Treat: Aged 12 hrs, 975 F
Properties: UTS 177 ksi
            TYS 166 ksi
            Elong 4.5 %
            Temp RT

Specimen: Kt=2.4
Stress Ratio: +0.1
Loading: Axial
Frequency: 340 Hz

Test Temperature: RT
Prior Exposure: None
Test Environment: Air
Source: Boeing (7)

Figure 28

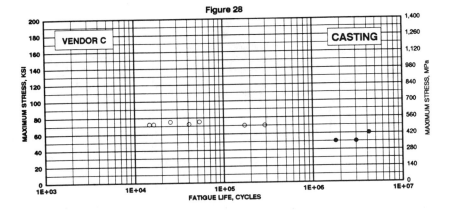

Product Form: 1-in thick Cast Plate
HIP: 1750 F, 2 hrs, 15 ksi
Heat Treat: Aged 12 hrs, 975 F
Properties: UTS 176 ksi
             TYS 165 ksi
             Elong 3.5 %
             Temp RT

Specimen: Kt=2.4
Stress Ratio: +0.1
Loading: Axial
Frequency: 30 Hz
Solid Marker = No Failure

Test Temperature: RT
Prior Exposure: None
Test Environment: Air
Source: Boeing (7)

Figure 29

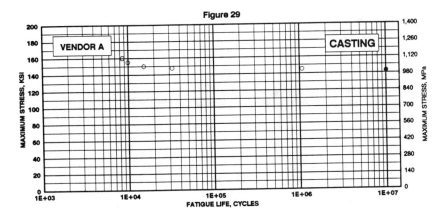

Product Form: 0.5-in diam. Cast Bar
HIP: 1750 F, 2 hrs, 15 ksi
Heat Treat: Aged 12 hrs, 975 F
Properties: UTS 181 ksi
             TYS 164 ksi
             Elong 6 %
             Temp RT

Specimen: Kt=1.0
Stress Ratio: +0.1
Loading: Axial

Frequency: 30 Hz
Solid Marker = No Failure

Test Temperature: RT
Prior Exposure: None
Test Environment: Air
Source: Boeing (7)

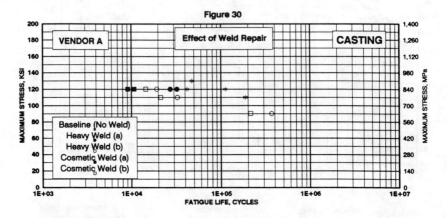

Figure 30

VENDOR A   Effect of Weld Repair   CASTING

Legend:
- Baseline (No Weld)
- Heavy Weld (a)
- Heavy Weld (b)
- Cosmetic Weld (a)
- Cosmetic Weld (b)

Product Form: 0.5-in diam. Cast Bar
HIP: 1750 F, 2 hrs, 15 ksi
Heat Treat: Aged 12 hrs, 975 F
Properties: UTS 180 ksi
[BASELINE] TYS 168 ksi
Elong 8 %
Temp RT

Specimen: Kt=1.0 (c)
Stress Ratio: +0.1
Loading: Axial
Frequency: 30 Hz

Test Temperature: RT
Prior Exposure: None
Test Environment: Air
Source: Boeing (8)

Properties: UTS 181 ksi
[WELD REPAIRED] TYS 170 ksi
Elong 5 %
Temp RT

NOTES:
(a) Aged at 975 F for 12 hrs after welding.
(b) Welded; Aged at 975 F for 12 hrs; Solution Treated at 1750 F for 1 hr,
Gas Fan Cooled; Aged at 975 F for 12 hrs.
(c) Weld repaired specimens consist of a 0.25-in radius notch in center
of gage length that has been filled-in by welding.

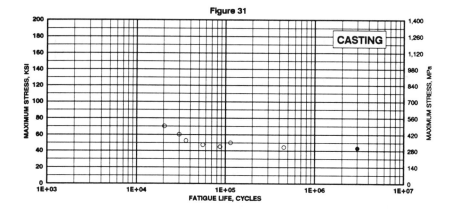

Figure 31

Product Form: Lockheed Flow Diverter
HIP: 1700 F, 2 hrs, 15 ksi
Heat Treat: Aged 12 hrs, 950 F
Properties: UTS 183 ksi
           TYS 170 ksi
           Elong 6 %
           Temp RT

Specimen: Kt=2.7
Stress Ratio: +0.1
Loading: Axial
Frequency: 25 Hz
Solid Marker = No Failure

Test Temperature: RT
Prior Exposure: None
Test Environment: Air
Source: TiTech (9)

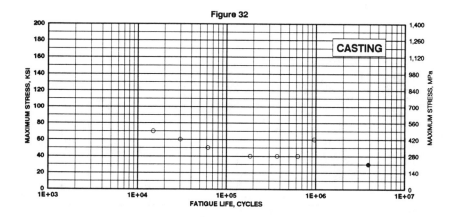

Figure 32

Product Form: Investment Casting
HIP: 1750 F, 2 hrs, 15 ksi
Heat Treat: Aged 12 hrs, 950 F
Properties: UTS 182 ksi
           TYS 167 ksi
           Elong 6 %
           Temp RT

Specimen: Kt=2.7
Stress Ratio: +0.1
Loading: Axial

Frequency: 30 Hz
Solid Marker = No Failure

Test Temperature: RT
Prior Exposure: None
Test Environment: Air
Source: TiTech
         and Lockheed (10)

Figure 33

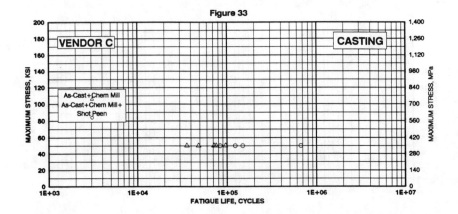

Product Form: 0.15-in Cast Plate
HIP: 1750 F, 2 hrs, 15 ksi
Heat Treat: Aged 12 hrs, 975 F
Properties: UTS 183 ksi
    TYS 163 ksi
    Elong 6 %
    Temp RT

Specimen: Kt=2.4
Stress Ratio: +0.1
Loading: Axial
Frequency: 30 Hz

Test Temperature: RT
Prior Exposure: None
Test Environment: Air
Source: Boeing (7)

Figure 34

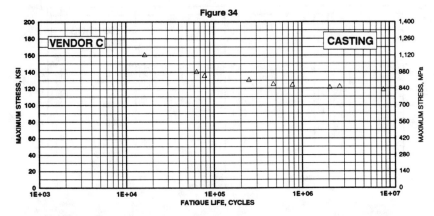

Product Form: 1-in thick Cast Plate
HIP: 1750 F, 2 hrs, 15 ksi
Heat Treat: Aged 12 hrs, 975 F
Properties: UTS 175 ksi
    TYS 165 ksi
    Elong 5 %
    Temp RT

Specimen: Kt=1.0
Stress Ratio: +0.1
Loading: Axial

Frequency: 30 Hz

Test Temperature: RT
Prior Exposure: None
Test Environment: Air

Source: Boeing (7)

460

Figure 35

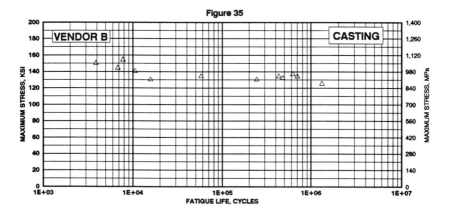

Product Form: 0.5-in Cast Test Bar
HIP: 1750 F, 2 hrs, 15 ksi
Heat Treat: Solution Treat 1750 F, 1 hr,
Gas Fan Cool; Then Aged 12 hrs, 975 F

Properties: UTS 175 ksi
            TYS 164 ksi
            Elong 5 %
            Temp RT

Specimen: Kt=1.0
Stress Ratio: +0.1
Loading: Axial
Frequency: 30 Hz

Test Temperature: RT
Prior Exposure: None
Test Environment: Air
Source: Eylon (11)

461

PROPERTIES OF BETA-C™ (Ti-3Al-8V-6Cr-4Mo-4Zr)

J. R. Wood, J. G. Ferrero, P. A. Russo, and R. W. Schutz

RMI Titanium Company
1000 Warren Avenue
Niles, Ohio 44446

Abstract

Beta-C™, a metastable beta titanium alloy developed by RMI Titanium Company, continues to be used in new applications due to its excellent mechanical properties and resistance to corrosion. New data in this paper include mechanical properties in large sections, cold drawn bar and welds. Fatigue crack growth and corrosion data are added to supplement previously published data.

Beta Titanium Alloys in the 1990's
Edited by D. Eylon, R.R. Boyer and D.A. Koss
The Minerals, Metals & Materials Society, 1993

General Description

Beta-C™, developed by RMI Titanium Company, is a metastable beta titanium alloy which can be heat treated to a variety of strength levels for various applications. It has good ductility and toughness, good hot and cold fabricability and a low elastic modulus. It has good resistance to general corrosion and resists hot, aggressive environments containing ferric chloride, sodium chloride, carbon dioxide and hydrogen sulfide. Its applications include coil springs, fasteners, rivets, downhole pipe and tooling for oil production equipment, and high strength structural components.

Chemical Composition (wt %)	
Aluminum	3.0 - 4.0
Vanadium	7.5 - 8.5
Chromium	5.5 - 6.5
Molybdenum	3.5 - 4.5
Zirconium	3.5 - 4.5
Iron	0.30 max.
Nitrogen	0.03 max.
Carbon	0.05 max.
Oxygen	0.12 max.
Hydrogen	0.020 max.
Others, total	0.40 max.
each	0.10 max.
Titanium	Remainder

Physical Properties	
Density, lb/cu in. (g/cu cm)	0.174 (4.82)
Beta transus, °F (°C)	1350 (732)
Thermal coef. expansion/°F (68-900°F) (STA*) /°C (20-482°C)	5.4×10^{-6} (3.0×10^{-6})
Modulus of elasticity, ‡psi (Tension) (Annealed) (MPa)	14.0×10^{6} (96.5×10^{6})
(Tension) (STA*) (MPa)	15.4×10^{6} (106.2×10^{6})
(Compression) (STA*) (MPa)	14.8×10^{6} (102.0×10^{6})
Modulus of rigidity, ‡psi (Torsion) (STA*) (MPa)	5.9×10^{6} (40.7×10^{6})

‡Values vary slightly depending on heat treatment.
*STA = Solution Treated and Aged.

Heat Treatment

Annealing or Solution Treatment: Heat to 1500-1700°F (816-927°C), hold 15-30 minutes at temperature, and cool in air or faster.

Aging: Reheat solution-treated material to 850-1000°F (454-538°C), hold at temperature for 4-24 hours and cool in air. Cold working prior to aging will increase the final aged tensile strength and maintain good ductility. The aging time and temperature are selected based on prior working history and strength requirements.

Thermal Stability of Beta-C™ Solution Annealed Sheet at 550°F(288°C)

Gage[1], in. (mm)	Exposure Time, Hours	YS, ksi (MPa)	UTS, ksi (MPa)	El, %
.040 (1)	0	122.1 (842)	125.8 (867)	10
	500	121.0 (834)	124.4 (858)	8
	1000	123.6 (852)	129.4 (892)	8
	1500	127.0 (876)	133.8 (923)	10
.063 (1.6)	0	120.7 (832)	124.2 (856)	9
	500	118.6 (818)	123.0 (848)	10
	1000	116.2 (801)	126.1 (869)	10
	1500	124.4 (858)	129.2 (891)	8

1. Solution Annealed 1450°F(788°C)-30Min-AC

Properties of Large Section Size Beta-C™

Heat Treatment	YS, ksi (MPa)	UTS, ksi (MPa)	El, %	RA, %	K_{1c}, ksi √in (MPa √in)
19" (48 cm) Diameter Billet					
1975°F(1079°C)-5Hr-AC +1500°F(816°C)-30Min-AC +975°F(534°C)-24Hr-AC	160.0[1] (1103)	170.1 (1173)	4.5	9.2	57.6 (63.4) (L-R)
1975°F(927°C)-5Hr-AC +1700°F(927°C)-1Hr-AC +950°F(510°C)-24Hr-AC	162.5[1] (1120)	170.8 (1178)	12.5	15.8	74.2 (81.6) (L-R)
18" (45.7 cm) Diameter x 9 3/8" (23.8 cm) ID Extrusion					
1500°F(816°C)-30Min-AC +1050°F(566°C)-24Hr-AC	158.1[2] (1090)	165.8 (1143)	5.5	8.6	65.6 (72.2) (C-R)
1700°F(927°C)-30Min-AC +1025°F(552°C)-24Hr-AC	159.8[2] (1102)	168.3 (1160)	9.5	20.6	91.0 (100.1) (C-R)

1. Longitudinal
2. Radial

Data Summary on Beta-C™ Electron Beam Welded Pipe

Sample	Heat Treatment	YS, ksi (MPa)	UTS, ksi (MPa)	El, %	RA, %	K_{Ie}, ksi √in (MPa √in)
Weld[1,2]	1700°F(927°C)-1Hr-AC + 1000°F(538°C)-24Hr-AC	157.4 (1067)	168.4 (1161)	7.0	19.0	75.8 (83.4)
Weld	1000°F(538°C)-24Hr-AC	156.7 (1080)	169.9 (1171)	6.3	13.8	73.2 (80.5)
Base Metal	1700°F(927°C)-1Hr-AC + 1000°F(538°C)-24Hr-AC	156.8 (1081)	162.2 (1118)	5.5	19.7	73.3 (80.6)
Base Metal	1000°F(538°C)-24Hr-AC	160.0 (1103)	171.5 (1182)	4.5	12.8	-

1. All material was solution annealed 1700°F(927°C)-1Hr-AC before welding.
2. 18" (45.7 cm) OD x 9 3/8" (23.8 cm) ID.

Strength-Toughness Trends in Large Section Beta-C™ Closed End Pipe

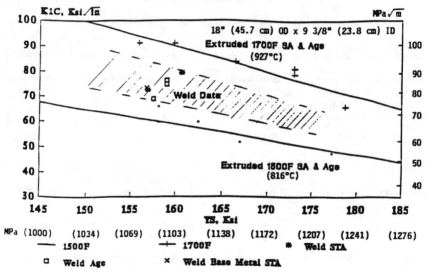

Products: 1) Extrusion
2) Trepanning of 18" (45.7 cm) billet and EB welding of end cap

Fatigue Crack Growth Rate as a Function of Stress Intensity
Range for Beta-C^{TM} Solutionized and Aged at 350 and 500°C

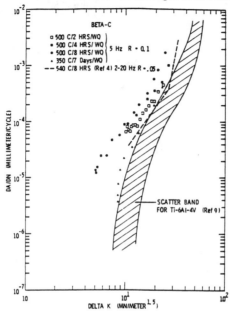

C. G. Rhodes and N. E. Paton "The Influence of Microstructure on Mechanical Properties in
Ti-3Al-8V-6Cr-4Mo-4Zr (Beta-C^{TM})", Met. Trans. A, Vol.8, November 1977, p. 1759.

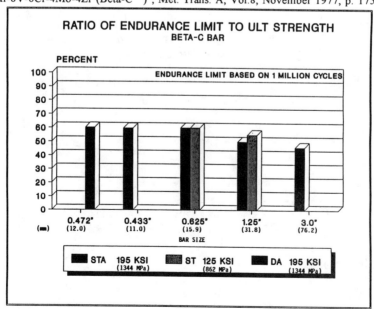

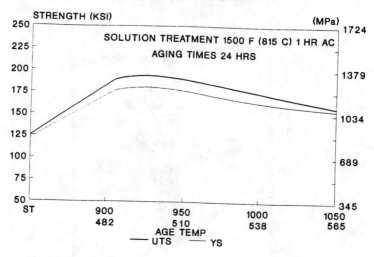

BETA-C BAR
STRENGTH vs AGE TEMP.

STRENGTH (KSI)

SOLUTION TREATMENT 1500 F (815 C) 1 HR AC
AGING TIMES 24 HRS

(MPa)

AGE TEMP
—— UTS —— YS

0.5 to 3" DIAMETER
(12.7 mm to 76.2 mm)

BETA-C BAR
DUCTILITY vs AGE TEMP.

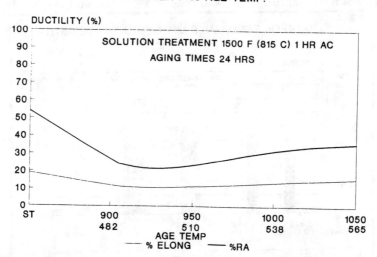

DUCTILITY (%)

SOLUTION TREATMENT 1500 F (815 C) 1 HR AC
AGING TIMES 24 HRS

AGE TEMP
—— % ELONG —— %RA

0.5 to 3" DIAMETER
(12.7 mm to 76.2 mm)

BETA-C BILLET
STRENGTH vs AGE TEMP.

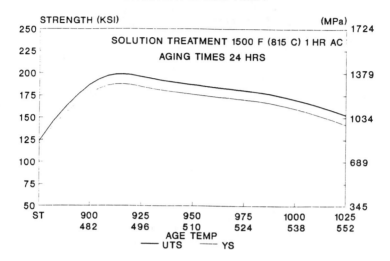

4 to 8" DIAMETER
(101.6 mm to 203.2 mm)

BETA-C BILLET
DUCTILITY vs AGE TEMP.

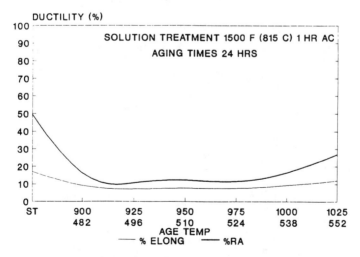

4 to 8" DIAMETER
(101.6 mm to 203.2 mm)

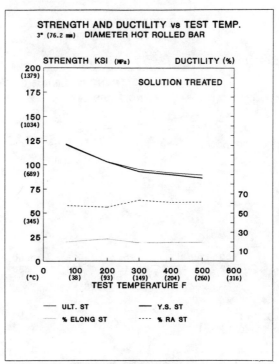

STRENGTH AND DUCTILITY vs TEST TEMP.
3" (76.2 mm) DIAMETER HOT ROLLED BAR

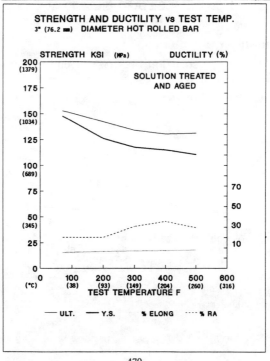

STRENGTH AND DUCTILITY vs TEST TEMP.
3" (76.2 mm) DIAMETER HOT ROLLED BAR

Effect of Aging Temp on Mechanical Properties of Cold Drawn Rod
Cold Drawn 40%

Size, in. (mm)	Condition	UTS, ksi (MPa)	YS, ksi (MPa)	% El	% RA	Double Shear
.063 (1.6)	As Drawn	169 (1165)	143 (986)	4	42	
	D.A. 900°F(482°C)-8Hrs	222 (1530)	207 (1427)	5	17	
.190 (4.8)	As Drawn	170 (1172)	165 (1137)	12	60	
	D.A. 925°F(496°C)-8Hrs	220 (1517)	208 (1434)	10	34	
.375 (9.5)	As Drawn	169 (1165)	160 (1103)	13	52	
	D.A. 925°F(496°C)-8Hrs	223 (1537)	210 (1448)	7.5	15	126
.472 (12)	As Drawn	166 (1144)	160 (1103)	11	48	101
	D.A. 900°F(482°C)-8Hrs	230 (1586)	215 (1482)	8	10	130
.530 (13.5)	As Drawn	156 (1075)	152 (1048)	10	35	100
	D.A. 950°F(510°C)-8Hrs	211 (1455)	191 (1317)	9	20	

COLD DRAWN BETA-C BAR
STRENGTH AND DUCTILITY vs TEST TEMP.

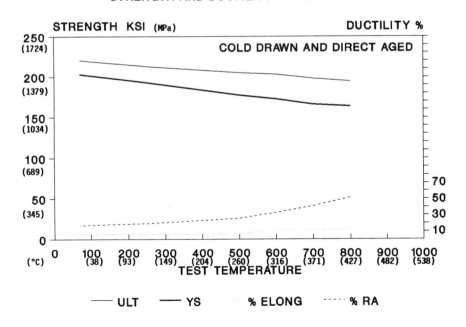

0.375" (9.5 mm) DIAMETER

General Corrosion Data for Beta-C™ in Various Media

Medium	Concentration, %	Temperature, °C	Corrosion Rate, mm/yr
Ferric chloride	10	Boiling	nil
Formic acid	50	Boiling	0.98
Hydrochloric acid	0.5	Boiling	0.003
	1.0	Boiling	0.058
	1.5	Boiling	0.26
Hydrochloric acid, aerated	pH 1	Boiling	nil
Hydrochloric acid, +0.1% FeCl₃	5	Boiling	0.018
Sulfuric acid, naturally aerated	1	Boiling	nil
	5	Boiling	1.85
Sulfuric acid + 3% Fe₂(SO₄)₃	50	Boiling	<0.03
Sulfuric acid + 1 g/L FeCl₃	10	Boiling	0.15
Sulfuric acid + 50 g/L FeCl₃	10	Boiling	0.05

Resistance of Beta Titanium Alloys to Boiling Oxidizing Acid Media

Alloy	Corrosion Rate (mm/yr)		
	Aerated HCl (pH = 1)	10% FeCl₃ (pH ~0.3)	10% HNO₃
Ti-Gr.2	0.64	0.01	0.08
Ti-6-4	0.60	0.00	0.33
Ti-8-8-2-3	0.00	0.01	--
Beta-C™*	0.00	0.01	0.18
Beta-C™/Pd* **	0.00	0.00	0.15
Beta 21S*	0.00	0.01	--

* ST and STA conditions
**Beta-C™/Pd contains .05% Pd

Comparison of Corrosion Rates of Beta-C^{TM} and CP Titanium
in Boiling Hydrochloric Acid

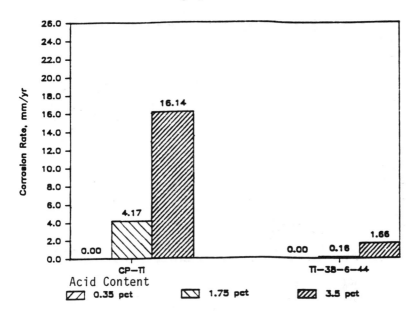

Comparison of Corrosion Rates of Beta-C^{TM} and CP Titanium
in Boiling Sulfuric Acid

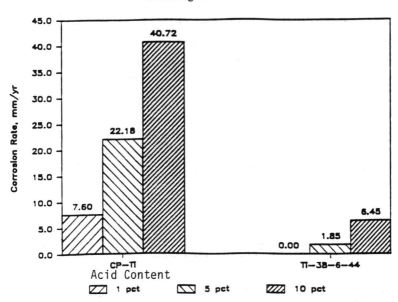

Comparison of Corrosion Rates of Beta-C™ and CP Titanium
in Boiling Nitric Acid

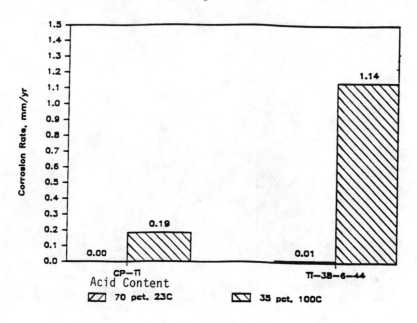

Comparison of Corrosion Rates of Beta-C™ and CP Titanium
in Boiling 10 Weight % Sulfuric Acid/Ferric Chloride

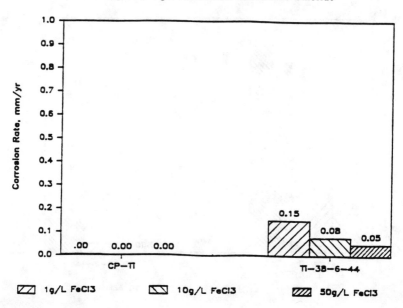

Corrosion Rate Profiles in Boiling HCl (Aged Alloys)

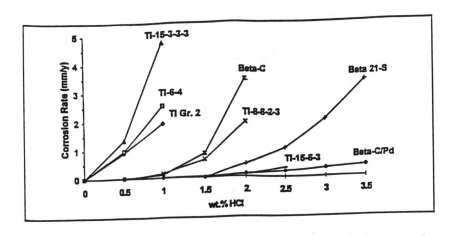

Crevice Corrosion Test Results for Standard and Pd-Enhanced Ti-38644 in Hot Brines

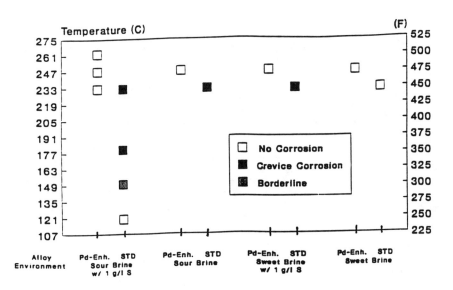

Temperature Thresholds for SCC of Beta-C^{TM} Titanium in a
Worst-Case Deep Gas Well Sour Brine

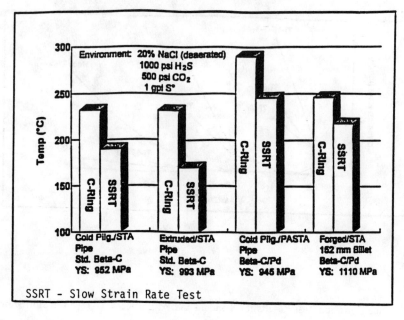

Reference

1.   <u>Beta Titanium Alloys in the 1980's</u>, edited by R. R. Boyer and H. W. Rosenberg,
     Metallurgical Society of AIME, Warrendale, PA, 1984.

ß-CEZ PROPERTIES

Y. Combres and B. Champin

Centre de Recherches de CEZUS
BP 33
73400 Ugine Cedex France

ABSTRACT

The ß-CEZ alloy has been designed by CEZUS in collaboration with the french DoD (STPA). The alloy is suitable for use in engine heavy section forgings where deep hardenability, strength, toughness, creep resistance, and low cycle fatigue are critical. It is one of those very few ß titanium alloys which offer creep strength up to 450 °C (842 °F). It has also been conceived for use in structural parts for aeronautics either under the form of forged parts where strength-toughness balance is sought : 1400 MPa (203 ksi) 0.2% YS-50 MPa.m$^{1/2}$ (45 ksi.in$^{1/2}$) K$_{Ic}$, or under the form of fasteners of the 1500 MPa (217 ksi)-15 % El category. At last, the alloy is capable of near net shape forging below the temperatures of the conventional $\alpha+\beta$ titanium alloys : superplasticity can be used as low as 700 °C (1292 °F) for instance.

The authors acknowledge Mr. Pichol, who compiled most of the data available to date.

Beta Titanium Alloys in the 1990's
Edited by D. Eylon, R.R. Boyer and D.A. Koss
The Minerals, Metals & Materials Society, 1993

INTRODUCTION

ß-CEZ can be defined as a metastable ß alloy because ß phase proportion at room temperature encompasses the 25-100 % range. Being predominantly ß, ß-CEZ is deep hardenable. By ageing, α phase precipitates out from the metastable phase ß and allows the achievement of high strength. The key to the successful application of ß-CEZ lies in the use of its excellent forgeability and high combinations of strength - ductility - toughness - fatigue - creep resistance [1-8].

CHEMICAL COMPOSITION

ß-CEZ is a metastable ß titanium alloy with 18 % addition elements. Mo, Cr and Fe strongly stabilize the ß phase insuring a large heat treatment response and improved formability.

Nominal : Ti-5Al-2Sn-4Zr-4Mo-2Cr-1Fe		
Chemistry specifications :		
Al	→	4,5-5,5
Sn	→	1,5-2,5
Zr	→	3,5-4,5
Mo	→	3,5-4,5
Cr	→	1,5-2,5
Fe	→	0,5-1,5
O between 800 and 1300 ppm		
H lower than 150 ppm		

Table I - Chemical composition of the ß-CEZ alloy.

PHYSICAL PROPERTIES

With this composition, the alloy exhibits the following thermal properties (table II) :

Tliquidus °C (°F)	1600 (2912)
Tsolidus °C (°F)	1550 (2822)
Tß °C (°F)	890 (1634)
density g/cm³ (lb/in³)	4.69 (0.169)
ST+A modulus GPa (ksi)	110-120 (15949-17398)
thermal expension* µm/m/K	9
specific heat* J/kg/K	580
thermal conductivity* W/m/K	6.7
emissivity*	0.7

* thermal properties at 20 °C

Table II - Thermal properties of ß-CEZ alloy.

PROCESSING

Hot working

The alloy can be hot worked either in the ß range or the α+ß range. A typical ß processing temperature is 920 °C (1688 °F), whereas α+ß deformation can take place in the 800-850 °C (1472-1562 °F) temperature range. Irrespective of the phase stability domain, the rheology of ß-CEZ is characterized by high strain rate sensitivity exponent m, insuring flow stability, and low apparent activation energies Q, insuring less sensitivity to temperature changes during the process. Superplastic deformation can be performed in the 700-800 °C (1292-1472 °F) temperature range acheiving 1000 % elongation. Table III summarizes these rheological parameters.

ß processing		
	temperature °C (°F)	920 (1688)
	strain rate range s^{-1}	10^{-2}-10
	mß	0.33
	Qß kJ/mole	175
α+ß processing		
	temperature range °C (°F)	800-850 (1472-1562)
	strain rate range s^{-1}	10^{-2}-10
	mα+ß	0.27
	Qα+ß kJ/mole	400
Superplasticity		
	temperature range °C (°F)	700-800 (1292-1472)
	strain rate range s^{-1}	10^{-4}-10^{-3}
	mspf	0.5
	Qspf kJ/mole	150
	El at 725 °C-8x10^{-4} s^{-1} %	1000
	El at 800 °C-3x10^{-4} s^{-1} %	600

Table III - Rheological parameters of ß-CEZ.

Heat treatments

They consist in solution treatments (ST) followed by ageing treatments (A). After different types of hot working followed by ST+A, the ß-CEZ microstructures are a combination of primary alpha (αI), secondary alpha (αII), stable ß (ßs) or metastable ß (ßm) or transformed ß (ßt). The morpohologies can be divided depending on the αI shape. They can be refered to as equiaxed, or lamellar (continuous α phase at the serrated ex-ß grains boundaries), or necklaced microstructures (fine α phase at the ex-ß grains boundaries) (fig. 1).

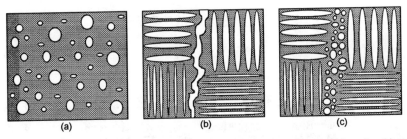

Figure 1 - Schematic of the different types of morphologies : (a) equiaxed, (b) lamellar, (c) necklaced

Solution treatment		
	temperature range °C (°F)	750-920 (1382-1688)
	duration h	1-4
Ageing		
	temperature range °C (°F)	500-650 (932-1202)
	duration h	1-8

Table IV - Typical conditions for solution and ageing treatments.

TENSILE PROPERTIES

Tables V to VII present some of the tensile properties.

Product	Heat treatment	UTS MPa (ksi)	0.2% YS MPa (ksi)	El %	RA %
	as forged	1040 (151)	960 (139)	18	45
Ø 150 mm	830 °C/1 h/WQ + 550 °C/8 h/AC	1601 (232)	1518 (220)	2	2.5
forged bar	830 °C/1 h/WQ + 600 °C/8 h/AC	1283 (186)	1208 (175)	11	15
(dia. 6")	860 °C/1 h/WQ + 550 °C/8 h/AC	1557 (226)	1478 (214)	2	6
	860 °C/1 h/WQ + 600 °C/8 h/AC	1370 (199)	1304 (189)	5	12
	as rolled	1157 (168)	1050 (152)	17	57
Ø 70 mm	830 °C/1 h/WQ + 550 °C/8 h/AC	1358 (197)	1272 (184)	11	36
rolled bar	830 °C/1 h/WQ + 600 °C/8 h/AC	1258 (182)	1194 (173)	14	48
(dia. 3")	860 °C/1 h/WQ + 550 °C/8 h/AC	1594 (231)	1533 (222)	8	18
	860 °C/1 h/WQ + 600 °C/8 h/AC	1491 (216)	1424 (206)	10	14
	as rolled	1490 (216)	1345 (195)	11	35
Ø 12.8 mm	830 °C/1 h/WQ + 550 °C/8 h/AC	1506 (218)	1460 (212)	13	50
rolled bar	830 °C/1 h/WQ + 600 °C/8 h/AC	1373 (199)	1349 (196)	15	52
(dia. 0.5")	860 °C/1 h/WQ + 550 °C/8 h/AC	1723 (250)	1683 (244)	7	12
	860 °C/1 h/WQ + 600 °C/8 h/AC	1540 (223)	1485 (215)	9	20
	as rolled L	1222 (177)	1124 (163)	15	-
	as rolled T	1260 (183)	1163 (169)	11	-
25^t mm	830 °C/1 h/WQ+600 °C/8 h/AC L	1334 (193)	1287 (187)	13	-
rolled plate	830 °C/1 h/WQ+600 °C/8 h/AC T	1351 (196)	1300 (188)	12	-
(1" thick)	860 °C/1 h/WQ+600 °C/8 h/AC L	1405 (204)	1338 (194)	10	-
	860 °C/1 h/WQ+600 °C/8 h/AC T	1418 (206)	1340 (194)	6	-
	as rolled L	1437 (208)	1385 (201)	2	-
	as rolled T	1330 (193)	1239 (180)	5	-
5^t mm	830 °C/1 h/ArC+600 °C/8 h/AC L	1270 (184)	1139 (165)	7	-
rolled plate	830 °C/1 h/ArC+600 °C/8 h/AC T	1292 (187)	1198 (174)	10	-
(1" thick)	860 °C/1 h/ArC+600 °C/8 h/AC L	1220 (177)	1143 (166)	8	-
	860 °C/1 h/ArC+600 °C/8 h/AC T	1280 (186)	1203 (174)	6	-

830 °C = 1526 °F ; 860 °C = 1580 °F ; 550 °C = 1022 °F ; 600 °C = 1112 °F

Table V - Selection of tensile properties on products with equiaxed microstructure.

Product	Heat treatment	UTS MPa (ksi)	0.2% YS MPa (ksi)	El %	RA %
Ø 300 mm	600 °C/8 h/AC	1608 (233)	1472 (213)	2	6
hot ß processed	830 °C/1 h/WQ + 570 °C/8 h/AC	1357 (197)	1171 (170)	5	9
pancake	830 °C/1 h/WQ + 600 °C/8 h/AC	1326 (192)	1188 (172)	6	9
hot dies at	860 °C/1 h/WQ + 570 °C/8 h/AC	1288 (187)	1111 (161)	5	12
830 °C (1526 °F)	860 °C/1 h/WQ + 600 °C/8 h/AC	1284 (186)	1142 (166)	7	13
Ø 300 mm	600 °C/8 h/AC	1631 (236)	1570 (228)	1	3
warm ß processed	830 °C/1 h/WQ + 570 °C/8 h/AC	1296 (188)	1148 (166)	7	13
pancake	830 °C/1 h/WQ + 600 °C/8 h/AC	1263 (183)	1145 (166)	7	15
warm dies at	860 °C/1 h/WQ + 570 °C/8 h/AC	1295 (188)	1135 (165)	6	15
750 °C (1382 °F)	860 °C/1 h/WQ + 600 °C/8 h/AC	1286 (186)	1143 (166)	8	18

830 °C = 1526 °F ; 860 °C = 1580 °F ; 570 °C = 1058 °F ; 600 °C = 1112 °F

Table VI - Selection of tensile properties on dia 12 " products with lamellar microstructure.

Product	Heat treatment	UTS MPa (ksi)	0.2% YS MPa (ksi)	El %	RA %
	as forged	1206 (175)	1137 (165)	15	30
Ø 80 mm	830 °C/1 h/AC + 580 °C/8 h/AC	1346 (195)	1287 (187)	10	16
forged bar	830 °C/1 h/AC + 600 °C/8 h/AC	1259 (183)	1222 (177)	14	23
(dia. 6")	860 °C/1 h/AC + 580 °C/8 h/AC	1341 (194)	1276 (185)	10	15
	860 °C/1 h/AC + 600 °C/8 h/AC	1348 (195)	1289 (187)	8	13
Ø 300 mm	600 °C/8 h/AC	1227 (178)	1138 (165)	10	23
through the	830 °C/1 h/WQ + 570 °C/8 h/AC	1314 (191)	1200 (174)	10	26
transus	830 °C/1 h/WQ + 600 °C/8 h/AC	1263 (183)	1170 (170)	11	25
processed	860 °C/1 h/WQ + 570 °C/8 h/AC	1295 (188)	1135 (165)	6	13
(dia. 12")	860 °C/1 h/WQ + 600 °C/8 h/AC	1286 (186)	1143 (166)	7	15

830 °C = 1526 °F ; 860 °C = 1580 °F ; 570 °C = 1058 °F ; 580 °C = 1076 °F ; 600 °C = 1112 °F

Table VII - Selection of tensile properties products with necklaced microstructure.

TOUGHNESS

Some of the obtained toughness values are given in table VIII.

Type of microstructure	Product	Heat treatment	Toughness MPa.m$^{1/2}$ (ksi.in$^{1/2}$)
		as forged	82 (74)
	Ø 150 mm	830 °C/1 h/WQ + 550 °C/8 h/AC	35 (32)
equiaxed	forged bar	830 °C/1 h/WQ + 600 °C/8 h/AC	48 (43)
	(dia. 6")	860 °C/1 h/WQ + 550 °C/8 h/AC	29 (26)
		860 °C/1 h/WQ + 600 °C/8 h/AC	39 (35)
	Ø 300 mm pancake	830 °C/1 h/WQ + 570 °C/8 h/AC	75 (69)
lamellar	830 °C (1526 °F) dies	830 °C/1 h/WQ + 630 °C/8 h/AC	90 (82)
	Ø 300 mm pancake	830 °C/1 h/WQ + 570 °C/8 h/AC	82 (76)
	750 °C (1382 °F) dies	830 °C/1 h/WQ + 630 °C/8 h/AC	98 (89)
		as forged	122 (110)
	Ø 80 mm	830 °C/1 h/AC + 580 °C/8 h/AC	73 (66)
	forged bar	830 °C/1 h/AC + 600 °C/8 h/AC	84 (76)
necklaced	(dia. 6")	860 °C/1 h/AC + 580 °C/8 h/AC	72 (65)
		860 °C/1 h/AC + 600 °C/8 h/AC	73 (66)
	Ø 300 mm through	830 °C/1 h/WQ + 570 °C/8 h/AC	72 (63)
	transus processed	830 °C/1 h/WQ + 630 °C/8 h/AC	96 (83)

830 °C = 1526 °F ; 860 °C = 1580 °F ; 550 °C = 1022 °F ; 570 °C = 1058 °F ; 580 °C = 1076 °F ; 600 °C = 1112 °F ; 630 °C = 1166 °F

Table VIII - Selection of toughness values on bars or Ø 300 mm (dia. 12") pancakes.

FATIGUE

Fatigue Crack Growth rates

Table IX gives the FCP characteristics at room temperature.

Type of microstructure	Heat treatment	da/dN mm/cycle (in/cycle)	ΔK at 10^{-6} mm/cycle MPa.m$^{1/2}$ (ksi.in$^{1/2}$)
lamellar	830 °C/1 h/WQ + 570 °C/8 h/AC	2x10^{-5} (8x10^{-7})	5 (4.5)
	830 °C/1 h/WQ + 630 °C/8 h/AC	2.5x10^{-5} (10^{-6})	5 (4.5)
necklaced	830 °C/1 h/WQ + 570 °C/8 h/AC	4x10^{-5} (1.6x10^{-6})	4 (3.6)
	830 °C/1 h/WQ + 630 °C/8 h/AC	2x10^{-5} (8x10^{-7})	5 (4.5)

830 °C = 1526 °F ; 570 °C = 1058 °F ; 630 °C = 1166 °F

Table IX - FCP characteristics at room temperature on Ø 300 mm (dia. 12") pancakes.

Low cycle fatigue

The LCF properties are presented in table X.

Type of microstructure	Heat treatment	Temperature °C (°F)	Stress for 10⁴ cycles MPa (ksi)
equiaxed	830 °C/1 h/WQ + 550 °C/8 h/AC	20 (68)	911 (132)
	830 °C/1 h/WQ + 600 °C/8 h/AC	20 (68)	900 (130)
	860 °C/1 h/WQ + 550 °C/8 h/AC	20 (68)	950 (138)
	860 °C/1 h/WQ + 600 °C/8 h/AC	20 (68)	913 (132)
lamellar	800 °C/4 h/WQ + 600 °C/8 h/AC	20 (68)	710 (103)
		300 (572)	640 (93)
		400 (752)	580 (84)
	770 °C/2 h/WQ + 600 °C/8 h/AC	20 (68)	720 (104)
		300 (572)	600 (87)
		400 (752)	550 (80)
necklaced	830 °C/1 h/AC + 560 °C/8 h/AC	20 (68)	910 (132)

770 °C = 1418 °F ; 800 °C = 1472 °F ; 830 °C = 1526 °F ; 860 °C = 1580 °F ; 550 °C = 1022 °F ; 560 °C = 1040 °F ; 600 °C = 1112 °F ;

Table X - LCF properties for Ø 150 mm (dia. 6") bar with equiaxed microstructure, Ø 300 mm (dia. 12") pancakes with lamellar microstructure, and Ø 80 mm (dia 3") bar with necklaced microstructure.

Fatigue strength

Table XI presents the fatigue strength on smooth specimens.

Type of microstructure	Heat treatment	Fatigue strength for 10⁵-10⁷ cycles MPa (ksi)
equiaxed	830 °C/1 h/WQ + 550 °C/8 h/AC	900 (130)
	830 °C/1 h/WQ + 600 °C/8 h/AC	900 (130)
	860 °C/1 h/WQ + 550 °C/8 h/AC	825 (120)
	860 °C/1 h/WQ + 600 °C/8 h/AC	825 (120)
lamellar	800 °C/4 h/WQ + 600 °C/8 h/AC	< 500 (< 72)
	770 °C/2 h/WQ + 600 °C/8 h/AC	< 500 (< 72)
necklaced	830 °C/1 h/AC + 560 °C/8 h/AC	900 (130)

770 °C = 1418 °F ; 800 °C = 1472 °F ; 830 °C = 1526 °F ; 860 °C = 1580 °F ; 550 °C = 1022 °F ; 560 °C = 1040 °F ; 600 °C = 1112 °F ;

Table XI - Fatigue strength for Ø 150 mm (dia. 6") bar with equiaxed microstructure, Ø 300 mm (dia. 12") pancakes with lamellar microstructure, and Ø 80 mm (dia 3") bar with necklaced microstructure.

CREEP STRENGTH

In table XII and XIII, the creep strength is given for the different microstructures.

Type of microstructure	Heat treatment	Temperature °C (°F)	Stress MPa (ksi)	Hours to 0.2 %
equiaxed	830 °C/1 h/WQ + 550 °C/8 h/AC	400 (752)	450 (65)	29
			600 (87)	13
	830 °C/1 h/WQ + 600 °C/8 h/AC	400 (752)	450 (65)	46
			600 (87)	15
	860 °C/1 h/WQ + 550 °C/8 h/AC	400 (752)	450 (65)	25-32
	860 °C/1 h/WQ + 600 °C/8 h/AC	400 (752)	450 (65)	25

830 °C = 1526 °F ; 860 °C = 1580 °F ; 550 °C = 1022 °F ; 600 °C = 1112 °F

Table XII - Creep resistance for a Ø 150 mm (dia. 6") bar with equiaxed microstructure.

Type of microstructure	Heat treatment	Temperature °C (°F)	Stress MPa (ksi)	Hours to 0.2 %
lamellar micro. hot die forged	800 °C/4 h/WQ + 600 °C/8 h/AC	400 (752)	500 (72)	95-153
		400 (752)	600 (87)	44-66
		425 (797)	300 (43)	82-83
		425 (797)	400 (58)	28-45
		425 (797)	500 (72)	17-33
		450 (842)	200 (29)	82-137
		450 (842)	300 (43)	15-29
	770 °C/2 h/WQ + 600 °C/8 h/AC	400 (752)	500 (72)	> 200
		400 (752)	600 (87)	66-68
		400 (752)	700 (101)	30-32
		425 (797)	300 (43)	120-168
		425 (797)	400 (58)	63-69
		425 (797)	500 (72)	24-33
		450 (842)	200 (29)	96-107
		450 (842)	300 (43)	31-37
lamellar micro. hot die forged	830 °C/1 h/WQ + 570 °C/8 h/AC	400 (752)	600 (87)	68-88
	830 °C/1 h/WQ + 630 °C/8 h/AC	400 (752)	600 (87)	82-88
lamellar micro. warm die forged	830 °C/1 h/WQ + 570 °C/8 h/AC	400 (752)	600 (87)	73-80
	830 °C/1 h/WQ + 630 °C/8 h/AC	400 (752)	600 (87)	58-82
necklaced micro. forged bar	830 °C/1 h/AC + 580 °C/8 h/AC	400 (752)	600 (87)	81
	830 °C/1 h/AC + 600 °C/8 h/AC	400 (752)	600 (87)	72
	860 °C/1 h/AC + 580 °C/8 h/AC	400 (752)	600 (87)	120
	860 °C/1 h/AC + 600 °C/8 h/AC	400 (752)	600 (87)	84
necklaced micro. pancake	830 °C/1 h/WQ + 570 °C/8 h/AC	400 (752)	600 (87)	88-101
	830 °C/1 h/WQ + 630 °C/8 h/AC	400 (752)	600 (87)	121-131

770 °C = 1418 °F ; 800 °C = 1472 °F ; 830 °C = 1526 °F ; 860 °C = 1580 °F ; 570 °C = 1042 °F ;
580 °C = 1076 °F ; 600 °C = 1112 °F ; 630 °C = 1166 °F ;

Table XIII - Creep resistance for Ø 300 mm (dia. 12") pancakes with lamellar or necklaced microstructures, and for Ø 80 mm (dia. 4") forged bar with necklaced microstructure.

References

[1] B. Prandi, J.-F. Wadier, F. Schwartz, P.-E. Mosser and A. vassel, Proc. 1990 TDA Int. Conf., TDA Ed. Dayton, OH, (1990) 150-159.
[2] B. Champin, B. Prandi, P.-E. Mosser and Y. Honorat, Proc. 1991 AAAF Meeting, (1991) 55-77.
[3] B. Prandi, E. Alheritiere, F. Schwarz and M. Thomas, Proc. 6th World Conf. on Titanium, Lacombe et al. Eds., Ed. Phys. Les Ulis, (1989) 811-816.
[4] A. Henri and A. Vassel, ONERA technical report n° 22/3578 M, (1992).
[5] Y. Combres, CEZUS technical report STPA contract n° 90-96-023, (1992).
[6] P. Paillere, CEZUS technical report STPA contract n° 90-96-014, (1991).
[7] A. Henri and A. Vassel, Proc. 7th World Conf. on Titanium, to be published.
[8] P.-E. Mosser, N. Marnier, and Y. Honorat., Proc. 7th World Conf. on Titanium, to be published.

ACLOA TITANIUM ALLOY Ti-10V-2Fe-3Al FORGINGS*

-DATA SHEETS-

G. W. Kuhlman
Alcoa Forged Products, 1600 Harvard Avenue, Cleveland, Ohio, 44105, USA

ABSTRACT

Ti-10V-2Fe-3Al is high strength, metastable beta titanium alloy developed by the Timet Division, Titanium Metals Corporation of America for enhanced forgeability and excellent mechanical property combinations, including deep hardening characteristics. Ti-10-2-3 has been the subject of intense thermomechanical processing (TMP) development and reduction to commercial practice for the full range of open die, closed-die and precision forged products utilized in aerospace and other applications. With commercially proven TMP, Ti-10-2-3 forgings provide a range of strength-fracture toughness combinations from 180 Ksi (1240 MPa) minimum U.T.S. to 140 Ksi (965 MPa) minimum U.T.S with commensurate fracture toughness. With optimal TMP and these strength-toughness combinations, Ti-10-2-3 forgings are found provide superior smooth and notched fatigue properties critical to aerospace applications over incumbent alpha-beta alloys. The alloy retains 80 percent of its room temperature strength at 600°F (315°C) and has been found highly fabricable in both conventional (warm) and hot die forging process technologies. Presented in this document is a summary of the critical thermomechanical and forging processing of Ti-10-2-3 forgings to achieve three strength levels specified by current AMS specifications. Presented as well is a compilaltion of first and second tier mechanical properties for Ti-10-2-3 forgings at all three strength levels and contrasts of Ti-10-2-3 properties with existing alpha-beta alloys. Mechanical properties are rationalized in terms of process-structure relationships. Finally, presented is information on post-forging processing of Ti-10-2-3 forgings to realize cost-efficient finished components for airframe and other structural applications.

* Originally Published As Alcoa Green Letter No. 224.

Beta Titanium Alloys in the 1990's
Edited by D. Eylon, R.R. Boyer and D.A. Koss
The Minerals, Metals & Materials Society, 1993

INTRODUCTION

Titanium alloy Ti-10V-2Fe-3Al (Ti-10-2-3) was developed by the Timet Divison of Titanium Metals Corporation of America in the early 1970's (1). This titanium alloy is classified as a metastable beta alloy and was specifically formulated to provide enhanced forgeability and excellent mechanical property combinations, including through hardening capability in relatively heavy sections, in the forging product form. Ti-10-2-3 forgings possess excellent combinations of strength, fracture toughness, and fatigue related properties that are important for aerospace applications. Further, Ti-10-2-3 retains 80 percent of its room temperature strength at 600°F (315°C). Forgings in Ti-10-2-3 are available in solution treated and aged (STA) or overaged (STOA) conditions and are typcially provided at three ultimate strength levels:

HIGH STRENGTH: 173-180 KSI (1193-1240 MPa)
INTERMEDIATE STRENGTH: 160 KSI (1103 MPa)
LOW STRENGTH: 140 KSI (965 MPa)

In addition to conventional open and closed-die forgings, Ti-10-2-3 is particularly well suited for Hot Die Forging techniques used in the manufacture of precision forgings.

TYPICAL APPLICATIONS

Ti-10-2-3 forgings are being used in many domestic and international aerospace applications, particularly airframe structural components. Further, Ti-10-2-3 forgings are being evaluated by turbine engine manufacturers for cold section components such as compressor/fan discs. Some typical aircraft applications include engine mounts, landing gear support parts, structural fittings, fuselage frames, bulkheads, ribs, and spars.

PHYSICAL METALLURGY

As is true for most titanium alloys, required mechanical property combinations of Ti-10-2-3 forgings are achieved by controlled thermal and working histories, e.g. thermomechanical processing (TMP), that creates the desired microstructural characteristics. Ti-10-2-3 is termed a metastable beta alloy, since, as contrasted to other beta alloys, alpha phase is retained during most TMP sequences. Alpha phase plays an important role in the final microstructure/mechanical properties achieved, particularly Ti-10-2-3's improved ductility and fracture toughness in relation to other beta alloys.

Further, unlike forgings in many competitive alpha-beta titanium alloys, Ti-10-2-3 forgings can be successfully thermomechanically processed to consistently achieve a wide range of strengths with commensurate fracture toughness. The major advantages of Ti-10-2-3 over available alpha-beta compositions of similar strength levels are its fracture toughness in air and salt water environment and its hardenability in heavy sections. Successful thermomechanical processing in Ti-10-2-3 controls the relative amounts of alpha and beta phases, phase morphologies and transformation mechanisms to achieve microstructural goals. More extensive treatment of the physical metallurgy of Ti-10-2-3 is contained in References 4 - 6.

CHEMICAL COMPOSITION

The nominal composition and the composition limits for Ti-10-2-3 are shown in Table 1. Vanadium and iron are the primary beta stabilizing elements. The nominal 2 percent iron content contributes significantly to Ti-10-2-3's ability to be heat treated in heavy sections and the alloy's forgeability, without introducing excessive microsegregation. Oxygen content is restricted to 0.13% maximum in order to maximize toughness. The nominal 3 percent aluminum stabilizes sufficient alpha such that the amount and morphology of this phase may be successfully controlled through TMP to achieve desired toughness and ductility combinations.

PHYSICAL PROPERTIES

Typical physical properties for Ti-10-2-3 are shown in Table 2.

TENSILE AND COMPRESSIVE MODULUS

Typical tensile stress-strain, compressive stress-strain and compressive tangent modulus curves for Ti-10-2-3 forgings at 173 KSI (1193 MPa) are shown in Figure 1(3). Figures 2 and 3 present tensile and compressive modulus as a function of temperature, from -65 to 465°F (-50 to 240°C) for all three strength levels of Ti-10-2-3 forgings. Supporting precision modulus data are presented in Table 3.

THERMOMECHANICAL PROCESSING

Essential to mechanical property capabilities of Ti-10-2-3 forgings is the design and execution of required thermomechanical processes, including forging procedures and heat treatments. Thermomechanical process development and resultant mechanical property verification for Ti-10-2-3 have been the subject of extensive efforts by a number of workers (7-20). The culmination of these efforts are commercially proven forging and thermal treatment processes that provide the optimum mechanical property combinations.

Ti-10-2-3 was designed to be successfully forged, under properly controlled conditions, above the alloy's beta transus (B_t). As a consequence, beta forging techniques are exploited in the design of forging processes. Further, subsequent heat treatments are designed to exploit the optimum transformation mechanisms of the alloy. Thus, the combination of beta and alpha-beta forging processes with solution treatment and aging provide thermomechanical processing sequences that achieve strength-toughness-fatigue related mechanical property combinations.

Figure 4 (4) presents the T-T-T diagram for Ti-10-2-3 and the transformation mechanisms as a function of post-solution treatment cooling rate. Two typical quench rates are presented. Note that these two cooling rates result in different transformation mechanisms which can, as shown in Figure 5 (28), provide different strength levels under equivalent aging temperature conditions. Further, Figure 6 (10) presents an aging curve for Ti-10-2-3 forgings when solution treated at 1385°F (752°C) and cold water quenched. Thus, the three available strength levels of Ti- 10-2-3 forgings are achieved by control of the aging temperature and by control of the post-solution treatment cooling rate.

The interaction of forging processes, e.g. beta and/or alpha-beta forging or combinations of these techniques, and subsquent heat treatments are controlled to meet specification mechanical property requirements and/or to tailor mechanical properties of this alloy to other specific user requirements for any given forging configuration. Illustrated in Figure 7a (17) and 7b (17) are two microstructural variants possible in Hot Die Forging of Ti-10-2-3 precision forgings. Alpha-beta Hot Die Forging, with the microstructure in Figure 7a, provides higher yield strength and lower fracture toughness than does beta Hot Die forging, with the microstructure in Figure 7b, while other mechanical properties, e.g. tensile strength and ductility, of these two forging process variants are essentially identical.

ROOM TEMPERATURE MECHANICAL PROPERTIES

Tables 4 to 6 present typical mechanical properties achieved in the three forging shapes depicted in Figure 8: a precision forging produced by Hot Die Forging; an intermediate size conventional, closed-die forging; and a very large, heavy section closed-die forging. Presented for the precision forging and intermediate size closed-die forging (Tables 4 and 5) are typical strength, ductility and fracture toughness properties obtained when these two forging shapes were processed to all three available strength levels for the alloy. In the case of the very large main landing gear beam forging (Table 6), mechanical property results presented are for the high strength level, UTS of 173 KSI (1193 MPa). Further, because of the extremely thick as-forged sections (Table 6), noted mechanical properties were achieved on coupons heat treated to noted thicknesses.

Tables 7 and 8 (3) present "S" basis design mechanical properties for the High Strength Level of Ti-10-2-3 precision and conventional forgings [173-180 KSI (1193-1240 MPa)] now incorporated into Mil-HDBK-5. "A" and "B" basis design mechanical properties for High Strength Level conventional Ti-10-2-3 forgings [173 KSI (1193 MPa)] are available and will be incorporated into Mil-HDBK-5 when analysis of required supporting production lot data, now in progress, is complete. Collection and analysis of design mechanical property data for the Intermediate Strength Level [160 KSI (1103 MPa)] and Low Strength Level [140 KSI (965 MPa)] are now in progress.

PROCUREMENT SPECIFICATIONS

The specification coverage for Ti-10-2-3 forgings is listed in Table 9. AMS specification required minimum mechanical properties are listed in Table 10. AMS 4983 and 4984 provide procurement specification requirements for High Strength Level Ti-10-2-3 forgings. AMS 4983 is used for precision forgings up to 1 in (25 mm) thick at minimum U.T.S. of 180 KSI (1240 MPa). AMS 4984 covers conventional, open-die and closed-die forgings up to 3 in (75 mm) thick at minimum U.T.S. of 173 KSI (1193 MPa). AMS 4986 and 4987 cover Intermediate Strength Level, minimum U.T.S. 160 KSI (1103 MPa) up to 3 in (75 mm) thick, and Low Strength Level, minimum U.T.S. 140 KSI (965 MPa) up to 4 in (100 mm) thick respectively. At a minimum U.T.S. of 140 KSI (965 MPa), Ti-10-2-3 forgings density compensated strength is equivalent to the widely used alpha-beta alloy Ti-6Al-4V.

Table 11 provides typical heavy section mechanical properties achieved in Ti-10-2-3 closed-die forgings, processed to all three available strength levels, where forged and heat treated section thicknesses exceed limits specified in AMS 4984, 4986 and 4987. These data illustrate the excellent hardenability of Ti-10-2-3 in very thick sections. Required minimum mechanical properties for forgings whose section thickness exceed the limits of the AMS specifications can be negotiated on an individual part basis.

ELEVATED TEMPERATURE MECHANICAL PROPERTIES

Typical tensile properties of Ti-10-2-3 forgings at elevated temperatures and at room temperature after exposure to elevated temperatures are shown in Table 12 and presented in Figure 9. Ti-10-2-3 forgings at all three available strength levels retain approximately 80 percent of room temperature properties at temperatures up to 600°F (315°C), and exposure to this temperature does not alter post-exposure room temperature strengths and ductility.

CREEP PROPERTIES

Figures 10a through 10e present Larson-Miller parameter data for 0.1, 0.2, 0.5, 1.0 and 2.0% creep for Ti-10-2-3 forgings at Intermediate [160 KSI (1103 MPa)] and Low [140 KSI (965 MPa)] strength levels. These two available strength levels are believed to be most pertinent to possible turbine engine applications for this alloy. Actual creep data are recorded in Table 13.

HIGH CYCLE FATIGUE PROPERTIES

Beta and metastable beta titanium alloys such as Ti-10-2-3, due to the highly refined nature of their microstructure, offer excellent smooth and/or notched high cycle fatigue lives. Alcoa (19, 21, 23) and others (10, 12, 13, 18, 22) have thoroughly evaluated the notched and smooth fatigue characteristics of Ti-10-2-3 forgings. Figures 11, 12 and 13 present a summary of axial-stress, high cycle, smooth and notched fatigue properties of Ti-10-2-3 forgings at all three available strength levels. Table 14 compares the high cycle fatigue capability of Ti-10-2-3 forgings with forgings in other available alpha-beta and beta titanium alloys in the forging product form. It is evident that Ti-10-2-3 forgings display superior high cycle smooth and notched fatigue lives in comparison to other aerospace titanium alloys. Reference 22 cites that the excellent fatigue performance of Ti-10-2-3 forgings is in part explained by the high proportion (e.g. ≥ 80 - 90 percent) of the total fatigue life consumed in fatigue crack initiation due to the highly refined microstructure typical of this alloy.

LOW CYCLE FATIGUE PROPERTIES

Figure 14 (21) presents strain controlled low cycle fatigue data for the Intermediate [160 KSI (1103 MPa)] and Low [140 KSI (965 MPa)] Strength levels in Ti-10-2-3 forgings. From these data, and data reported in Reference 18, the low cycle fatigue characteristics of Ti-10-2-3 are at least comparable to other titanium alloys used in aerospace applications.

FATIGUE CRACK GROWTH RATE PROPERTIES

Constant amplitude, fatigue crack growth rate (FCGR) data for Ti-10-2-3 forgings at all three available strength levels are shown in Figures 15, 16 and 17 (21). The fatigue crack growth rates of Ti-10-2-3 forgings, when processed to achieve all three available strength levels, are comparable. Therefore FCGR of Ti-10-2-3 is independent of strength and from References 6, 19 and 21 is insensitive to microstructure and thermomechanical processing variations. Table 14 compares the threshold stress intensity (ΔK_{th}) of Ti-10-2-3 forgings with other alpha-beta and beta titanium alloys in the forging product form, using ΔK_{th} data from Figures 15 to 17(21) and Reference 23. The FCGR of Ti-10-2-3 forgings is comparable to other high strength alpha-beta alloys, such as Ti-6Al-2Sn-4Zr-6Mo and Ti-6Al-6V-2Sn, but may be inferior to certain compositional and thermomechanical processing variations of the alpha-beta alloy Ti-6Al-4V.

FRACTURE TOUGHNESS

Ti-10-2-3 forgings have been found by Alcoa (7, 16, 17, 19, 21) and others (8 - 13, 15, 20) to display excellent combinations of strength and fracture toughness. Specification guaranteed minima for fracture toughness of all three available strength levels are shown in Table 10. Figure 18(16) presents a compilation of strength-fracture toughness data from a wide variety of forging shapes, including those in Tables 4 to 6, and illustrates the strength-fracture toughness relationship achieved in Ti-10-2-3 forgings. Table 14 compares the critical crack length, a, for unstable fracture of Ti-10-2-3 forgings at each available strength level with other alpha-beta and beta titanium alloys in the forging product form. The critical crack length in Ti-10-2-3 is superior to many high strength alpha-beta and beta titanium alloys and comparable or superior to all compositional and thermomechanical processing variants of Ti-6Al-4V, except Ti-6Al-4V Eli, beta annealed.

STRESS CORROSION CRACKING RESISTANCE

Ti-10-2-3 forgings display excellent stress corrosion cracking resistance in salt water, as measured by subjecting a fatigue crack to a static load in the K_{1scc} test method. From References 1, 9, 10,12, K_{1scc} is > 0.7 K_{1c} for Ti-10-2-3 forgings processed to all three available strength levels. Also, from Reference 9, the fatigue crack growth resistance of Ti-10-2-3 forgings is essentially unaffected by exposure to sea water. In this respect Ti-10-2-3 may be superior to the alpha-beta alloy Ti-6Al-4V for which a salt water FCGR debit exists at lower test frequencies (9).

GENERAL CORROSION RESISTANCE

Like most titanium alloys, forgings in Ti-10-2-3 display excellent general corrosion resistance under exposure conditions prevalent in most aerospace and marine applications for this alloy in forgings. Ti-10-2-3 forgings are typically used in aerospace and marine applications without special surface treatments. Like all beta titanium alloys, however, Ti-10-2-3 has a greater affinity for hydrogen than is typical of alpha or alpha-beta alloys. Thus if Ti-10-2-3 forgings, in service, are to be exposed to environmental conditions that may create significant hydrogen potential, it may be necessary to apply surface finishes, such as conversion treatments, which retard hydrogen pick-up.

FORGEABILITY

Ti-10-2-3 is a highly forgeable titanium alloy that is well suited to both conventional (warm die) and Hot Die closed-die forging techniques. Alcoa (7,16, 26 - 28) and others (8, 9, 11, 13, 24, 25) have extensively evaluated the forging and deformation characteristics of this alloy. Figure 19 (8) and Figure 20 (9) illustrate the deformation behavior and flow stresses of Ti-10-2-3 in comparison to the widely used alpha-beta alloy Ti-6Al-4V. Note that Ti-10-2-3 is successfully forged at temperatures at least 200 - 300°F (110 - 167°C) lower than Ti-6Al-4V. Further, the forging pressure requirements and/or flow stresses of Ti-10-2-3 are lower than Ti-6Al-4V under most forging process conditions.

Thus, Ti-10-2-3 is readily fabricable into desired conventional hand and closed-die forging shapes using all types of forging equipment. Further, Ti-10-2-3 can be successfully Hot Die or Isothermally forged using relatively low die temperatures in comparison to the die temperatures required for Hot Die or Isothermal Forging of alpha-beta alloys (24, 25). Ti-10-2-3 has therefore been found to be particularly amenable to commercial exploitation of Hot Die Forging techniques in the manufacture of precision forging shapes (26 - 28). For more information on precision forgings in Ti-10-2-3, the Alcoa Precision Forging Design Manual.

THERMAL TREATMENTS

Ti-10-2-3 forgings are typically supplied in the final heat treatment condition, either STA or STOA, as is outlined in AMS 4983, 4984, 4986, or 4987. Recommended heat treatment practices are outlined in these AMS specifications and Mil-H-81200, "Heat Treatment of Titanium Alloys." The nominal practices listed below are taken from these documents.

Solution Treatment and Quench

The recommended temperature for solution heat treating Ti-10-2-3 forgings is heating to a temperature 60 to 100°F (33 to 55°C) below the beta transus, e.g. typically 1370 to 1410°F (745 to 765°C). The recommended time at temperature is a 30 minutes minimum. Required quench is as follows:

AMS 4983: Air cool or faster
AMS 4984/4986/4987: Water quench

For conventional hand and closed-die forgings supplied to the High Strength level of AMS 4984, an optional solution anneal at 60 to 100°F (33 to 55°C) below the beta transus followed by furnace or air cooling, prior to solution treatment is permitted. Further, all specifications permit alternate heat treatments as agreed between purchaser and supplier. For example, it may be necessary to employ alternate quench rates as illustrated in Figure 5.

Aging

Recommended aging practices for the three available strength levels of Ti-10-2-3 forgings are also outlined in AMS 4983, 4984, 4986 and 4987 and Mil-H-81200. Nominal practices are as follows:

AMS 4983: 900 - 975°F (480 - 525°C), 8 hours at temperature, followed by air cool.
AMS 4984: 900 - 950°F (480 - 510°C), 8 hours at temperature, followed by air cool.
AMS 4986: 950 - 1000°F (510 - 540°C), 8 hours at temperature, followed by air cool.
AMS 4987: 1050 - 1100°F (565 - 620°C), 8 hours at temperature, followed by air cool.

MACHINABILITY

Ti-10-2-3 is somewhat more difficult to machine than the alpha-beta alloy Ti-6Al-4V but less difficult to machine than all other beta alloys (1). When water quenching is used during heat treatment, there can be some distortion resulting from redistribution of residual stresses during machining. But because of the alloy's low creep strength at the aging temperatures employed [900 to 1100F (480 - 620C)], distortion is much less severe. Typical hardness levels demonstrated by the three available strength levels are as follows:

High Strength Level (AMS 4983/4984): Rc 38 - 42
Intermediate Strength Level (AMS 4986): Rc 34 - 38
Low Strength Level (AMS 4987): Rc 30 - 34

JOINING

Ti-10-2-3 forgings are quite weldable by several techniques including tungsten inert gas (TIG), metal inert gas (MIG) and electron beam or plasma welding (1). The latter techniques are preferred for fabrication and joining of critical structure. The excellent weldability of Ti-10-2-3 has been attributed to the relatively fine weldment grain size (1). Ti-10-2-3 weldments, particularly using electron beam techniques, have been found to display excellent ductility and toughness (1).

FINISHING, CLEANING AND CHEMICAL MILLING

Ti-10-2-3 forgings are typically used in current aerospace and mar- ine applications without surface finishes. TI-10-2-3 forgings are cleaned, descaled and pickled, using identical methods employed for other titanium alloys. Either mechanical or chemical (molten caustic or sodium hydride baths) methods of scale removal are accepted processes, followed by pickling in HNO_3 - HF baths at approximately a 7:1 ratio.

As with forgings in other titanium alloys, most specifications require removal of all alpha case from Ti-10-2-3 alloys. As a beta alloy, the depth of alpha case formation in Ti-10-2-3 during typical forging and heat treatment thermal exposures has been found to be shallower than typical of similarly processed alpha and alpha-beta alloys. Ti-10-2-3 has a greater affinity for hydrogen than typical of alpha and alpha-beta alloys and thus pickling and/or chemical milling processes must be adequately controlled to prevent unnecessary hydrogen pick-up. Acid strength and bath temperature are effective in controlling pickling rate.

Ti-10-2-3 can be successfully chemically milled by available commercial chemical milling processes and baths employed for other titanium alloys.

CONCLUSION

Ti-10-2-3 forgings in the three available strength levels should be considered for applications requiring strength levels in the range of precipitation hardening (PH) stainless steels (High Strength Level) and/or other high strength alpha-beta titanium alloys (Intermediate and Low Strenth Level) coupled with good fracture toughness and excellent fatigue lives.

References

1. Timet Division of Titanium Metals Corporation of America, Timet Titanium Data (Developmental), "Metallurgical and Mechanical Properties of an Advanced High Toughness Alloy, Ti-10V-2Fe-3Al," 1977 (under revision).
2. Society of Automotive Engineers, Aerospace Materials Specification, AMS 4983A, 4984, 4986, 4987, 87-01-01.
3. Military Standardization Handbook, Metallic Materials and Elements for Aerospace Vehicle Structures, Mil-HDBK-5, Rev. E, Change Notice No. 1, March/April 1988.
4. T.W. Duerig, G.T. Terlinde, J.C. Williams, "Phase Transformations and Tensile Properties of Ti-10V-2Fe-3Al," Metallurgical Transactions, Vol.11A, 1980, pp.1987 - 1998.
5. G.T. Terlinde, T.W. Duerig, J.C. Williams, "Microstructure, Tensile Deformation and Fracture in Aged Ti-10V-2Fe-3Al," Metallurgical Transactions A, Vol. 14A, 1983, pp. 2101 - 2115.
6. T.W. Duerig, J.E. Allison, J.C. Williams, "Microstructural Influences on Fatigue Crack Propagation in Ti-10V-2Fe-3Al", Metallurgical Transactions, Volume 16A, 1985, pp. 739-752.
7. G.W. Kuhlman, T.B. Gurganus, "Optimizing Thermomechanical Processing of Titanium Ti-10V-2Fe-3Al Forgings," Metal Progress, Vol. 118, No. 2, July 1980, pp. 30 - 38.
8. C.C. Chen, R.R. Boyer, "Practical Considerations for Manufacturing High Strength Ti-10V-2Fe-3Al Forgings," Journal of Metals, Vol. 36, No. 7, July 1979, pp. 33 - 39.
9. C.C. Chen, J.A. Hall, R.R. Boyer, "HIgh Strength Beta Titanium Alloy Forgings for Aircraft Structural Applications," Titanium 80, Vol. 1, pp. 457 - 466.
10. R.R. Boyer, "Design Properties of High Strength Titanium Alloy, Ti-10V-2Fe-3Al," Journal of Metals, Vol. 37, No. 3, March 1980, pp. 61 - 65.
11. C.C. Chen, "Forgeability, Structures, and Properties of Hot-Die Processed Ti-10V-2Fe-3Al Thin Section Forgings," Wyman-Gordon Co., RD74-120, November 1974 and RD75-118, November 1975.
12. Mechanical Property Data Ti-10V-2Fe-3Al Alloy, AFWAl/Battelle, F33615-80-C-5168, July 1982.
13. D.J. Moracz, Isothermal Forging Beta Titanium, AFWAL-TR-80-4169, December 1980.
14. E.A. Starke, Jr., J.C. Williams,"The Role of Thermomechanical Processing In Tailoring the Properties of Aluminum and Titanium Alloys," Deformation Processing and Structure, ASM, 1986, pp. 279 - 305.
15. W.D. Gooden, J.C. Williams, Y. Kohsaka, "Structure-Property Relationships in Ti-10V-2Fe-3Al," Senior Thesis, Carnegie-Mellon University, 1986.
16. G.W. Kuhlman, R. Pishko, J.R. Kahrs, J.W. Nelson, "Isothermal Forging of Beta and Near-Beta Titanium Alloys," Beta Titanium Alloys in the 1980's, TMS-AIME, 1984, pp. 255 - 280.
17. G.W. Kuhlman, R. Pishko, "Processing-Property Relationships in Hot Die Forged Alpha-Beta, Beta and Near-Beta Titanium Alloys," Titanium Science and Technology, Deutsche Gesellschaft fur Metallkunde, e.V., Vol. 1, 1985 pp. 469 - 476.
18. R.S. Carey, R.R. Boyer, H.W. Rosenberg, "Fatigue Properties of Ti-10V-2Fe-3Al," IBID, Vol. 2, pp. 1261 - 1268.
19. R.R. Boyer, G.W. Kuhlman, "Processing-Properties Relationships of Ti-10V-2Fe-3Al," TMS-AIME Annual Meeting, New Orleans, La., February 1986, in publication.
20. G. Terlinde, H.J. Rathjen, K.H. Schwalbe, G.W. Kuhlman, "Effect of Thermomechanical Processing on Fracture Toughness and Ductility of the Beta-Ti Alloy Ti-10V-2Fe-3Al," Proceedings, 6th European Conference on Fracture, Amsterdam, Holland, June 1986.
21. G.W. Kuhlman, A.K. Chakrabarti, R. Pshko, T.L. Yu, G. Terlinde, "LCF, Fracture Toughness, and Fatigue/Fatigue Crack Propagation Resistance Optimization of Ti-10V-2Fe-3Al Alloy Through Microstructural Modification," Microstructure Fracture Toughness and Fatigue Crack Growth Rate in Titanium Alloys, A.K. Chakrabarit and J.C. Chestnutt, Editors, TMS-AIME, 1987, pp. 171-192.

491

22. G.R. Yoder, L.A. Cooley, R.R. Boyer, "Microstructure/Crack Tolerance Aspects of Notched Fatigue Life in Ti-10V-2Fe-3Al Alloy," IBID, pp. 209 - 231.
23. G.W. Kuhlman, A.K. Chakrabarti, "Room Temperature Fatigue Crack Propagation in Beta Titanium Alloys," IBID, pp. 3 - 19.
24. S.L. Semiatan, T. Althan, Isothermal and Hot Die Forging of High Temperature Alloys, MCIC-83-47, MCIC, 1983.
25. S.N. Shah, J.D. McKeogh, Status of Near-net Shape Forging For Major Aerospace Applications, MF83-908, SME, 1983.
26. G.W. Kuhlman, J.W. Nelson, Precision Forging Technology: A Change in the State-of-the-Art for Aluminum and Titanium Alloy Net Forging Shapes, MF84-256, SME, 1984.
27. G.W. Kuhlman, R. Pishko, "Practical Considerations for the Fabrication of Cost Effect Net and Near-Net Titanium Forging Shapes Through Hot Die and Isothermal Forging Technology," Titanium Science and Technology, Vol. 1, Deutsche Gesellschaft fur Metallkunde e.V., 1985, pp. 609 - 616.
28. R. Pishko, T.L. Yu, G.W. Kuhlman, "Precision Forging of Titanium Alloys," Proceedings, 1986 Internation Conference on Titanium Products and Applications, Titanium Development Association, 1987, pp. 376 - 406.

Table 1: Chemical Composition, Ti-10V-2Fe-3Al
(AMS 4983/4984/4986/4987)

Element	Wt. Per Cent
Vanadium	9.0 - 11.0
Aluminum	2.6 - 3.4
Iron	1.6 - 2.2
Oxygen	0.13 Max
Carbon	0.05 Max
Nitrogen	0.05 Max (500 ppm)
Hydrogen	0.015 Max (150 ppm)
Yttrium	0.005 Max (50 ppm)
Others, each	0.10
Others, total	0.30
Titanium	Remainder

Table 2: Physical Properties of Ti-10V-2Fe-3Al

Property	Value
Tranformation Temperature, °F (°C):	1470 (800)
Density, lbs/in^3 (g/cm^3):	0.168 (4.65)
Tensile Modulus, -60 to 465F (-50 to 240C): KSI x 10^3 (MPa x 10^3)	15.0 - 15.9 (103 - 110)
Compressive Modulus, -60 to 465F (-50 to 240C) KSI x 10^3 (MPa x 10^3)	14.9 - 16.1 (103 - 111)
Coefficient of Thermal Expansion, 68 - 800F(20 - 425C): in/in/°F x 10^{-6} (m/m/°C X 10^{-6})	5.4 (9.72)

Table 3: Precision Modulus Data*, Ti-10V-2Fe-3Al Forgings
Processed To Three Strength Levels

Temp °F(°C)	Modulus of Elasticity, 10^3 Ksi (10^3 MPa)		
	180 Ksi(1240 MPa) Minimum U.T.S.	160 Ksi(1103 MPa) Mininmum U.T.S.	140 Ksi (965 MPa) Mininimum U.T.S.

I. TENSILE MODULUS:

Temp °F(°C)	180 Ksi		160 Ksi		140 Ksi	
-58 (-50)	16.26	(110.7)	15.60	(107.6)	15.76	(108.8)
72 (22)	15.87	(109.4)	15.39	(106.1)	15.47	(106.7)
122 (50)	15.73	(108.4)	15.25	(105.1)	15.38	(106.0)
212 (100)	15.48	(106.7)	15.01	(103.5)	15.22	(104.9)
302 (150)	15.56	(107.3)	15.16	(104.5)	15.34	(105.8)
482 (250)	14.99	(103.4)	15.17	(104.6)	14.88	(102.6)

II. COMPRESSIVE MODULUS:

Temp °F(°C)	180 Ksi		160 Ksi		140 Ksi	
-58 (-50)	16.40	(113.1)	15.92	(109.8)	15.86	(109.4)
72 (22)	16.00	(110.3)	15.51	(106.9)	15.45	(106.5)
122 (50)	15.82	(109.1)	15.37	(106.0)	15.30	(105.5)
212 (100)	15.55	(107.2)	15.15	(104.4)	15.06	(103.8)
302 (150)	15.51	(106.9)	15.15	(104.4)	14.97	(103.2)
482 (250)	15.08	(104.0)	15.10	(104.1)	14.59	(100.6)

*Average of three determinations

Table 4: Typical Tensile Mechanical Properties, Ti-10V-2Fe-3Al
Splice Angle Precision Forging (Figure 8)

Spec No	Dir	Yld Str Ksi(MPa)	Ult Ten Str Ksi(MPa)	El (%)	RA (%)	K1c Ksi√in(MPa√m)
HIGH STRENGTH LEVEL (AMS 4983A, Ti-10V-2Fe-3Al STA)						
1	L	172 (1185)	181 (1248)	8	18	47 (52) T-L
2	L	168 (1158)	182 (1255)	6	20	
3	T	167 (1151)	186 (1282)	9	33	
4	T	174 (1200)	184 (1269)	8	20	
5	L	178 (1227)	189 (1303)	9	31	
INTERMEDIATE STRENGTH LEVEL(AMS 4986, Ti-10V-2Fe-3Al STA)						
1	L	157 (1082)	168 (1158)	16	44	66 (73) T-L
2	L	155 (1069)	167 (1151)	11	30	
3	T	153 (1055)	163 (1124)	18	55	
4	T	158 (1089)	169 (1165)	14	33	
5	L	160 (1103)	169 (1165)	14	39	
LOW STRENGTH LEVEL (AMS 4987, Ti-10V-2Fe-3Al STOA)						
1	L	136 (938)	148 (1020)	22	61	93 (102) T-L
2	L	137 (945)	149 (1027)	20	61	
3	T	134 (924)	146 (1007)	18	60	
4	T	137 (945)	155 (1069)	12	51	
5	L	140 (965)	146 (1007)	16	57	

Table 5: **Typical Tensile Mechanical Properties, Ti-10V-2Fe-3Al Rear Support Conventional Closed-Die Forging (Figure 8)**

Spec No	Dir	Thk in(mm)	Yld Str Ksi(MPa)		Ult Ten Str Ksi(MPa)		El (%)	RA (%)	Klc Ksi√in(Mpa√m)
HIGH STRENGTH LEVEL (AMS 4984, Ti-10V-2Fe-3Al STA)									
1	L	1 (25)	173	(1193)	184	(1269)	8	17	63 (69) L-T
2	L	1 (25)	177	(1220)	185	(1276)	6	28	60 (66) T-L
3	T	2 (50)	169	(1165)	179	(1234)	11	32	50 (55) S-L
4	T	3 (75)	167	(1151)	178	(1227)	9	38	
5	T	2 (50)	172	(1186)	180	(1240)	11	39	
6	L	2 (50)	169	(1165)	178	(1227)	9	33	
7	T	1 (25)	173	(1192)	183	(1262)	8	23	
INTERMEDIATE STRENGTH LEVEL(AMS 4986, Ti-10V-2Fe-3Al STA)									
1	L	1 (25)	152	(1048)	168	(1158)	15	35	76 (84) L-T
2	L	1 (25)	158	(1089)	165	(1138)	12	25	84 (92) T-L
3	T	2 (50)	155	(1069)	162	(1117)	10	24	77 (85) S-L
4	T	3 (75)	157	(1082)	161	(1110)	12	30	
5	T	2 (50)	162	(1117)	166	(1145)	12	29	
6	L	2 (50)	156	(1075)	164	(1131)	13	31	
7	T	1 (25)	153	(1055)	163	(1124)	12	26	
LOW STRENGTH LEVEL (AMS 4987, Ti-10V-2Fe-3Al STOA)									
1	L	1 (25)	134	(924)	147	(1013)	21	48	106 (117) L-T
2	L	1 (25)	137	(945)	149	(1027)	20	45	88 (97) T-L
3	T	2 (50)	135	(931)	148	(1020)	18	44	108 (119) S-L
4	T	3 (75)	132	(910)	143	(986)	19	48	
5	T	2 (50)	135	(931)	151	(1041)	19	43	
6	L	2 (50)	137	(945)	152	(1048)	20	48	
7	T	1 (25)	136	(938)	148	(1020)	21	40	

Table 6: **Typical Tensile Mechanical Properties, Ti-10V-2Fe-3Al Main Landing Gear Beam Conventional Forging (Figure 8)**

HIGH STRENGTH LEVEL (AMS 4984, Ti-10V-2Fe-3Al STA)

Spec No	Dir	Frg Thk in(mm)	H.T. Thk in(mm)	Yld Str Ksi(Mpa)		Ult Ten Str Ksi(MPa)		El (%)	RA (%)	Klc Ksi√in(MPa√m)
1	T	4 (100)	1 (25)	171	(1179)	175	(1207)	7	6	60 (66) L-S
2	T	4 (100)	1 (25)	161	(1110)	179	(1234)	5	5	56 (62) S-L
3	L	3 (75)	1 (25)	170	(1172)	179	(1234)	4	4	66 (73) T-L
4	T	9 (229)	3 (75)	170	(1172)	176	(1213)	4	4	65 (72) S-L
5	T	5 (127)	3 (75)	165	(1138)	174	(1200)	5	6	65 (72) T-L
6	T	7 (178)	3 (75)	169	(1165)	176	(1213)	5	5	
7	L	4 (100)	1 (25)	171	(1179)	176	(1213)	4	4	
8	T	9 (229)	3 (75)	170	(1172)	176	(1213)	4	4	
9	L	4 (100)	1 (25)	173	(1193)	178	(1227)	4	4	
10	T	5 (127)	1 (25)	172	(1186)	180	(1240)	5	5	
11	T	4 (100)	1 (25)	176	(1213)	181	(1248)	6	6	
12	T	5 (127)	3 (75)	176	(1213)	179	(1234)	5	5	
13	T	3 (75)	1 (25)	173	(1192)	183	(1262)	4	4	
14	T	5 (127)	3 (75)	171	(1179)	177	(1220)	4	7	
15	L	4 (100)	1 (25)	176	(1213)	179	(1234)	4	4	
16	L	4 (100)	1 (25)	178	(1227)	187	(1289)	5	7	
17	T	5 (127)	1 (25)	173	(1192)	181	(1248)	5	7	
18	T	5 (127)	1 (25)	173	(1192)	179	(1234)	4	4	

494

Tables 7 and 8: Design Mechanical Properties of Ti-10V-2Fe-Al Die Forgings *

	Table 7 Precision Forgings (≤ 1 in., 25 mm)	Table 8 Conventional Forgings (≤3 in., 75 mm)
Specification	AMS 4983	AMS 4984
Form	Conventional Die Forgings	
Temper	Solution Treated and Aged (900 - 950°F)	
Thickness or Diameter, in.	< 1.000	< 3.000
Basis	S	S
Mechanical Properties:		
F_{tu}, ksi:		
L......................	180	173
LT...................	180[b]	173[b]
ST..................	. . . .	173[b]
F_{ty}, ksi:		
L......................	160	160
LT...................	160[b]	160[b]
ST..................	. . . .	160[b]
F_{cy}, ksi		
L......................	168	168
LT...................	166	166
ST..................	. . . .	166
F_{su}, ksi..............	101	97
F_{bru},[a] ksi:		
(e/D = 1.5).....	244	234
(e/D = 2.0).....	295	284
e, percent		
L......................	4	4
LT...................	4[b]	4[b]
ST..................	. . . .	4[b]
E, 10^3ksi...........	15.9	
E_c, 10^3ksi.........	16.3	
G, 10^3ksi...........	. . . .	
μ...........................	. . . .	
Physical Properties:		
ω, lbs./in^3..........	0.168	
α, 10^{-6} in./in./°F	5.4 (68-800F)	
C and K...............	. . . .	

a - Bearing Values are "dry pin" valudes per section 1.4.7.1
b - Applicable providing LT or ST dimension is > 2.500 in.
* - Excerpted From Mil-Hdbk-5, Revision E, Change Notice 1.

TI-10V-2FE-3AL FORGINGS SPECIFICATION COVERAGE

FORGING PRODUCT   TEMPER   SPECIFICATION   MIL-HDBK-5

Precision Forgings   STA   AMS 4983A   "S" VALUES#
Up to 1in (25mm) 180 Ksi (1240 MPa) Ult.Tens.Str.

Conventional Hand &   STA   AMS 4984   "S" VALUES#
Closed-Die Forgings Up to 3in (75mm)173 Ksi (1193 MPa) Ult.Ten.Str.

Precision, Hand &   STA   AMS 4986   "S" VALUES#
Closed-Die Forgings Up to 4in (100mm)160 Ksi (1103 MPa)Ult.Ten.Str.

Precision, Hand &   STOA   AMS 4987   "S" VALUES#
Closed-Die Forgings Up to 4in (100mm) 140 Ksi (965 MPa) Ult.Ten.Str.

NOTES:
*   "S" Values, outlined in Table 7, were calculated from actual design allow
able test results on forgings conforming with AMS 4984, publish-ed in Change
Notice 1 to Revision E.
"S" Values, contained in Table 8, are published in Change Notice 1 to
Revision E.   Design allowable tests for "A" and "B" values have been complet
ed for ten lots, with publication of these values in Mil-HDBK-5 pending re
quired accumulation and   analysis of two hundred quality assurance/let re
lease tests now in progress.

Table 10: Guaranteed Minimum Mechanical Properties, Ti-10V-2Fe-3Al Precision, Hand and Closed Die Forgings

Specification	Dir	Yld Str Ksi(MPa)	Ult Ten Str Ksi(MPa)	El (%)	RA (%)	Klc Ksi√in(MPa√m)
AMS 4983A	L	160 (1103)	180 (1240)	4	-*	40 (44)
(STA)	T	160 (1103)	180 (1240)	4	-*	
		Precision Forgings Up to 1 in (25 mm) Thick.				
AMS 4984	L	160 (1103)	173 (1193)	4	-*	40 (44)
(STA)	T	160 (1103)	173 (1193)	4	-*	
		Hand and Closed-Die Forgings Up to 3 in (75 mm) Thick.				
AMS 4986	L	145 (1000)	160 (1103)	6	10	55 (60)
(STA)	T	145 (1000)	160 (1103)	6	10	
		Precison, Hand and Closed Die Forgings Up to 4 in (100 mm) Thick.				
AMS 4987	L	130 (896)	140 (965)	8	20	80 (88)
(STOA)	T	130 (896)	140 (965)	8	20	
		Precision, Hand and Closed Die Forgings Up to 4 in (100 mm) Thick.				

NOTE: * Reduction in Area to be reported only.

Table 11: Heavy Section Mechanical Properties, Ti-10V-2Fe-3Al Forgings. Section Thickness Exceeds Specification Limits

Sect.Thk. in (mm)	Dir	Yld Str Ksi(MPa)	Ult Ten Str Ksi(MPa)	El (%)	RA (%)	Klc Ksi√in(MPa√m)
CLOSED-DIE FORGING PROCESSED PER AMS 4984:						
4 (100)	L	161 (1110)	171 (1179)	8	11	52 (57)
4 (100)	L	159 (1096)	173 (1193)	11	21	
4 (100)	T	163 (1123)	169 (1165)	10	18	
5 (127)	L	158 (1179)	171 (1089)	6	9	56 (62)
5 (127)	T	156 (1076)	168 (1158)	8	10	
CLOSED-DIE FORGING PROCESSED PER AMS 4986:						
5 (127)	L	148 (1020)	158 (1089)	13	31	81 (89)
5 (127)	T	144 (993)	157 (1082)	10	23	
5 (127)	T	142 (979)	155 (1069)	12	30	
6 (150)	L	141 (972)	155 (1069)	14	29	78 (86)
6 (150)	T	143 (986)	156 (1076)	11	27	
CLOSED-DIE FORGING PROCESSED PER AMS 4987:						
5 (127)	L	132 (910)	139 (958)	16	44	101 (111)
5 (127)	T	129 (889)	138 (951)	15	41	
6 (150)	L	131 (903)	137 (945)	16	39	97 (107)
6 (150)	T	127 (876)	136 (938)	13	34	

Table 12: Typical Elevated Temperature Tensile Properties,* Ti-10V-2Fe-3Al Closed-Die Forgings

Exposure + Temp °F(°C)	Test Temp °F(°C)	Yld Str Ksi(MPa)	Ult Ten Str Ksi(MPa)	El (%)	RA (%)	
PROCESSED PER AMS 4984 FOR HIGH STRENGTH LEVEL:						
75 (20)	75 (20)	162 (1116)	175 (1207)	9	31	
400 (204)	400 (204)	121 (834)	147 (1014)	13	45	
500 (260)	500 (260)	111 (765)	142 (979)	12	44	
600 (315)	600 (315)	103 (710)	136 (938)	12	39	
700 (371)	700 (371)	92 (634)	144 (993)	14	53	
600# (315)	75 (20)	164 (1131)	180 (1240)	11	28	
PROCESSED PER AMS 4986 FOR INTERMEDIATE STRENGTH LEVEL:						
75 (20)	75 (20)	148 (1020)	163 (1123)	11	19	
400 (204)	400 (204)	112 (772)	132 (910)	16	52	
500 (260)	500 (260)	106 (731)	131 (903)	14	56	
600 (315)	600 (315)	95 (655)	128 (882)	14	65	
700 (371)	700 (371)	92 (634)	123 (848)	19	71	
600# (315)	75 (20)	149 (1027)	165 (1138)	12	45	
PROCESSED PER AMS 4987 FOR LOW STRENGTH LEVEL:						
75 (20)	75 (20)	139 (958)	151 (1041)	13	47	
400 (204)	400 (204)	110 (758)	130 (896)	17	53	500
(260)	500 (260)	106 (731)	131 (903)	16	57	
600 (315)	600 (315)	98 (676)	134 (924)	16	55	
700 (371)	700 (371)	92 (634)	121 (834)	17	71	
600# (315)	75 (20)	141 (972)	151 (1041)	14	52	

Notes: * Average of two test results.
+ Exposure time for all tests was 100 hours.
# Tested at room temperature after noted exposure.

Table 13: **Creep Data For Ti-10V-2Fe-3Al Forgings Processed To Two Strength Levels**

Temp °F(°C)	Stress Ksi (MPa)	0.1	0.2	0.5	1.0	2.0
I. 160 Ksi (1103 MPa) MINIMUM ULTIMATE TENSILE STRENGTH:						
650 (343)	120 (827)	0.49	1.2	5.6	19.3	61
650 (343)	100 (690)	1.5	5.1	27	94	347
650 (343)	80 (552)	10	36	189	--	--
750 (399)	95 (655)	0.12	0.24	0.64	1.6	4.4
750 (399)	70 (483)	0.9	1.9	4.7	13	42
750 (399)	50 (345)	0.7	2.2	12	50	238
850 (454)	65 (448)	0.04	0.08	0.26	0.66	1.6
850 (454)	40 (276)	0.35	0.7	1.9	4.8	13
850 (454)	20 (138)	0.06	1.8	17	80	321
950 (510)	30 (207)	0.02	0.06	0.2	0.61	1.6
950 (510)	20 (138)	0.19	0.38	1.7	5.1	14
950 (510)	10 (69)	1.0	2.9	13	41	127
II. 140 Ksi (965 MPa) MINIMUM ULTIMATE TENSILE STRENGTH:						
650 (343)	120 (827)	1.2	5	22	62	--
650 (343)	85 (586)	3.7	12	52	181	--
650 (343	70 (483)	8.9	23	91	392	--
750 (399)	95 (655)	0.04	0.12	0.37	0.86	1.9
750 (399)	70 (483)	0.3	0.3	1.6	4.9	18
750 (399)	50 (345)	0.6	1.9	9.2	36	212
850 (454)	65 (448)	0.04	0.08	0.25	0.55	1.2
850 (454)	40 (276)	0.3	0.6	1.8	5.2	14
850 (454)	20 (138)	0.8	2.5	16	73	294
950 (510)	30 (207)	0.04	0.09	0.27	0.62	1.4
950 (510)	20 (138)	0.2	0.3	1.5	4.3	11
950 (510)	10 (69)	1.1	3.2	16	54	156

Table 14: Comparison of Critical First and Second Tier Mechanical Properties of Ti-10V-2Fe-3Al Forgings With Other Alpha-Beta and Beta Titanium Alloys

ALLOY/ COND	Min Y.S. Ksi(MPa)	Min U.T.S. Ksi(MPa)	Min K_{Ic} Ksi√in (MPa√m)	a[+] in (mm)	ΔK_{th}[++] Ksi√in (MPa√m)	Sm. Fat.[+++] @10[7]Cycles Ksi(MPa)
Ti-10V-2Fe-3Al STA (AMS 4984)	160 (1103)	173 (1193)	40 (44)	0.07 (1.8)	3.6 (4.0)	130 (896)
Ti-10V-2Fe-3Al STA (AMS 4986)	145 (1000)	160 (1103)	55 (60)	0.16 (4.1)	3.7 (4.1)	125 (862)
Ti-10V-2Fe-3Al STOA(AMS 4987)	130 (896)	140 (965)	80 (88)	0.42 (10.7)	3.9 (4.3)	120 (827)

ALPHA-BETA ALLOYS:

ALLOY/ COND	Min Y.S. Ksi(MPa)	Min U.T.S. Ksi(MPa)	Min K_{Ic} Ksi√in (MPa√m)	a[+] in (mm)	ΔK_{th}[++] Ksi√in (MPa√m)	Sm. Fat.[+++] @10[7]Cycles Ksi(MPa)
Ti-6Al-4V Std Cond A	120 (827)	130 (896)	50* (55)	0.19 (4.8)	5.3 (5.8)	65 (448)
Ti-6Al-4V Std Cond STA	130 (896)	140 (965)	40* (44)	0.10 (2.5)	---	75 (517)
Ti-6Al-4V Ell Cond RA	115 (792)	125 (862)	75 (82)	0.47 (11.9)	5.9 (6.5)	70 (483)
Ti-6Al-4V Ell Cond BA	110 (758)	120 (827)	85 (94)	0.66 (16.8)	6.2 (6.8)	70 (483)
Ti-6Al-6V-2Sn Cond A	125 (862)	135 (931)	35* (40)	0.07 (1.8)	---	---
Ti-6Al-2Sn-4Zr-6Mo Cond STA	150 (1034)	160 (1103)	30* (33)	0.04 (1.0)	<4.0 (<4.4)	65 (448)
Ti-6Al-2Sn-4Zr-6Mo Cond B/STA	150 (1034)	160 (1103)	45 (50)	0.10 (2.5)	<4.0 (<4.4)	80 (552)

BETA ALLOYS:

ALLOY/ COND	Min Y.S. Ksi(MPa)	Min U.T.S. Ksi(MPa)	Min K_{Ic} Ksi√in (MPa√m)	a[+] in (mm)	ΔK_{th}[++] Ksi√in (MPa√m)	Sm. Fat.[+++] @10[7]Cycles Ksi(MPa)
Beta C (Ti-3Al-8V-6Cr-4Mo-4Zr) Cond STA	150** (1034)	160** (1103)	40** (44)	0.08 (2.0)	<4.0 (<4.4)	100 (690)
Ti-15V-3Cr-3Sn-3Al Cond STA	150** (1034)	160** (1103)	50** (55)	0.12 (3.0)	---	120 (827)

Std = Std. Al. and O Contents   A = Annealed or Sol'n Treat plus Anneal
Ell = Controlled Al and O Contents   RA = Recrystallized Annealed
STOA = Solution Treat & Overaged   STA = Solution Treated and Aged
BA = Beta Annealed or Quenched   B/STA = Beta Hot Die Frg'd + STA

Notes:
+ - a is critical crack length calculated as, $a \approx 1.1[Kic/Y.S.]^2$
++ - ΔK_{th} threshold stress intensity in fatigue crack growth tests.

+++ - Smooth fatigue stress at 10^7 cycles, tests at R = 0.1 to 0.3.
F = 30 - 125 Hz. Ti-6Al-4V may be underestimated, for strengths were low in these forgings due to low oxygen content,
* - Kic is not guaranteed for this alloy/condition but note is estimated minimum Kic if it were guaranteed.
** - Estimated minima in forgings based upon limited data.

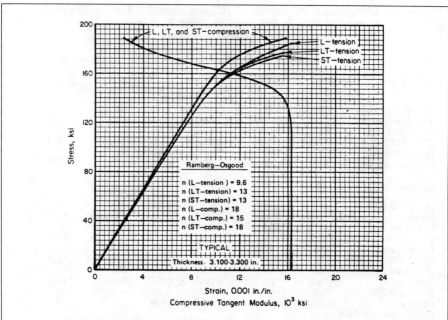

Figure 1: Typical Tensile Stress-Strain, Compressive Sress-Strain, and Compressive Tangent Modulus Curves For Solution Treated and Aged (900-950°F) Ti-10V-2Fe-3Al Forgings [Excerpted Mil-HDBK-5, Rev. E, CN 1]

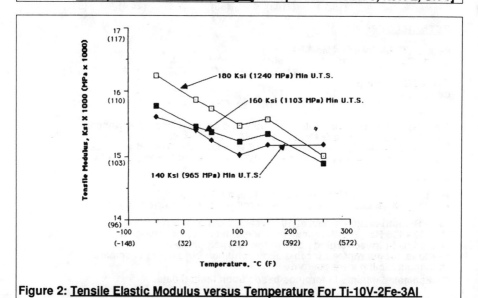

Figure 2: Tensile Elastic Modulus versus Temperature For Ti-10V-2Fe-3Al Forgings, Processed to Three Strength Levels

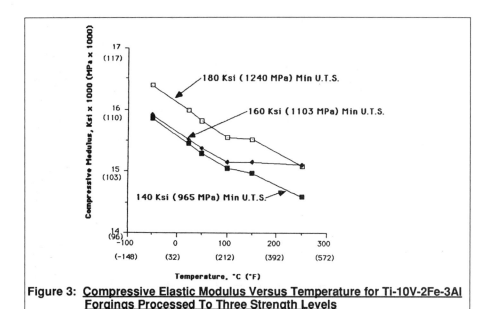

Figure 3: Compressive Elastic Modulus Versus Temperature for Ti-10V-2Fe-3Al Forgings Processed To Three Strength Levels

GA-18521.22

Figure 4: Qualitative TTT Diagram for β-ST Ti-10V-2Fe-3Al (From Duerig, et.al. Met. Trans., Dec. 1980)

501

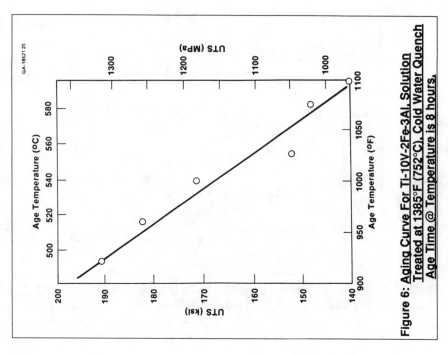

Figure 6: Aging Curve For Ti-10V-2Fe-3Al. Solution Treated at 1385°F (752°C). Cold Water Quench Age Time @ Temperature is 8 hours.

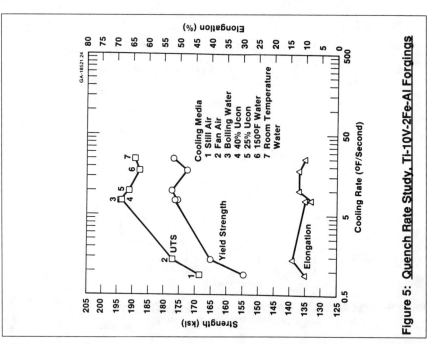

Figure 5: Quench Rate Study. Ti-10V-2Fe-Al Forgings

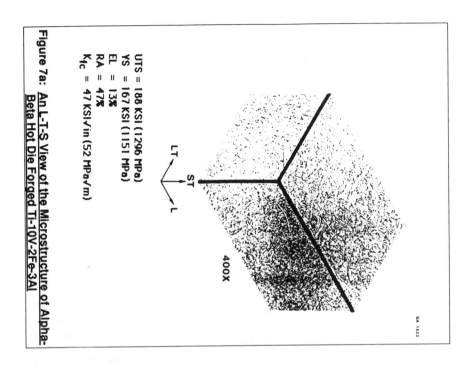

UTS = 188 KSI (1296 MPa)
YS = 167 KSI (1151 MPa)
EL = 13%
RA = 47%
K_{Ic} = 47 KSI√in (52 MPa√m)

Figure 7a: <u>An L-T-S View of the Microstructure of Alpha-Beta Hot Die Forged Ti-10V-2Fe-3Al</u>

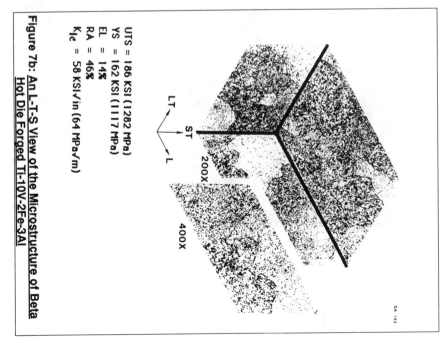

UTS = 186 KSI (1282 MPa)
YS = 162 KSI (1117 MPa)
EL = 14%
RA = 46%
K_{Ic} = 58 KSI√in (64 MPa√m)

Figure 7b: <u>An L-T-S View of the Microstructure of Beta Hot Die Forged Ti-10V-2Fe-3Al</u>

503

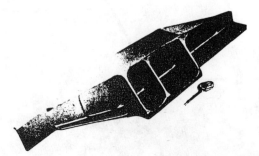

Splice Angle
Precision Forging

13.79 in (349 mm) Long
21 sq in (135 sq cm) PVA
Weight 1.5 lbs (0.7 Kg)

Intermediate Size
Closed-Die Forging
Rear Support
29 in (737 mm) Long
145 sq in (935 sq cm) PVA
Weight 75 lbs (34 Kg)

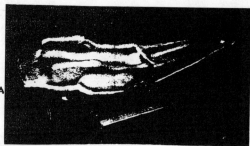

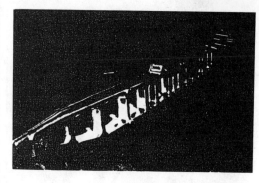

Heavy Section
Closed-Die Forging
Main Landing Gear Beam
176 in (447 cm) Long
2,650 sq in (17,100 sq cm) PVA
Weight 3,000 lbs (1365 Kg)

Figure 8: Representative Ti-10V-2Fe-3Al Precision and Conventional Closed-Die Forgings

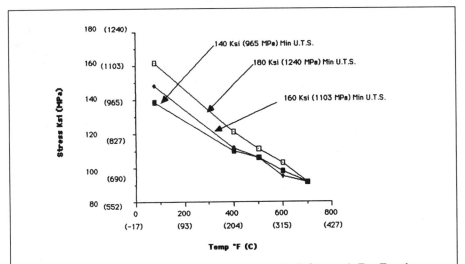

Figure 9: <u>Ti-10V-2Fe-3Al Elevated Temperature Yield Strength For Forgings Processed To Three Strength Levels</u>

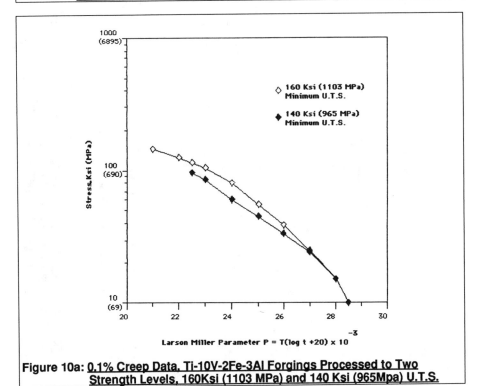

Figure 10a: <u>0.1% Creep Data, Ti-10V-2Fe-3Al Forgings Processed to Two Strength Levels, 160Ksi (1103 MPa) and 140 Ksi (965Mpa) U.T.S.</u>

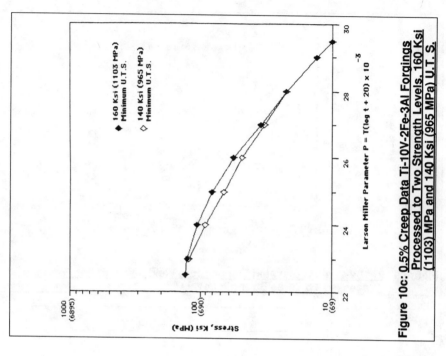

Figure 10c: <u>0.5% Creep Data Ti-10V-2Fe-3Al Forgings Processed to Two Strength Levels, 160 Ksi (1103) MPa and 140 Ksi (965 MPa) U.T.S.</u>

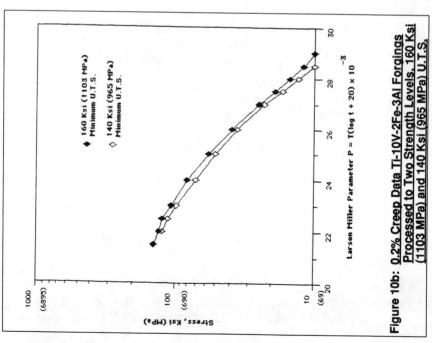

Figure 10b: <u>0.2% Creep Data Ti-10V-2Fe-3Al Forgings Processed to Two Strength Levels, 160 Ksi (1103 MPa) and 140 Ksi (965 MPa) U.T.S.</u>

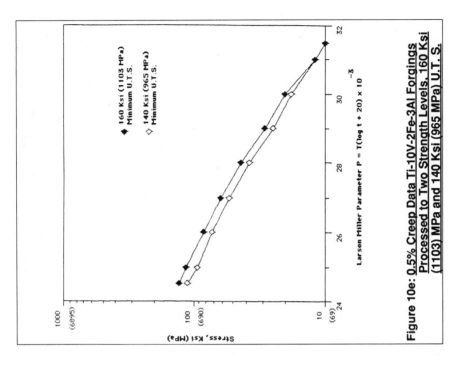

Figure 10e: 0.5% Creep Data Ti-10V-2Fe-3Al Forgings Processed to Two Strength Levels, 160 Ksi (1103) MPa and 140 Ksi (965 MPa) U.T.S.

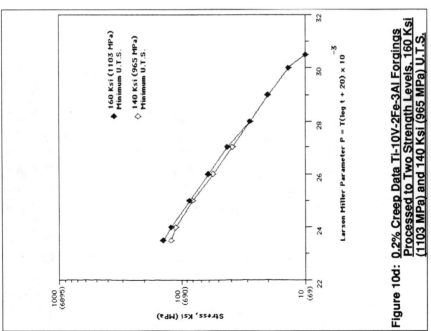

Figure 10d: 0.2% Creep Data Ti-10V-2Fe-3Al Forgings Processed to Two Strength Levels, 160 Ksi (1103 MPa) and 140 Ksi (965 MPa) U.T.S.

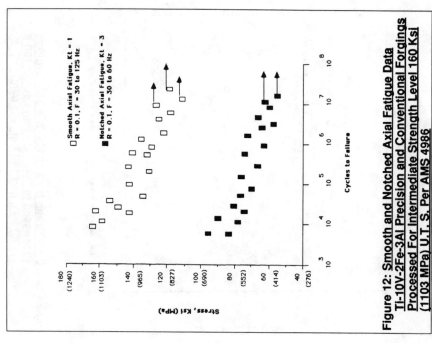

Figure 12: Smooth and Notched Axial Fatigue Data
Ti-10V-2Fe-3Al Precision and Conventional Forgings
Processed For Intermediate Strength Level 160 Ksi
(1103 MPa) U.T.S. Per AMS 4986

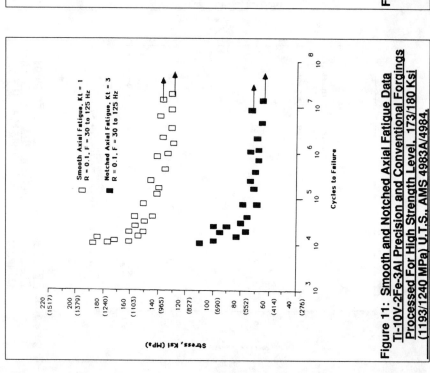

Figure 11: Smooth and Notched Axial Fatigue Data
Ti-10V-2Fe-3Al Precision and Conventional Forgings
Processed For High Strength Level, 173/180 Ksi
(1193/1240 MPa) U.T.S., AMS 4983A/4984,

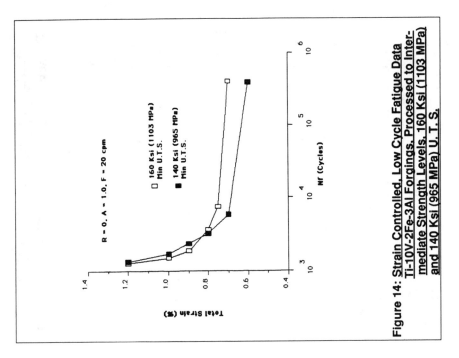

Figure 14: Strain Controlled, Low Cycle Fatigue Data Ti-10V-2Fe-3Al Forgings, Processed to Intermediate Strength Levels, 160 Ksi (1103 MPa) and 140 Ksi (965 MPa) U.T.S.

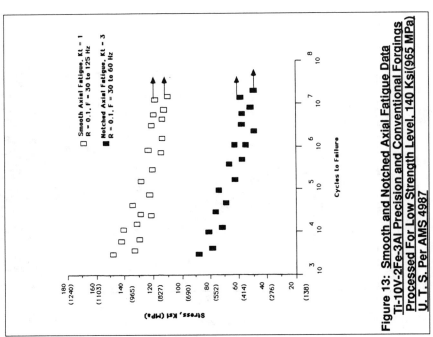

Figure 13: Smooth and Notched Axial Fatigue Data Ti-10V-2Fe-3Al Precision and Conventional Forgings Processed For Low Strength Level, 140 Ksi(965 MPa) U.T.S. Per AMS 4987

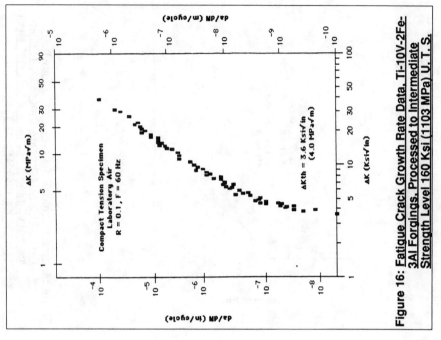

Figure 16: Fatigue Crack Growth Rate Data, Ti-10V-2Fe-3Al Forgings, Processed to Intermediate Strength Level 160 Ksi (1103 MPa) U.T.S.

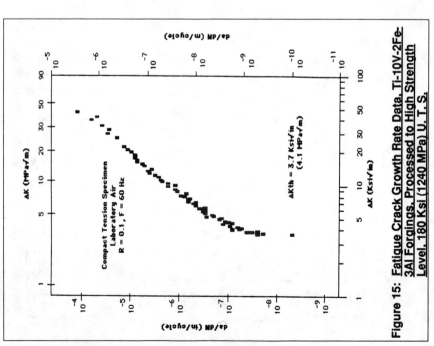

Figure 15: Fatigue Crack Growth Rate Data, Ti-10V-2Fe-3Al Forgings, Processed to High Strength Level, 180 Ksi (1240 MPa) U.T.S.

510

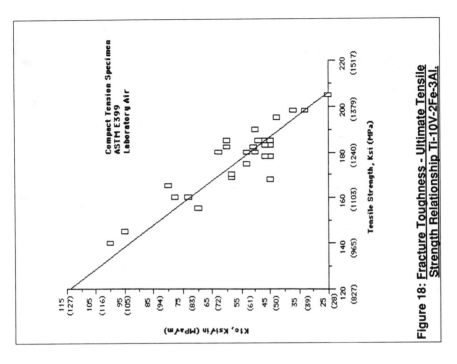

Figure 18: Fracture Toughness - Ultimate Tensile Strength Relationship Ti-10V-2Fe-3Al.

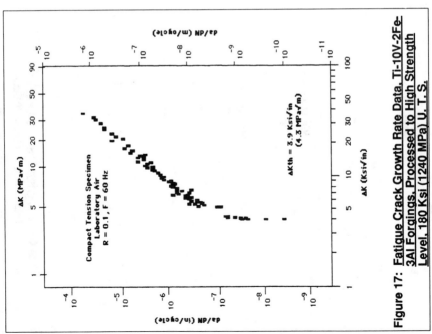

Figure 17: Fatigue Crack Growth Rate Data, Ti-10V-2Fe-3Al Forgings, Processed to High Strength Level, 180 Ksi (1240 MPa) U.T.S.

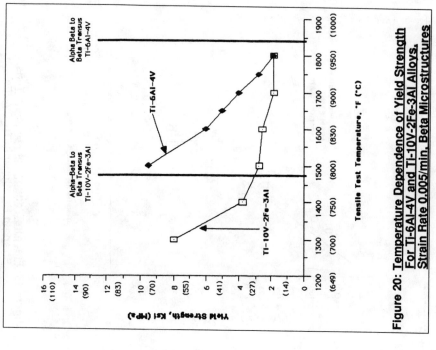

Figure 20: Temperature Dependence of Yield Strength For Ti-6Al-4V and Ti-10V-2Fe-3Al Alloys, Strain Rate 0.005/min, Beta Microstructures

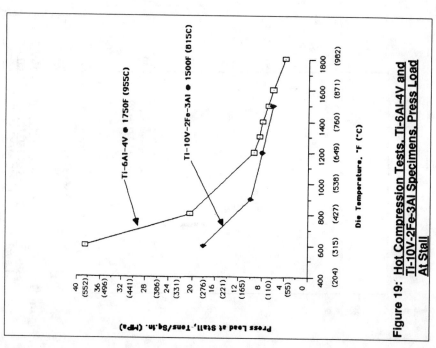

Figure 19: Hot Compression Tests, Ti-6Al-4V and Ti-10V-2Fe-3Al Specimens, Press Load At Stall

512

SP-700 TITANIUM ALLOY DATA SHEETS

A.Ogawa, H.Fukai, K.Minakawa, and C.Ouchi

Materials and Processing Research Center
NKK Corporation
Kawasaki, JAPAN

Abstract

SP-700, an emerging β-rich α+β titanium alloy, is designed to improve superplastic formability as well as mechanical properties over Ti-6Al-4V alloy. Owing to its fine microstructure and low β-transus temperature, it is superplastic-formable at temperature below 1073K(800°C) with low flow stress. Remarkable workability of this alloy is also retained in conventional manufacturing processes. Another advantage of SP-700 is heat treatment response which includes deep hardenability and quick aging kinetics. Corrosion resistance and machinability are equivalent to or better than Ti-6Al-4V alloy.

Beta Titanium Alloys in the 1990's
Edited by D. Eylon, R.R. Boyer and D.A. Koss
The Minerals, Metals & Materials Society, 1993

Introduction

SP-700 [1,2] is designed to offer excellent superplasticity by the additions of molybdenum and iron to the Ti-Al-V alloy. As well as in superplastic forming, remarkable workability is achieved in conventional manufacturing processes due to its fine microstructure and low β-transus temperature. Also improved in SP-700 are heat treatment response and mechanical properties over the Ti-6Al-4V alloy. These data sheets present the microstructures, physical and mechanical properties and corrosion resistance of SP-700 .

Chemical Composition

Table 1 - Nominal chemical composition
(unit:wt%)

Al	V	Mo	Fe
4.5	3.0	2.0	2.0

Physical Properties

Table 2 - Physical properties of mill annealed SP-700

Density	4.54 g/cm^3 at RT	Electric Resistivity	1.62 μΩm
Melting Point	1593 ± 5 °C	Magnetic Susceptibility	3.51 x 10^{-6} cm^3/g
Beta Transus	900 ± 5°C	Magnetic Permeability	1.0020
Thermal Conductivity	6.77 W/m·°C at RT	Young's Modulus	110 GPa at RT
Thermal Diffusivity	305.3 m^2/s at RT	Poisson's Rate	0.33 at RT
Specific Heat	493.8 J/kg °C at RT	Internal Friction (AQ^{-1})	9.3 x 10^{-6} at 400Hz
Thermal Expansion	8.90 x 10^{-6}/°C (25 - 600°C)		

Microstructure

Figure 1 depicts typical microstructures of SP-700 in mill annealed, recrystallization annealed, β annealed, and solution treated and aged conditions. Effects of solution treatment temperature and cooling rate on microstructure are given in Figures 2 and 3, respectively. Phase maps as a function of solution treatment temperature and cooling rate are illustrated in Figure 4.

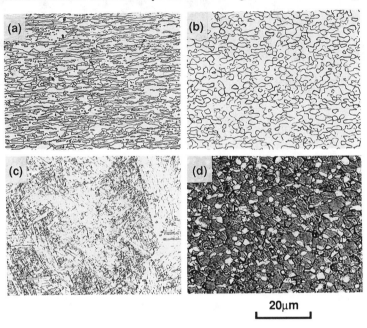

20μm

Figure 1 - Typical microstructure of SP-700 in (a)mill annealed, (b)recrystallization annealed, (c)β annealed , and (d)solution treated and aged conditions.

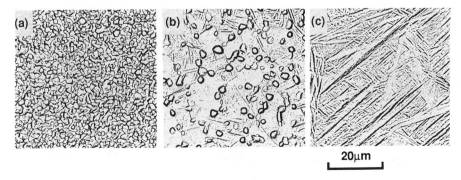

20μm

Figure 2 - Microstructure of SP-700 in solution treated at (a)800°C, (b)875°C, and (c)900°C.

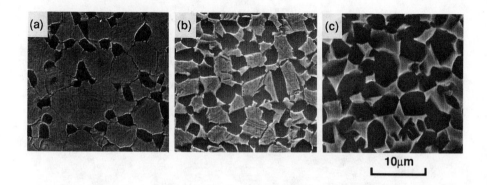

Figure 3 - Effect of cooling rate on microstructure of SP-700. Cooling rates are (a)10°C/s, (b)1°C/s, and (c)0.1°C/s, respectively.

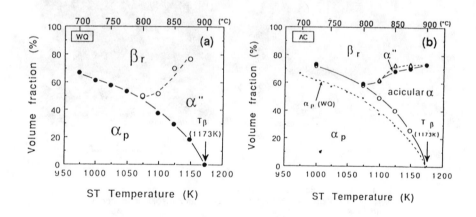

Figure 4 - Phase maps of SP-700 in conditions of (a)water quenching and (b)air cooling. Dotted line with αp(WQ) in (b) indicates percentage of primary α in condition of water quenching.

Mechanical Properties

Hardenability

SP-700 is less sensitive to slow or delayed quenching than Ti-6Al-4V owing to its more stable β phase(see Figures 5 and 6).

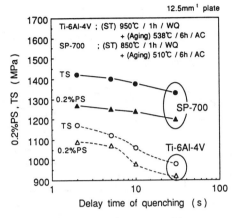

12.5mm¹ plate

Figure 5 - Effect of delay in quenching on tensile properties of SP-700 in solution treated and aged condition.

Figure 6 - Effect of cooling rate on tensile properties of SP-700 in solution treated and aged condition.

Hardness

Mill annealed SP-700 has a hardness of 300 HV. Recrystallization annealed SP-700 has a hardness of 280 to 320 HV. Hardness of solution treated and aged SP-700 ranges from 350 to 510 HV depending on solution treating and aging conditions(see Figure 7).

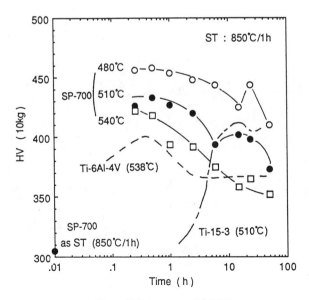

Figure 7. Aging curve of SP-700.

Tensile Properties

Table 3 - Typical tensile properties of SP-700 plate and sheet in mill annealed condition.

Thickness / mm	Direction	0.2%PS / MPa	UTS / MPa	El / %
0.8	L	1023	1073	10.4
	T	1023	1073	10.2
2.0	L	953	1014	13.2
	T	924	996	15.0
3.8	L	949	1025	22.8
	T	929	1015	21.0

Table 4 - Typical tensile properties of mill annealed SP-700

Product Form	0.2%PS / MPa	UTS / MPa	El / %
Plate (15.0mmt)	990	1028	16.8
Sheet(3.8mmt)	949	1025	22.8
Bar (ϕ22.0mm)	936	1007	18.4

Table 5 - Typical tensile properties of SP-700 in various heat treatment conditions.

Heat Treatment	0.2%PS / MPa	TS / MPa	El / %	RA / %
Mill Annealing 720°C/1hr/AC	972	1023	19.0	61.9
Recry. Annealing 800°C/1hr/AC	917	966	20.8	61.6
ST(WQ)+Aging 850°C/1hr/WQ + 560°C/6hr/AC	1240	1377	11.6	28.0
ST(AC)+Aging 850°C/1hr/AC + 510°C/6hr/AC	1114	1213	14.4	39.6

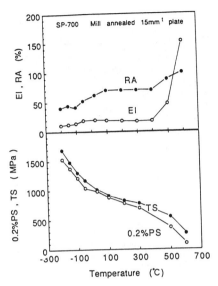

Figure 8 - Temperature dependence of tensile properties of SP-700.

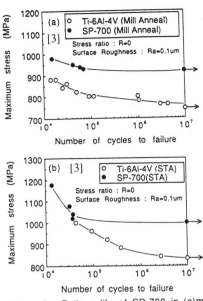

Figure 9 - Fatigue life of SP-700 in (a)mill annealed and (b)solution treated and aged conditions with cylindrical specimens.

Fatigue Properties

Fatigue strength of SP-700 in the mill annealed condition or in the solution treated and aged condition is higher than that of Ti-6Al-4V because of its higher strength and finer microstructure(see Figures 9 and 10). Fatigue crack propagation rate of SP-700 in the mill annealed condition is shown in Figure 11.

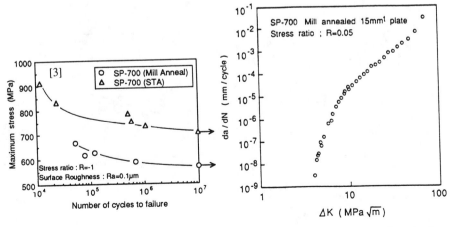

Figure 10 - Fatigue life of SP-700(R=-1).

figure 11 - Fatigue crack propagation rate of SP-700 in mill annealed condition.

Fracture Toughness

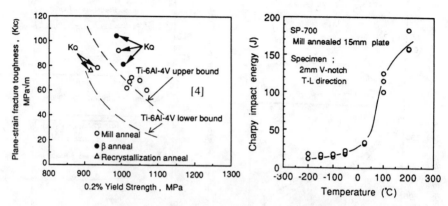

Figure 12 - Fracture toughness of SP-700.

Figure 13 - Temperature dependence of Charpy impact energy of SP-700 in mill annealed condition.

Superplasticity

SP-700 shows excellent superplastic formability at temperatures around 775°C, which is over 100°C lower than Ti-6Al-4V(see Figures 14 and 15). SP-700 gives much lower flow stress at temperatures between 600 and 900°C(see Figure 14). Room temperature tensile properties after superplastic deformation are better retained in SP-700 owing to its finer microstructure (see Figures 16 and 17, and Table 6).

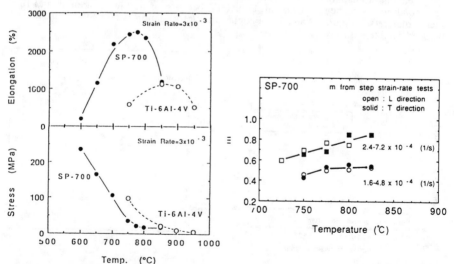

Figure 14 - Superplastic elongation and flow stress of SP-700.

Figure 15 - Strain rate sensitivity index, m-value, of SP-700.

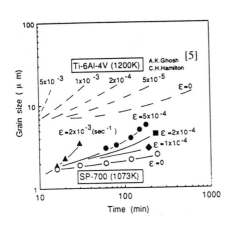

Figure 16 - Grain growth of SP-700 and Ti-6Al-4V during superplastic deformation.

Figure 17 - Microstructure of pre- and post-superplastic deformation.

Table 6 - Tensile properties of SP-700 after SPF

	Tensile Properties			Grain Size / μm
	0.2%PS / MPa	UTS / MPa	El / %	
Before SPF	1029	1038	19.7	1.6
After SPF	933	986	14.4	2.6

SPF condition   SPF temperature:775°C
Strain rate:1 x 10^{-4}/s
SPF period:3hr

Bending

SP-700 shows much better bendability at room temperature as compared to Ti-6Al-4V(see Table 7).

Table 7. Bend factor and limit of cold rolling reduction

Alloy	Direction	Bend Factor (R/t)	Cold Rolling Reduction Limit
SP-700	L	2.1	69%
	T	2.1	58%
Ti-6Al-4V	L, T	4	20%

Forgeability

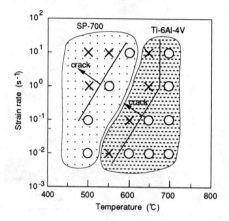

Figure 18. Forgeability of SP-700 measured after 70% reduction in compression of notched cylindrical specimens(6mm in dia. x 10 mm in height). "X" symbol denotes cracks were observed and "O" denotes no cracks observed.

Corrosion resistance

General Corrosion

The corrosion resistance of SP-700 depends on the formation of a protective oxide layer, the same as in commercially pure titanium. SP-700 resists corrosion under a salt environment and has slightly higher corrosion resistance in hot or concentrated solution of the reducing acids(hydrochloric, sulphuric acid) than pure titanium and Ti-6Al-4V(see Table 8).

Table 8. Corrosion rate of SP-700, Ti-6Al-4V, and CP-2 Ti
(unit:mm/year)

Environment	Concentration	Temperature	SP-700	Ti-6Al-4V	CP-2 Ti
NaCl	25%	Boiling	0	0	0
HCl	2%	20°C	0	0	0
		Boiling	4.8	5.6	4.3
	4%	20°C	0	0	0
		Boiling	52.8	68.3	78.7
H_2SO_4	2%	20°C	0	0	0
		Boiling	20.5	23.5	20.3
	4%	20°C	0	0	0.1
		Boiling	45.9	46.4	47.8

Crevice Corrosion

SP-700 resists crevice corrosion in boiling 5% sodium chloride aqueous solution(see Table 9).

Hot Corrosion

SP-700 shows excellent resistance to hot salt cracking. Specimen with 15.5 g/m^2-thickness salt coating did not crack under a stress of 245 MPa at 350°C(see Table 9).

Table 9. Results of crevice corrosion test and hot salt cracking test of SP-700, Ti-6Al-4V, and CP-2 Ti.

		SP-700	Ti-6Al-4V	CP-2 Ti
Crevice Corrosion Boiling 5% NaCl Testing Period:720hr	Surface Observation	no corrosion	no corrosion	no corrosion
Hot Salt Cracking Test Applied Stress: 245MPa Salt Coating:15.1g/m thickness Testing Temperature :350°C		not failed	not failed	

Diffusion Bonding

SP-700 can be diffusion-bonded. Bonding strength is shown in Figure 19, and microstructures at interface of diffusion bonding are shown in Figure 20.

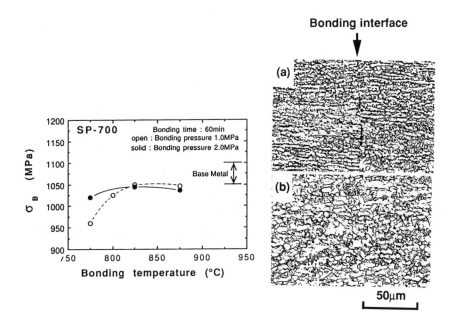

Figure 19. Strength of diffusion bonded SP-700.

Figure 20. Microstructure of SP-700 at interface of (a) 3 minutes and (b)1hour diffusion bonding.

523

Tool Life

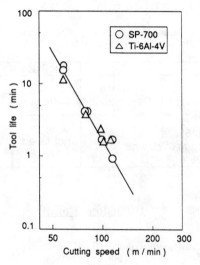

Figure 21. Tool lives of cemented carbide bits on mill annealed SP-700 with 60-200mm/min of speed and 0.1mm/rev of feed. No lubricant was used.

References

1. C.Ouchi *et al*.,"Development of β-rich α+β Titanium Alloy : SP-700," <u>NKK Technical Review</u>, No.65 (1992), 61-67
2. M.Ishikawa *et al*.,"Microstructure and Mechanical Properties Relationship of β-rich α+β Titanium Alloy : SP-700 (Paper presented at the Seventh World Conference on Titanium, San Diego, California, 29 June 1992), 8.
3. T.Fujita *et al*.,"Fatigue and Fracture Toughness Properties in the Beta-Rich α+β Titanium Alloy SP-700 (Paper presented at the Beta Titanium Alloys Symposium, 1993 TMS annual meeting, Denver Colorado, 22 February 1993),11.
4. A.W.Bowen and C.A.Stubbington, <u>Titanium and Titanium Alloys</u>, ed. J.C.Williams and A.F.Belov (1981), 1989-2001.
5. A.K.Ghosh, C.H.Hamilton, and J.A.Werty, <u>Met.Forum</u>, vol.8, No.4, 172-190 (1985)

SUBJECT INDEX

The **bold type** page number(s) denote locations in which the subject is discussed in greater detail. Complete alloy listing and alloy chemistries are under *Titanium alloys* and are listed by elemental alphabetical order followed by numerical order of weight or atomic percent.

AUTHOR INDEX